ECOHYDROLOGY: PROCESSES, MODELS AND CASE STUDIES

An approach to the sustainable management of water resources

Ecohydrology: Processes, Models and Case Studies

An approach to the sustainable management of water resources

Edited by

David Harper

Maciej Zalewski

and

Nic Pacini

www.cabi.org

CABI is a trading name of CAB International

CABI Head Office
Nosworthy Way
Wallingford
Oxfordshire OX10 8DE
UK

Tel: +44 (0)1491 832111
Fax: +44 (0)1491 833508
E-mail: cabi@cabi.org
Website: www.cabi.org

CABI North American Office
875 Massachusetts Avenue
7th Floor
Cambridge, MA 02139
USA

Tel: +1 617 395 4056
Fax: +1 617 354 6875
E-mail: cabi-nao@cabi.org

A catalogue record for this book is available from the British Library, London, UK.

Library of Congress Cataloging-in-Publication Data
Ecohydrology : processes, models and case studies : an approach to the
sustainable management of water resources / editors, David Harper,
Maciej Zalewski & Nic Pacini.
      p. cm.
   Includes bibliographical references and index.
   ISBN 978-1-84593-002-8 (alk. paper)
  1. Ecohydrology. 2. Water-supply–Management. I. Harper, David M.
II. Zalewski, Maciej. III. Pacini, Nic.

   QH541.15.E19E263 2008
   551.48–dc22

                        2008007715

ISBN-13: 978 1 84593 002 8

Typeset by AMA Dataset, Preston.
Printed and bound in the UK by Cromwell Press, Trowbridge.

The paper used for the text pages in this book is FSC certified.
The FSC (Forest Stewardship Council) is an international network to promote
responsible management of the world's forests.

Contents

Contributors

István Bíró, [Dr], *Environmental Protection and Water Management Research Institute (VITUKI), H-1453 Budapest Pf. 27, Hungary.*

Giuseppe Crosa, [Professor], *Department of Biotechnology and Molecular Sciences, University of Insubria, Via Dunant 3, I-21100 Varese, Italy. Email: Giuseppe.Crosa@uninsubria.it*

Benoit Demars, [Dr], *Macaulay Institute, Macaulay Drive, Craigiebuckler, Aberdeen AB15 8QH, UK. Email: B.Demars@macaulay.ac.uk*

Boris Fashchevsky, [Professor], *International Sakharov Environmental University, 48–99 Kalinovskogo Street, 220086 Minsk, Republic of Belarus. Email: borisf2@yandex.ru*

Emmanuel Gereta, [Dr], *Tanzania Wildlife Research Institute, PO Box 661, Arusha, Tanzania. Email: e.gereta@habari.co.tz*

James A. Gore, [Professor], *Program in Environmental Science, Policy and Geography, University of South Florida, 140 Seventh Avenue South, St Petersburg, FL 33701, USA. Email: jagore@stpt.usf.edu*

Dieter Gutknecht, [Professor], *Austrian Academy of Sciences, Vienna University of Technology, Dr. Ignaz Seipel-Platz 2, A-1010 Vienna, Austria. Email: gutknecht@hydro.tuwien.ac.at*

David M. Harper, [Dr], *Department of Biology, University of Leicester, Leicester LE1 7RH, UK. Email: dmh@le.ac.uk*

Bert Higler, [Professor], *ALTERRA, PO Box 47, 6700 AA Wageningen, The Netherlands. Email: L.W.G.Higler@Alterra.wag-ur.nl*

Anna Hillbricht-Ilkowska, [Professor], *Centre for Ecological Research, Polish Academy of Sciences, Dziekanow Lesny, 05-092 Lomianki, Poland. Email: anna.ilkowska@ neostrada.pl*

Christoph Humborg, [Dr], *Department of Systems Ecology, Stockholm University, S-10691 Stockholm, Sweden. Email: christop@system.ecology.su.se*

Venugopalan Ittekkot, [Dr], *Centre for Marine Tropical Ecology – Bremen, Fahrenheitstrasse 6, D-28359 Bremen, Germany. Email: ittekkot@uni-bremen.de*

Georg A. Janauer, [Professor], *Department für Limnologie und Hydrobotanik, Faculty of Life Sciences, University of Vienna, Althanstrasse 14, A-1091 Vienna, Austria. Email: georg.janauer@univie.ac.at*

Géza Jolánki, [Professor], *Environmental Protection and Water Management Research Institute (VITUKI), H-1453 Budapest Pf. 27, Hungary. Email: jolankai@vituki.hu*

Joanna Kemp, [Dr], *SEPA Aberdeen, Greyhope House, Greyhope Road, Torry, Aberdeen AB11 9RD, UK. Email: Joanna.Kemp@sepa.org.uk*

Michael E. McClain, [Dr], *Department of Environmental Studies, Institute for Sustainability Science, Florida International University, Miami, FL 33199, USA. Email: michael.mcclain@fiu.edu*

Jim Mead, [Dr], *Division of Water Resources, North Carolina Department of Environment and Natural Resources, PO Box 27687, Raleigh, NC 27611, USA. Email: jim_mead@mail.ehnr.state.nc.us*

Olga Oksiyuk, [Professor], *Institute of Hydrobiology, Ukrainian Academy of Sciences, Geroyev Stalingrada av. 12, 04210 Kyiv, Ukraine.*

Nic Pacini, [Dr], *Department of Ecology, University of Calabria, I-87036 Arcavacata di Rende (Cosenza), Italy. Email: kilapacini@hotmail.com*

Tadeusz Penczak, [Professor], *Department of Ecology and Vertebrate Zoology, University of Lodz, 12/16 Banacha Str., 90-237 Lodz, Poland. Email: penczakt@biol.uni.lodz.pl*

Lars Rahm, [Dr], *Department of Water and Environmental Studies, Linköpings Universitet, S-58183 Linköping, Sweden. Email: larra@tema.liu.se*

Han Runhaar, [Dr], *Kiwa Water Research, Groningenhaven 7, 3433 PE Nieuwegein, The Netherlands. Email: Han.Runhaar@kiwa.nl*

Kevin Skinner, [Dr], *Jacobs Ltd, Sheldon Court, Wagon Lane, Coventry Road, Sheldon, Birmingham B26 3DU, UK. Email: kevin.skinner@jacobs.com*

Leszek Starkel, [Professor], *Department of Geomorphology and Hydrology, Institute of Geography, Polish Academy of Sciences, 31-018 Kraków, sw. Jana 22, Poland. Email: starkel@zg.pan.krakow.pl*

Vladimir Timchenko, [Professor], *Institute of Hydrobiology, Ukrainian Academy of Sciences, Geroyev Stalingrada av. 12, 04210 Kyiv, Ukraine. Email: timol@iptelecom.net.ua*

R. Van Ek, [Dr], *De Rondom 1, 5612 AP Eindhoven, The Netherlands. Email: remco.vanek@tno.nl*

Iwona Wagner, [Dr], *European Regional Centre for Ecology u/a UNESCO, 3 Tylna Street, 91-849 Lodz, Poland. Email: iwwag@biol.uni.lodz.pl*

Jan-Philip M. Witte, [Dr], *Kiwa Water Research and Vrije Universiteit, Institute of Ecological Science, De Boelelaan 1085, 1081 HV Amsterdam, The Netherlands. Email: Flip.Witte@kiwa.nl*

Eric Wolanski, [Professor], *Australian Institute of Marine Science, PMB No. 3, Townsville MC, Queensland 4810, Australia. Email: e.wolanski@aims.gov.au*

Maciej Zaleweski, [Professor], *European Regional Centre for Ecology u/a UNESCO, 3 Tylna Street, 91-849 Lodz, Poland. Email: mzal@biol.uni.lodz.pl*

Preface

The River Taw flows from its source in the high ground of north Dartmoor to the north Devon coast near Barnstaple. Its journey is one of change, growth, maturation and dispersal. It is a long transparency embodying an infinite number of liquid states – I think of it as a living entity reflecting a human microcosm.

Susan Derges, 1997

Scientific discourse is absolutely loaded with metaphors, and it must be so in order for it to have meaning. It's not a bad thing, it's a good thing, but of course, you have got to have the right metaphors. The dominant metaphors in biology at the moment are metaphors of survival, competition, selfish genes etc., which has its origins in Darwinism and 19th-century economics, but it seems to me that a better metaphor, one that is truer to nature, is that of creation and transformation, and therefore inevitably participation. Participation involves subjectivity, which is a very tricky area for most scientists, because subjectivity has no obvious rules. . .but I believe that subjectivity, and intuition, are in fact the source of understanding wholes and the reality of nature. You can understand the parts with your analytical mind, but to understand the whole you have to use your intuition, and that involves participation.

Susan Derges in dialogue with Brian Goodwin, 1997

Overleaf: Susan Derges photogram *The River Taw (Hawthorn), 25 May 1998*

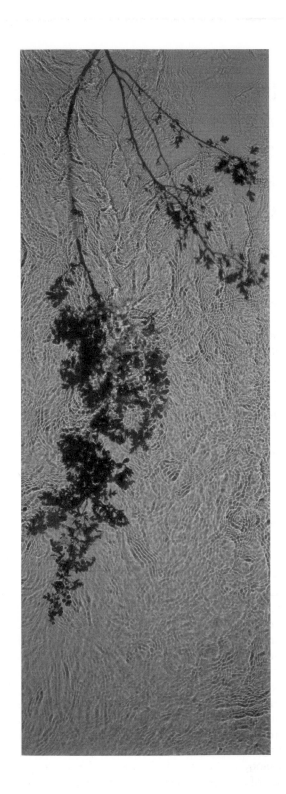

1 Linking Biological and Physical Processes at the River Basin Scale: the Origins, Scientific Background and Scope of Ecohydrology

M. Zalewski[1], D.M. Harper[2], B. Demars[3], G. Jolánkai[4], G. Crosa[5], G.A. Janauer[6] and N. Pacini[7]

[1]European Regional Centre for Ecology u/a UNESCO, Lodz, Poland; [2]Department of Biology, University of Leicester, Leicester, UK; [3]Macaulay Institute, Aberdeen, UK; [4]Environmental Protection and Water Management Research Institute (VITUKI), Budapest, Hungary; [5]Department of Biotechnology and Molecular Sciences, University of Insubria, Varese, Italy; [6]Department für Limnologie und Hydrobotanik, Faculty of Life Sciences, University of Vienna, Vienna, Austria; [7]Department of Ecology, University of Calabria, Arcavacata di Rende (Cosenza), Italy

What is Ecohydrology?

Ecohydrology, as used in this book, is a new term (used since the late 1990s) to describe a new scientific way of managing the water cycle in order to achieve the sustainable use of water by societies. It is an understanding of how hydrological processes integrate with ecological ones (e.g. the discharge regimes of rivers, lakes, wetlands and reservoirs influence the populations and interactions between them, superimposed on dynamics of their physical performance in ecosystems) and conversely, how ecological ones may subsequently regulate hydrological ones (e.g. how distribution of vegetation in a catchment affects the hydrological cycle by modification of evapotranspiration and runoff at basin scale, riparian vegetation and debris dams in headwaters and floodplain wetlands in lower reaches of rivers regulate discharge timing). It then integrates the knowledge of those two processes and uses it to find innovative solutions to the problems of river basin degradation caused by our society.

Ecohydrology is thus a sub-discipline of hydrology, dealing with the ecological aspects of the water cycle. It is based on the assumption that hydrological processes are the major abiotic drivers of natural ecosystems – the main templet (Southwood, 1977) of lotic systems for example (Zalewski and Naiman, 1985; Poff and Ward, 1989). Thus the authors in this book use the term 'Ecohydrology' as opposed to 'Hydroecology', because they see the need to understand how hydrology regulates ecology (and how biota affect local hydrology), in order to utilize that knowledge to achieve the optimal management of ecosystem capacity.

Eighty-three per cent of the land surface of this planet has been modified by the engineering activities of human beings (Meybeck, 2003), in most cases without any understanding of how the impacts of those modifications have changed processes in natural ecosystems, such as biogeochemical cycles and energy flows. Virtually all those modifications have had direct and indirect impacts upon disturbances of the water cycles, sometimes both locally and globally. The technical benefits that many of these modifications brought (e.g. power and drinking water from dams, food from irrigation, treatment of sewage) have not been shared equitably around the world and, as a result, access to safe water is a major one of the UN Millennium Development Goals (United Nations, 2005).

Ecohydrology addresses solutions to the global problems of the 21st century through enabling scientists to identify how they can use the ecological regulatory processes to increase the assimilative capacity of ecosystems (Zalewski, 2000), particularly their change in resilience and adaptation in the face of global changes (e.g. Chapin *et al.*, 2006).

Ecohydrology is a thus a new way of thinking and acting for scientists, most of whom are, even today, usually sectorially-educated. It offers insights into environmental processes, integrating water resources and ecosystem sciences, based not only on hydrology and ecology but also considering molecular biology, genetics, chemistry, as well as socio-economic and legal sciences. It proposes a new scientific paradigm, or way of thinking and seeking solutions to the many new problems that humankind's development has created. Most of the problems have been caused by the way in which society has sought to utilize and 'tame' nature over the past few centuries. The utilization has now frequently become over-utilization (e.g. over-fishing, soil erosion) and the 'taming' (e.g. rivers for flood control, dams for hydropower and water supply) has resulted in simplification for single goals, which has now led to major losses elsewhere (e.g. loss of floodplain assimilative capacity for water quality and quality after its disconnection from the river channel) (Fig 1.1).

Practicing science in an interdisciplinary fashion is necessary for achieving the acceptance of a regulatory science by society, which itself is the only way to achieve sustainable development goals. It has been seen as the necessity for maintenance of scientific research funding; is expressly stated within the Millenium Development Goals; is essential for ecosystem rehabilitation and restoration (Hulse and Gregory, 2004) and has been identified as a need in a critical review of ecohydrological and related concepts (Hannah *et al.*, 2004). It is a natural stage in the evolution of a new paradigm, as will be shown in this book. The International Council for Science (ICSU, 2005) has identified key elements of 21st century sciences to be integrative, problem-solving and interdisciplinary. Ecohydrology provides these three elements.

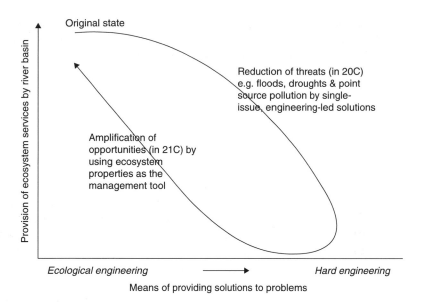

Fig. 1.1. Ecohydrology principles in the context of decision-making theory. Modified from Zalewski *et al.* (1997).

The Definition(s) of Ecohydrology

In its development, ecohydrology has been given several different definitions; none of them yet in hard-copy dictionaries. Its origins are clear - it comes from *eco*, *hydro* and *logy*, derived from the Greek, *oikos*, *hudôr* and *logos*, respectively meaning house, water and science. Therefore eco-hydro-logy is the science of water and ecology. It has not yet though achieved a single definition established by agreement or by common usage. Its range of definitions is most easily found through an internet search-engine, or a keyword search on a science database. Three subject areas appear. The first is focused upon plant–water dynamics on land – ranging from a single species, through vegetation type to a landscape and its (micro)climate (Baird & Wilby, 2001). The second is more connected with quantities in the water cycle and the impact of changes in quantity upon ecology in rivers (Acreman, 2001). The third meaning is inclusive of the subjects addressed by the first two and advocates an integrated vision of physical and biotic processes driving the dynamic evolution of river basins (Zalewski, 2006); it is this one that has driven the production of this book.

Two terms have been commonly involved in these three areas of development – ecohydrology and hydroecology. These and the three contexts in which they had been used, were given a thorough review by Kundzewicz (2002a). He made the point that any new concept passes through a phase of multiple uses of its descriptive terms. In describing ecohydrology, he wrote:

> It is indeed being *in statu nascendi*, offering scientific challenges galore, room for excitement and dynamic development. It will take some time before the notion ripens and a broad consensus is reached.

The value of the term was also debated in an issue of the *Hydrological Sciences Journal* in the same year (Kundzewicz, 2002b). Bonnell (2002) concluded his discussion with the comments, 'Thus the ecohydrology umbrella is providing a means of integrating landscape hydrology with freshwater biology, and this is the important paradigm shift.'

The term, as it is defined in this book, is the most widely used in the scientific world. It was first defined by Zalewski *et al.*, (1997) in the technical manual introducing it as a sub-programme of the 5th UNESCO International Hydrological Programme (IHP) initiative. As a consequence of the Programme's development, a more appropriate definition now is:

> The quantification and modelling of the dual regulation of biota by hydrology and vice versa within a basin, understanding their modification and synergistic integration in order to buffer man-made impacts with the ultimate goal of preserving, enhancing or restoring the capacity of the basin's aquatic ecosystems for sustainable use.

The concept that this definition espouses, falls into four linked parts:

- It is first of all *a scientific approach,* which can generate testable hypotheses and hence offers transparent rigour about the *dual regulation.* It represents a key to the interpretation of empirical data and models.
- Second, the concept is used to *promote integrative science* – hydrology and ecology – *for problem-solving* between professional scientists, decision-makers and stakeholders, at a river basin scale.
- Third, the integration directs the management of natural processes in the water cycle within the basin to try to *enhance the resilience, resistance and adaptation of aquatic ecosystems.*
- The ultimate goal of this management is to provide *sustainable use of a river basin's natural resources* for the benefit of nature and its human population.

The Need for Ecohydrology

Every country in the world has problems of water allocation because there is too little, of the right quality, in the right place. The government of every country in the world recognizes this, though few are taking action, many recognize it but are taking limited action, and too many recognize it but are doing nothing because the timescale of the solution exceeds the timescale of government rule. UNESCO (the lead agency in science) and UNEP (the lead agency in environment), among the major inter-governmental organisations addressing environmental issues, had recognized the problems of water scarcity and allocation in their support for the ecohydrological approach. Implicit in this support is the recognition that past water management approaches are no longer appropriate or effective for the 21st century, and new approaches are needed.

Former water management consisted of capital-intensive, high-technology, engineering-based schemes (collectively known as 'hydro-technology'). Construction of dams, water diversion, flood relief, agricultural development implying drainage and irrigation, and sewage treatment schemes all fit this description.

The schemes almost always tried to control, or at best to exploit, the natural elements of the water cycles rather than work with them. They almost always resulted in more widespread and unpredicted deterioration of the natural components of the water cycle. Individually-small perturbations can be seen almost everywhere on the globe; for example on a very small scale, the eutrophication of Esthwaite Water in the English Lake District was begun by the construction of the sewage treatment works in the village of Hawkshead, when a piped water supply was first laid into the village in 1923 and flush toilets replaced earth closets (Pennington, 1981). Individually-large and catastrophic examples are not so ubiquitous and not so obvious to the 'ordinary person', implying that such mistakes continue to be repeated; for example the damming of rivers in Africa or the salinization and drying up of the Aral Sea in Central Asia (shared by Uzbekistan and Kashakstan) through the diversion of Amu Darya and Syr Darya tributaries to help grow cotton. The Aral Sea has shrunk by more than half of its surface (68,329 km^2) and by 75 per cent of its volume after 90 per cent of the natural inflows were diverted to irrigate massive cotton monocultures, originally extending for more than 7 million ha in the USSR (Kindler and Matthews, 1997; SIWI, 2001).

Most 'hydro-technical' schemes fall closer to the Esthwaite Lake example than the Aral Sea one, so most people are not aware of their individual negative impacts. In most cases too, the proponents and the constructors were not, at the time, aware of the extent to which their activities would disrupt the homeostasis of the natural ecosystems. Collectively though, small impacts have added up and cause both environmental and health hazards. There is no longer a single lowland river in England unaffected by nutrient enrichment due to diffuse pollution and treated sewage effluents (Muscutt and Withers 1996; Demars and Harper, 2005a). Neither is there one un-impacted by discharge regulation through direct abstraction or by land drainage schemes (Brookes, 1995). The 'vicious circle', set by technologically-intensive environmental solutions, producing unpredicted environmental damage is drawing to an end however, with the growing realisation that the human population will exceed the total carrying capacity of the planet very soon (Cohen *et al.*, 1995). Every small decision, at every governmental level, affecting land, water and air, now needs to be taken with the true application of the principles of 'sustainability' in mind – using the natural properties of an ecosystem's capacity to absorb human impact and to mitigate damage. The effort directed at environmental change for maximising nature and human benefit must be turned on to a more 'precise' understanding of natural processes and a better perception of the risk engendered by resource exploitation.

The Genesis of Ecohydrology

Ecohydrology was born in the UNESCO stable as part of the 5th IHP (International Hydrological Programme, 1996-2001). This was in one sense a response to the formal statements arising from the Dublin Conference on Water and promotion of Integrated Water Management (Solanes and Gonzalez-Villarreal, 1999) but in another sense it represented the intellectual development of the

UNESCO Man and the Biosphere (MAB) Programme. MAB, launched in 1971, had quickly realized the importance of human impact on aquatic systems, as it was reflected, among others, in the *Land Use Impacts on Aquatic Systems* project of the MAB programme (Jolánkai and Roberts, 1984). This importance was translated into a determination to understand the potential role of sub-systems on buffering the worst effects of human impact, resulting in the 5-year MAB programme *Role of Land/Inland Water Ecotones in Landscape Management and Restoration* (Naiman *et al.*, 1989; Naiman and Decamps 1990; Zalewski, *et al.*, 2001). The final meeting of the Ecotone project in 1994 concluded that an important development would be the integration of ecology and other sciences, a natural development of earlier integrative initiatives within UNESCO. Consequently, the lessons from the Ecotone project, which had laid emphasis on ecological issues, became incorporated into the needs of the new IHP-V project, which saw the close cooperation between freshwater ecology, geomorphology, hydrology and water engineering as a central component to make ecohydrology the holistic 'tool for the sustainable management of aquatic resources' (Zalewski *et al.*, 1997) that is the subject of this book. Throughout this period, key staff members of the UNESCO IHP programme had the vision and the intellect to drive the evolution of the concept forward, by providing support for conferences, meetings and subsequent publications.

Development of integrated thinking about water resources occurred concurrently with the recognition of the importance of the meaning of 'sustainability' before and after the 1992 United Nations Conference on Environment and Development (UNCED, known as the 'Rio Convention') (Membratu, 1998). Although recognition of our general environmental deterioration had started before Rio – it can be traced back to varied sources such as 'Silent Spring' by Rachel Carson in 1963 and the 1972 Stockholm Conference (Meadows *et al.*, 1992) – the specific impact of this deterioration on water resource availability was highlighted by the Dublin Conference on *Water & the Environment* early in 1992 that helped to prepare for UNCED in Rio. Dublin was followed by the Paris conference, 6 years later – *Water, a Looming Crisis?* (Zebidi, 1998).

The community of aquatic scientists had also been moving along several parallel routes towards an integrated approach to aquatic management, for a decade prior to the first use of the term ecohydrology in this context. Academic conference titles since the early 1990s show this: for example *Hydrological, Chemical and Biological Processes of Contaminant Transformation and Transport in River and Lake Systems* (Jolánkai, 1992), *Habitat Hydraulics* (LeClerc *et al.*, 1996), *Hydro-ecology* (Acreman, 2001), *Environmental Flows for River Systems* (Petts, 2003), *Aquatic Habitats: Analysis and Restoration* (Garcia and Martinez, 2005). These meetings and proceedings encouraged aquatic scientists to work with their neighbouring disciplines. Proceedings showed the beginnings of integration of ecology with hydrology, through 'habitat hydraulics' or 'ecohydraulics' or 'environmental flows' – the terms given to the allocation of flows in rivers or releases from reservoirs for the maintenance of aquatic habitats and life.

Over the same time period, policy was moving in the same direction. The new democratic South Africa enshrined the concepts of 'water for people' and 'water for nature' in one of the first pieces of legislation in 1996. The European Union's Water Framework Directive (WFD), by inviting a strategy for dealing with cumulative basin impacts rather than focusing on pre-determined discharge-point limits, instituted a strong regulatory principle calling for integrated river basin management. The management of fluvial hydrology is at the heart of the WFD, as stated in Article 1, which addresses the hydrological needs of aquatic, terrestrial and wetland ecosystems (1a) as well as floods and droughts (1e). Several guidance documents, defined under the Common Implementation Strategy phase of the WFD, such as the 'Wetlands Horizontal Guidance' and 'Impacts and Pressures' have highlighted that. Even though the WFD does not cite ecohydrology explicitly, it clearly expresses a concrete scope for its implementation. The development of ecohydrology as a management tool for guiding sustainable water management in the UK has moved on from just being a tool for dealing with flow regulation issues ('Hydro-Ecology') into one fully addressing the need to achieve 'good ecosystem quality' under the WFD (Environment Agency, 2004).

Thus integrated ecological and hydrological thinking, regardless of its name, was moving in the same direction – the need to understand human impacts and find ways of mitigating or reversing them. Integration is a precondition necessary for an ecohydrological approach now enshrined in modern water management terms – the *Catchment Management Planning* (CMP) concept and particularly the *Integrated Water (or River) Basin Management* (IWBM) concept, which is being promoted by the major international environmental NGOs such as IUCN (as *Water and Nature initiative*) and WWF (as *Living Waters Programme*). Ecohydrology is the major means of achieving Integrated Water Basin Management, within what is now called *The Ecosystem Approach* (IUCN, 2008).

The Scientific Evolution of Ecohydrology

The principles of ecohydrology have evolved since the late 1990s in such publications, promoted by UNESCO support, as: Zalewski (2000); Zalewski and Harper, (2002); Zalewski *et al.*, (2004). They have focused upon the links between the disciplines and the use of low-cost ecological technologies for the management of wetland, instream and riparian plant communities. These were first given the term *Ecological Engineering* (Mitsch and Jørgensen, 1989) and have now become an integral part of ecohydrology, termed 'phytotechnologies'. The linkage gained support from the United Nations Environment Programme (UNEP), in recognition of phytotechnologies' widespread global value as a low-cost sustainable solution to mitigating pollution on land and water (Zalewski, 2002; Zalewski *et al.*, 2003; Zalewski, 2004). The range of techniques that can be enhanced through ecological engineering to increase the ecosystem services that river basins can provide is illustrated in Fig. 1.2 and 1.3.

The scientific development of ecohydrology can be traced to two major theories at different scales – that of the 'ecosystem' and that of 'Gaia'. Both of them describe the emergent properties of groups of interacting living organisms. Both have sought the analogy of a 'super-organism' to aid understanding.

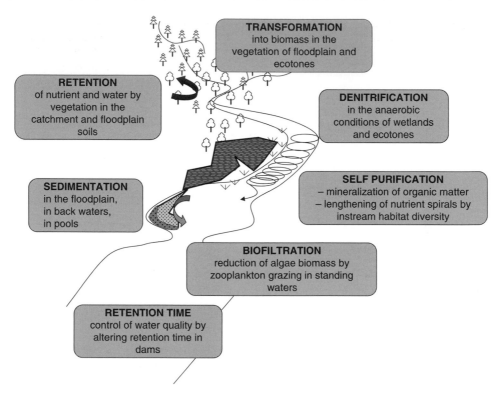

Fig. 1.2. Diagramatic representation of ecohydrological processes, which, in different parts of the catchment, can be used to control hydrological and hydrochemical ones.

The ecosystem

The smaller scale and earlier concept, that of the ecosystem, comes from a term coined by Arthur Tansley (1935) but reflecting 40 or so years of earlier thinking about the linkage between biological systems and their environment. The ecosystem was proposed as an alternative to the super-organism concepts of vegetation that had been proposed by Clements (1916). The simile of a super-organism has been used repeatedly during the 20th century to describe a group of organisms having the physiological properties of a single organism (Lincoln *et al.*, 1982). It has difficulties if it is used literally, because an individual organism is the template for natural selection to operate upon; but it does have value if used metaphorically to assist the understanding of a concept with emergent properties, such as Ecosystem or River Basin.

An important theory in support of the earliest thinking about ecosystems, was the apparent homeostasis in lakes created by its feedback loops in, for example, nutrient cycling. This can be traced back to the last few decades of the 19th century, when Forbes (1887) envisaged a lake as a 'microcosm', where 'a balance between building up and breaking down [occurs], in which the struggle for existence and natural selection have produced an equilibrium'. The

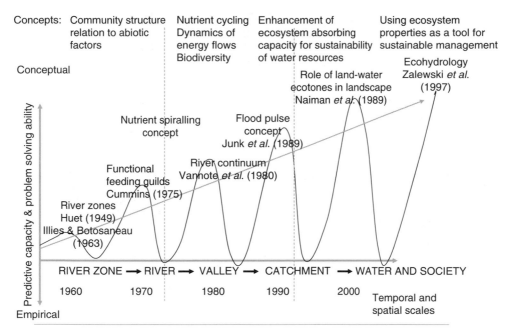

Fig. 1.3. Diagrammatic representation of the evolution of ecohydrology as an integrating concept in water science and management from earlier aquatic science concepts, which enabled the increase in scales and predictive capacity.

understanding of nutrient cycling brought ideas from geochemistry into ecological thinking, such as those of Lotka (1925), who envisaged the earth as a single system whose parts, linked by chemical changes, were all driven by solar input. Important progress was made by Lindeman (1942), who clearly showed the abiotic-biotic linkages in a bog-lake ecosystem, through interpretation of the food web in terms of energy flow through the different trophic levels. He interpreted ecological succession in terms of energy transfer efficiencies. Odum (1953) subsequently placed the ecosystem concept and the link with biogeochemical cycles firmly in the forefront of ecological thinking in his textbook, *Fundamentals of Ecology*, which had a major influence on the discipline of ecology over the following 30 years. Thomas Odum (1971, 1983) developed Lotka's physical approach much further, making the link with physical laws clear by showing how ecosystems operate under thermodynamic laws. This thinking has attracted continuous interest (e.g. Jorgensen and Kay, 2000; Jorgensen and Svirezhev, 2004)

Odum (1969) established ecosystem characteristics (covering community energetics and structure, life history, nutrient cycling, selection pressure and overall homeostasis). He made predictions of the internal trends to be expected in the succession of an ecosystem through its development stages, mature stages. He later included predictions for stressed ecosystems (Odum 1985).

Both Eugene and Thomas Odum had always integrated man into their concept of the ecosystem (Odum, 1971, 1983,). Eugene Odum's predictions of anthropogenic impacts (Odum, 1985) were tested by Schindler (1987, 1990) in whole lake experiments, which found that either the lake's recovery from stress (nutrient enrichment or acidification) was much slower than its initial degradation or that the ecosystem's dynamic equilibrium state had changed.

The importance of the energy that does not enter the biotic component, in structuring a lake ecosystem through its effect upon stratification and mixing, had already been recognized by Juday (1940). The importance of biotic influence upon energy flow was given impetus by the treatise of Gates (1962), a physicist. Other physical scientists were able to understand larger scale processes of river basin development as well as just individual lakes, through the application of thermodynamic principles (Leopold and Langbein, 1962). Several ecologists had argued for the integrity of the basin for much of the 20th century as part of the intellectual debate about the ecosystem (Golley, 1993).

Ecological theory was taking longer timescales into its frame of thinking as well as incorporating the physical driving laws, linking the processes of ecosystem succession with complexity and stability (Margalef, 1960, 1963). Spatial scales were seen to be important influences both upwards as well as downwards; Connell (1978) showed that instability on the point scale may result in a high stability in the total system. Natural selection, operating at the level of the individual, had earlier appeared to contradict these theories of ecosystem properties, but Southwood (1977) developed a new concept of how natural selection operates on an organism through its habitat. This Habitat Templet concept (Southwood 1988) hypothesized that the spatial and temporal gradients provide the frame, upon which evolution forges characteristic assemblages of species traits. The habitat of the species can be characterized by two gradients: the spatial heterogeneity (adversity gradient) and the temporal variability (disturbance gradient) of the environment (Townsend and Hildrew, 1994). Both gradients can be characterized by three components defining the realized niche (Hutchinson, 1970) of the species: biotic (competitors, consumers), chemical (alkalinity, nutrients, pollutants) and physical (geomorphological).

O'Neill (2001) recently questioned the ecosystem concept and suggested that, although the metaphor of the super-organism has been replaced by the machine analogy to facilitate communication of ecology to the public, the scientific paradigm needed to be rejuvenated. He saw the most serious scientific gap to bridge was an explanation of the stability of the ecosystem (self-regulation) without using the old concept of 'Balance in Nature' and with the integration of more evolutionary biology. His last two paragraphs about ecosystem theory are important for ecohydrological thinking:

> Perhaps the most important implication involves our view of human society. *Homo sapiens* is not an external disturbance, it is a keystone species within the system. In the long term, it may not be the magnitude of extracted goods and services that will determine sustainability. It may well be our disruption of ecological recovery and stability mechanisms that determines system collapse.
>
> Certainly, we don't want to dismiss the current theory prematurely. But we must understand that the machine analogy is critically limited. In so far as the local system maximizes environmental potential, it necessarily sacrifices stability when that potential changes. The challenge to the ecological system is optimization to a

moving target. Optimize too rapidly and the system is trapped in a local attractor and, like an overspecialized species, cannot adapt when conditions change. So it would not be wise to send the old dobbin to the glue factory before we determine how well the new one takes the bit. But it certainly seems to be time to start shopping for a new colt.

(Robert V. O'Neill, 2001)

Ecosystem and super-organism concepts have now been incorporated into ecological risk assessment guidelines, which developed in the United States during the early 1990s (USEPA, 1998). Even if several basic differences do exist, in many ways ecological risk assessment is perceived as an extension of risk analysis methodologies developed to protect human health. A super-organism analogy can be perceived in the definition of essential ecosystem functions, 'ecosystem physiology', and in the environmental destiny of multiple stressors being transferred between different matrices (air, water, soil, sediment) in a similar way to substances transferred between the organs of a single organism. Risk-based thinking has become a central tenet of modern environmental management (as in, for example the EU Water Framework Directive) to which ecohydrology is contributing its vision of river basin unity and its attention to habitat integrity.

Gaia

The higher theory within which ecohydrology fits, is the Gaia theory of planetary self-regulation. This, first proposed by Lovelock (1972, 1979, 1988), explained that the homeostasis of the Earth's atmosphere was maintained by the negative feedback activities of the biosphere, because it was far from thermodynamic equilibrium. The theory was at first heavily criticised for three reasons, because:

- Lovelock and Margulis (1974) had ignored much earlier scientific works putting forward the idea of the influence of the biota on its environment (e.g. as early as Spencer 1844, Huxley 1877),
- they did not fully foresee the homeostatic, teleonomic and optimising implications of their theory (Kirchner 1989), and
- the metaphor of the Earth (Gaia) seen as a living entity, which did not reproduce, was not compatible with neo-Darwinism, even if the metaphor was illustrating the second law of thermodynamics (Lotkla 1925; Schrödinger 1944).

Patten and Odum (1981) tackled the teleonomic epistemological gap to defend the view that ecosystems are cybernetic systems (c.f. Margalef 1968, Odum and Odum 1971) and rejected the idea of the super-organism, in response to Engelberg and Boyarsky (1979). Lovelock was himself heavily criticized at the American Geophysical Union's Annual Chapman Conference, in March 1988, dedicated to Gaia (Kirchner 1989; Schneider & Boston 1991). His theory then had to integrate the neo-Darwinist criticisms to survive and try to demonstrate that emergent properties may arise from biota at the global level and regulate the atmosphere. This it did (Watson and Lovelock 1983; Lenton 1998; Lenton and Lovelock, 2000).

Many books have been written about Gaia. This has strengthened its credibility (e.g. Williams 1996, Volk 1998) and applicability (Bunyard 1996). Recently Dagg (2002) suggested a common ground for *The Extended Phenotype* (Dawkins 1982) and *Gaia* (Lenton 1998) as a result of emergentism.

The Role of Aquatic Sciences

Lake ecology had made major contributions to the development of ecological theory in the early–mid-20th century, and running water ecology soon caught up. Thirty years ago, the beginnings of understanding of the consequences of flow for riverine ecosystem structure (Cummins, 1974), led to a major step forward in integrating physical and ecological processes in running water (Vannote *et al.*, 1980), the River Continuum Concept (RCC). This also made sense of an earlier, more descriptive phase of aquatic ecology, river zonation, which had been based on fish communities (Huet, 1954) and the typology of river stretches (Illies, 1961; Illies and Botoseanu, 1963) summarized by Hynes (1970) and Hawkes (1975). The RCC did this by integrating the physical driving forces from source to mouth with these biological zones in a river:

> 'in natural river systems biological communities form a temporal continuum of synchronized species replacements following the flow from the spring to the river mouth'
>
> (Vannote *et al.*, 1980)

Some aspects of the continuum, particularly the spiralling behaviour of nutrients in rivers, had already been suggested (Webster and Pattern, 1979) and were elaborated shortly after the RCC (Newbold *et al.*, 1983). Subsequently the RCC was reshaped and extended to encompass broader spatial and temporal scales (Cummins *et al.*, 1984; Minshall *et al.*, 1985). The role of woody debris in holding back river discharge (Triska, 1984) and influencing floodplain structure on temporal scales for up to hundreds of years, was the most important ecological regulatory process highlighted. This triggered further research and many investigations have been devoted to this subject since then (Harmon *et al.*, 1986; Gurnell *et al.*, 1999; Robertson and Augspurger, 1999). The RCC had not explicitly initially addressed human impact, but the Serial Discontinuity Concept (SDC) by Ward and Stanford (1983) was a way of understanding the magnitude of disruption, initially by dams. In larger rivers the influence of the river floodplain on the main channel in the lower reaches was not fully included in the RCC and as a result the Flood Pulse Concept (FPC), was formulated – initially for the largest flood plain system of the world, the Amazon and its basin (Junk, 1982; Junk *et al.*, 1989). More recently, its ideas were extended to include temperate floodplains (Tockner *et al.*, 2000). Although flooding has major consequences for floodplain functioning and productivity at all latitudes (Bayley, 1995; Zalewski, 2006), the flood pulse effect is particularly relevant in the tropics due to temperature coupling, i.e. the seasonal convergence of high discharge and high temperatures, which maximise biotic processes such as fish development and biomass accumulation (Junk *et al.*, 1989; Junk, 2000; Junk and Wantzen, 2003). The Riverine Productivity Model (RPM)

(Thorp and Delong, 1994) identified the different sources of organic carbon in: (i) local autochthonous production; (ii) direct inputs from the riparian zone, and; (iii) instream primary productivity, so placing the floodplain's influence in material cycling in the context of its whole catchment.

Attention was increasingly directed to the ecological problems in rivers that were heavily modified by regulation and human use of water resources, which highlighted the loss of connectivity between the main river channel and the aquatic habitats in the flood plain. The way connectivity was impaired by regulation measures was highlighted by Ward and Stanford (1995) and the impact of engineering works shown by Bravard *et al.* (1986). The role of deposition resulting from geomorphic processes and ecological succession in disconnected side channels was described by Petts and Amoros (1996). Bornette *et al.* (1998) had described in detail how the diversity of aquatic plant life is linked to the level of connectivity in flood plain water bodies. Most recently Demars and Harper (2005b) found that hydrological connectivity along heavily-modified lowland rivers, and isolation between rivers, better explains aquatic plant distribution than did local (artificial) environmental conditions such as impoundments created by water mills.

Many hydrologists and environmental engineers had initially moved towards ecology through recognition of the wider importance of nutrient and sediment loading (Vollenweider, 1970; Wischmeier and Smith, 1978; Vollenveider and Kerekes, 1981; OECD, 1982). Even in the 19th century however, Hungarian water engineers had termed the obligatory release flow from dams into rivers as the 'living water', indicating, at least in terminology, that they wished to preserve life in the rivers. Modelling of freshwater systems was triggered by Odum´s (1957) study on the energy budget of a large spring system and quickly followed the development of computing power (Imboden and Gachter, 1975; Lorensen, 1975; Jörgensen, 1983), leading to an holistic approach to the processes of the entire basin (Jolánkai, 1983; Thornton *et al.*, 1999).

The largest challenge that integration of ecology and hydrology faces is that of scale. Numerical dimensions of sampling sites, expressed by terms like point, catchment, meso-, macro-, or mega-scale are not at all defined within each field, and even more numerical deviation will be found if projects in ecology and hydrology are compared. Hydrological elements used by Sloane *et al.* (1997) were about 100 km^2 in size. It is rare for limnological investigations to deal with minimum element areas of this size. Even by trying to come closer to biological dimensions, Sloane *et al.* (1997) remained short of a real biological view of the environment. The same applies to the meso-habitat scale of Bovee (1996) who tried to reduce the need for 'high precision sampling techniques', since useful ecological information, like species composition, is very often only available by direct observation on the spot and cannot be worked out by more remote techniques. Structural ecological elements like ecotones are usually below the size of typical hydrological point scale investigations. Highly aggregated biological structures in a fluvial system like corridors, ecotone complexes or a mosaic of habitat patches may just reach the meso-scale of hydrology.

The linking of hydrological variables with biological properties at the appropriate scale is the key to the uptake and the success of ecohydrological thinking.

Evaluating the ecological health of a river basin, in terms of its deviation from a supposed 'natural state', has been an important goal for water science and management since the late 1970s. Methods have been developed which, on their own, evaluated a part of river health, but in combination and through eco-hydrological thinking, can now provide a greater value than the sum of the parts. Within a hierarchy of scales, the added value of ecohydrology is in its emphasis upon the range of scales within which the biota can cause feedback regulation of the hydrological processes. This integration constitutes a fundamental forward step from understanding the more obvious regulation of ecological processes by hydrology. It is illustrated diagrammatically in Fig. 1.4 by the range of ecological and technological measures being implemented and planned on the Pilica river in Poland to mitigate high nutrient loads, which have rendered a downstream reservoir unfit for human domestic consumption.

The most appropriate scale for management activities is a middle scale since, at this spatial scale, it is possible to know the appropriate morphological or flow schemes that will control the local hydraulic heterogeneity, and thus the river's ecological processes. In fluvial systems, this approach allows a shift of observational targets from instream biological characteristics (i.e. diversity, trophic guild ratios) or processes (i.e. production/respiration) to physical characteristics that can be easily measured and with high precision. The establishment of such relationships provides the optimum opportunity for the ecohydrological

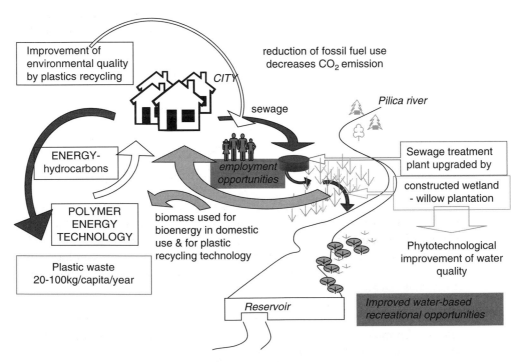

Fig. 1.4. Diagramatic representation of the development of ecohydrological methods in the Pilica River basin, central Poland, to reduce nutrient load to the downstream reservoir, reduce carbon emissions and increase employment opportunities.

management of water courses; mesohabitats (or their key hydraulic variables), can serve as monitoring tools within ecological assessment procedures. Biological assets can be improved by restoring the physical heterogeneity through direct substrate re-naturalisation or through the design of adequate flow management schemes.

At catchment scale, the integration of hydrological regimes within an ecosystem context is at the heart of the Continum Concept of Vannote *et al.* (1980). In certain river basin types, advances have been made in understanding and quantifying this integration. The best examples of processes in near-natural condition river basins, probably come from studies of undisturbed basins in Boreal latitudes (Naiman *et al.*, 1986, Helfield and Naiman, 2001; Naiman *et al.*, 2002; Helfield and Naiman, 2006), while the heavily modified rivers of Eurasia provide an example of the opposite extreme. There are no explicit procedures relating hydrological regime descriptors to anthropogenic changes in an ecosystem (Richter *et al.*, 1996), but different practical methods for evaluating aspects of ecological health fit within a hierarchy of physical scales.

Ecohydrological principles can be implemented at a range of different scales through the development of integrative scientific methods, enabling new directions of research and management as well as a reinterpretation of former concepts. The value of ecohydrology is surfacing in regional and national legislation across the world. In Europe it is occurring through the development of methods designed to achieve 'good ecological status' as required by the EU Water Framework Directive (2000/60/EC); in the USA through methods designed to achieve 'biotic integrity' in implementing the Clean Water Act and the Water Quality Act; in South Africa through methods designed to achieve 'the ecological reserve' in implementing the National Water Act (Mackay, 1999) and in Australia through methods designed to achieve 'ecological flows' in implementing the Water Act and Landcare programmes.

The Purpose of This Book

It is the goal of this volume to promote the integration of ecology with hydrology to readers of either disciplines and to provide enough theoretical basis to demonstrate to the reader that ecohydrology not only works scientifically, but is also the concept most likely to deliver a 'concrete advancement', out of the many buzz-words in the first 5 years of the 21st century (Kundzewicz, 2002a). This volume provides support for the more practical Manual of Ecohydrology, published by UNESCO and UNEP (Zalewski and Wagner-Lotkowska, 2004) and its Guidelines (Zalewski, 2002).

What does all this imply? First of all, that ecologists understand the hydrological and chemical drivers at the basis of the structure and function of aquatic systems (Chapters 2–7). Equally, it means that hydrologists understand the ecological consequences of natural and modified flows, both under their quantity and quality aspects (Chapters 8–10).

Scientists of all disciplines nowadays rely on modelling for understanding the present and predicting the future, but both scale and accuracy must be appropriate to their needs. This is particularly so at the 'sharp-end' of ecohydrology, which is the practical use of ecosystem processes to regulate habitat properties through hydrology. The ecohydrological concept only works when the spatial scale of a catchment is firmly kept in mind, even where the practical application may be larger or smaller, in order to achieve the regulatory feedback desired (Chapters 11–14). The temporal scale is arguably the most important for our society: increasing numbers of scientists and planners, using increasingly sophisticated models, are trying to predict future global changes. Understanding ecohydrological processes from the past is an important pointer to interpreting the future (Chapters 14–15).

Concluding Remarks

The ecohydrology concept is held together by a number of scientific threads, such as the ecosystem definition, the emergent properties across spatial and temporal scales and the metaphor of a super-organism. This super-organism metaphor, which had been used repetitively in the past – for example by the Greek philosopher Plato (c.429-c.347 BC), the polymath Leonardo da Vinci (1452–1519), and the geologist James Hutton (Hutton 1788) – was revisited during the latter half of the 20th century with the ecosystem and the Gaia concepts (Odum and Odum, 1959; Odum, 1969; Lovelock and Lodge, 1972; Lovelock, 1972). It might now be the key to unlock society's understanding of the urgent need for ecohydrological solutions in the crowded, warmer, world of the 21st century. So too might the recently-discredited term 'The Balance of Nature'. This concept came from the 19th century (Humboldt, 1845, 1847), lost favour but re-appeared in scientific literature in the second half of the 20th, when nature became likened to a living entity.

Metaphors are still used successfully today in science for example, *The Red Queen* (van Valen, 1973) and *The Selfish Gene* (Dawkins, 1976), but science is not only about metaphors. Ecohydrology seeks to bridge the sciences, arts and society to achieve its ultimate goal: the sustainable management of river basins. In doing this it has to follow the pace of the Red Queen (Carroll, 1871, van Valen, 1973), so that its principles are continually revised as sciences, arts and society move on.

It is clear that the next phase of the development of ecohydrology must be to better engage with people ('stakeholders') and policy and politics ('decision-makers'). It has been suggested that, in order to do so, ecohydrology must develop into a unified science from sociology, hydrology and ecology (Hiwasaki and Arico, 2007). That is not the view taken by the authors of this book. We all seek to engage with people in nations, river basins and local communities, in order to achieve meaningful and sustainable management of their water resources. But to do so, it is not necessary to reconstruct disciplines that already integrate well due to their common physical science base. However, it is very necessary to use a common language and to seek new ways of communicating about problems (such as films, see the UN University virtual field course web site), which coherently link disciplines, and in doing so provide solutions. This is illustrated in Fig 1.5.

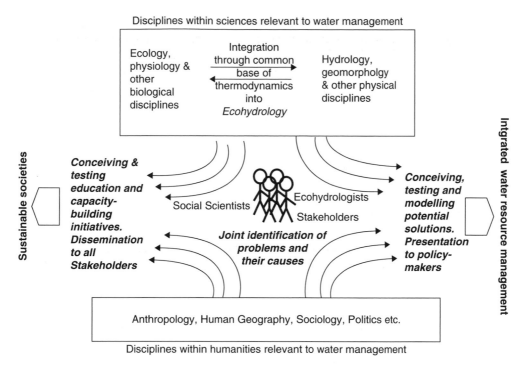

Fig. 1.5. A conceptual diagram showing the way in which ecohydrology must develop over the next decade in order to become a globally effective tool for sustainable water management, by engaging social science disciplines and educationists in a trans-disciplinary fashion to produce unified, integrated solutions to water resource problems and coherent educational tools.

The Preface of the book presents a Cibachrome photogram, which was made without a camera by directly immersing, at night, large-scale sheets of positive photographic paper beneath the flowing water surface and with exposition to a flashlight. The River Taw (Devon, England) was used as a negative and the landscape as the dark room. Ambient light in the sky added colour. This photogram is only one moment extracted out of a whole time series of photograms capturing daily and seasonal changes from source to sea, capturing the interplay of the river and its environment (Derges, 1997, 1999). Susan Derges' work on the River Taw is only part of a whole continually evolving body of experiments with nature on wave and particles, the observer and the observed, growth and forms (Derges, 1985, 1999). This work is firmly rooted in ancient philosophy (e.g. Zen philosophy, Watchmann and Kruse, 2004) and natural history (e.g. Reclus, 1869), yet remains intuitive, creative and communicative. It represents one artist's vision of an aquatic ecosystem in a way that can be shared with many. Ecohydrology seeks to do similar things – it creates a vision of a sustainable aquatic ecosystem, which if shared with enough people, can be achieved in reality. We hope that you will enjoy this book and share our vision.

2 Patterns and Processes in the Catchment

D. GUTKNECHT[1], G. JOLÁNKAI[2] AND K. SKINNER[3]

[1]Austrian Academy of Sciences, Vienna University of Technology, Vienna, Austria; [2]Environmental Protection and Water Management Research Institute (VITUKI), Budapest, Hungary; [3]Jacobs Ltd, Sheldon, Birmingham, UK

Hydrological Patterns – a Link between Catchment Characteristics and the River Ecosystem

River ecosystems evolve under the influence of hydrological patterns that develop by the interplay of climatic and meteorological factors with the geomorphology, soils and vegetation characteristics of the catchment. This pattern depicts, in an integrated way, the overall effect of catchment-scale water dynamics. One example of the manifestation of these effects is the regime type (Gustard, 1992), defined as the temporal pattern characterizing the annual stream hydrograph. In an Alpine environment, regime types include snowmelt-dominated regimes with high flows in early summer, rainfall-dominated regimes with uniform flows with peaks throughout the year, and groundwater-dominated regimes with smoothly varying streamflow. From a stream ecology perspective, these characteristics are reflected in specific values of low flows, mean runoff and the level and frequency of flood plain inundation (Fig. 2.1). Insight into underlying processes explaining characteristic hydrological patterns can be gained by examining the ecohydrology of river catchments.

The interaction of ecological processes with hydrological ones produces relevant feedback loops. Among these, particular relevance is attributed to the dominating role of plant nutrients, especially phosphorus, which is well known but still not fully understood (Jolánkai, 1992; Tiessen, 1995; Haygarth, 1997), and to a series of hydrological processes that are decisive for the fate of solutes in aquatic systems.

The major processes driving the hydrological cycle, their effects on the fate of pollutants and thus on the evolution of ecosystems, can be summarized as follows (Fig. 2.2).

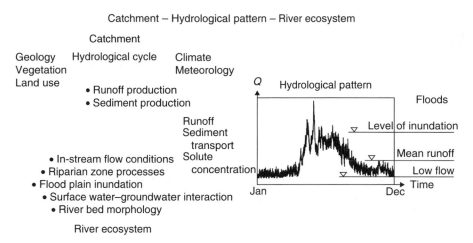

Fig. 2.1. Links between problems of ecohydrological interest, features of hydrological patterns and catchment processes.

Precipitation

Water, by evaporating from the ocean, from other water surfaces and from the land surface or by transpiring through vegetation, condenses into clouds that are displaced by the wind. Clouds release the water vapour as precipitation: rain, snow and hail.

Precipitation affects solute transport in one or more of the following ways:

- Rainfall and hail provide the kinetic energy for the detachment of sediment particles and sediment-borne contaminants.
- Precipitation determines the quantity of water available for further transport over and below the ground surface.
- Precipitation causes the wet deposition of airborne pollutants and generates the largest flux of total atmospheric pollution (Jolánkai, 1983, 1992).

Experimental evidence (Jolánkai, 1986; Wu *et al.*, 1989; Razavian, 1990), as well as theoretical models of particle detachment, proves that the intensity and duration of rainstorms determines the magnitude of solid particles exported per unit area of drainage basin. The higher the intensity of rainfall, the more kinetic energy is available; the longer the duration of rainfall events, the more water is available for further particle transport.

The amount and the spatial and temporal distribution of precipitation depend to a large extent on transpiration and evaporation fluxes associated with the terrestrial vegetation. The major global role played by vast tropical rainforests is familiar even to the general public; however, it is not as well appreciated that smaller-scale regional and even local effects might also be determined by vegetation. A more refined quantification of such processes (i.e. what amount of atmospheric

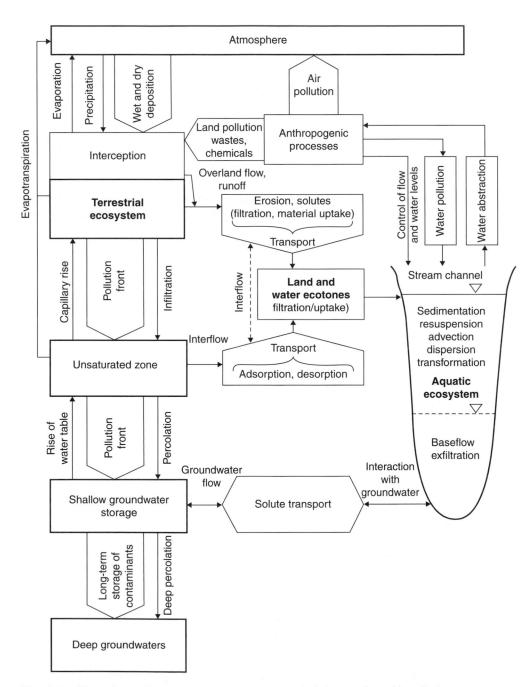

Fig. 2.2. Flow chart of hydrological processes and their interaction with pollution processes and ecology. (From Jolánkai & Bíró, 2001.)

moisture is produced by given plant communities under different substrates and climatic conditions) is still needed.

Runoff

All processes that generate runoff following precipitation events will generate transport phenomena. Three processes are the major ones:

- *Interception*. This is the portion of the precipitation flux that becomes intercepted and/or retained by the vegetative canopy. That portion not intercepted, either throughflow or stemflow, reaches the ground surface and becomes available for runoff and runoff-induced particle and solute export. Atmospheric pollutants are intercepted by vegetation under the form of both dry as well as wet deposition. This process creates significant feedback effects, since intercepted pollutants cause a deterioration of the vegetation resulting in increased runoff and hence erosion.
- *Evapotranspiration*. This is a combined term for the portion of precipitation that is returned to the atmosphere by evaporation from wet plant surfaces and by transpiration from soil–plant–water systems. On a global scale, evapotranspiration accounts for 70% of the annual precipitation flux, and thus it represents the dominant mechanism driving the hydrological cycle and runoff-induced transport processes. Transpiration is also a primary biological process through which plants regulate the uptake of water and nutrients.

Interception, evaporation and transpiration demonstrate the key role of vegetation as the most important factor governing and controlling hydrological cycling processes. Following rain events, by physical detention and biochemical uptake, the vegetation forms a barrier to surface and subsurface fluxes of nutrients. Recognizing these properties, most diffuse pollution control strategies are based on characteristics of the vegetation component of the terrestrial ecosystem.

- *Infiltration*. The portion of precipitation that does not evaporate becomes surface runoff or infiltrates into the ground. Infiltration water is then either returned to the atmosphere through transpiration or contributes to soil moisture storage, at near-surface levels as interflow, or enters the groundwater compartment. Infiltration determines the rate at which pollutants are leached into the ground, travel downwards to the water table or move with interflow towards further surface or subsurface recipients. The rate of infiltration defines the time available for chemical and biological processes that govern the composition of the soil leachate and of surface runoff.

Water storage capacities, characterizing a given portion of a drainage area within a certain period in time, will affect local runoff water quantities. Storage compartments can be subdivided into temporary interception stores, surface depressions and soil moisture.

Water Pathways Approach – a Hypothesis for Catchment Studies

For a selected spatial unit, the relationship between hydrological elements can be expressed mathematically in a time-varying or steady-state mode. The basic water balance equation establishes that:

$$P - ET - S - R = \Delta V_1 + \Delta V_2 \tag{2.1}$$

where

P = precipitation,
ET = evapotranspiration,
S = subsurface runoff,
R = surface runoff,
ΔV_1 = change in surface storage,
ΔV_2 = change in subsurface storage.

Given the interplay of the myriad of complex and interwoven processes and process controls, one may ask what could be the best way to describe and quantify this system. The hypothesis put forward in this chapter is that the water pathway is a useful pivotal point for describing and interpreting catchment processes. Consider as a starting point the local runoff generation processes (Fig. 2.3) where precipitation, vegetation, soil and bedrock characteristics are the key controls of runoff generation (Dunne, 1978). The dominant factors and

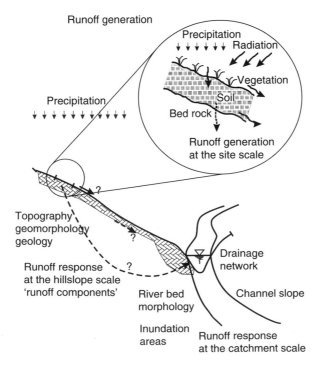

Fig. 2.3. A 'water pathways' perspective to catchment processes.

processes change as scale increases. At the hillslope scale, runoff moves laterally through the subsurface and the penetration depth controls the temporal characteristics of the runoff response (Gutknecht, 1996a). In hydrology, the importance of temporal characteristics (such as transit times, travel times, response times and hydrograph rise) led to the development of the concept of runoff components, which include direct and delayed storm runoff. Moving up to the catchment scale, stream-wide processes need to be considered. Here, the conformation of the drainage network, the channel slope, the flood plain morphology and the extension of inundation areas are important controls of hydrological patterns. Temporal streamflow patterns are scale-dependent and overall processes at different scales are dominated by different factors (Blöschl and Sivapalan, 1995).

Factor Groups

From a more quantitative perspective, there is a need for developing process-oriented modelling approaches that consider the interplay of all relevant factors leading to runoff response. These issues can be addressed by defining appropriate 'factor groups' (Fig. 2.4). A first factor group is centred on soil and vegetation processes (Fig. 2.4, upper right). This group involves both atmospheric and soil moisture forcing. The landscape exerts an important influence through

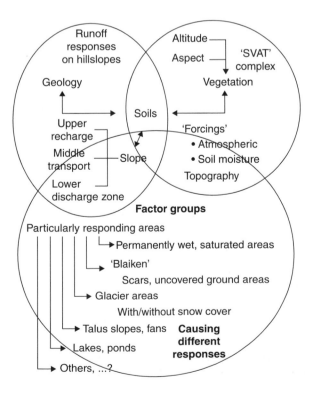

Fig. 2.4. Factor groups governing runoff generation.

altitude and aspect dependencies. These elements are common to the SVAT (soil–vegetation–atmosphere–transfer) complex that has recently gained increasing importance in the atmospheric sciences in the context of global climate modelling (e.g. Salvucci and Entekhabi, 1995).

A second factor group is centred on soils and geologic controls (Fig. 2.4, upper left) (e.g. Rulon *et al.*, 1985; Genereux *et al.*, 1993; Jenkins *et al.*, 1994). Here, landscape properties are interpreted by considering the relative position of a given site on a hillslope (e.g. upper hillslope involving recharge; middle hillslope involving lateral runoff; lower hillslope involving discharge into a waterbody). These landscape functions have a long tradition in pedology and have been associated with the catena concept according to which different positions within the landscape lead to the evolution of different soil types. Overall, the second factor group is related to runoff responses at the hillslope scale.

The third factor group is related to areas exhibiting a hydraulic behaviour that may be different from the rest of the catchment (Fig. 2.4, lower). Depending on the climate and hydrology of the catchment under study, these particular conditions may have widely different physical causes. Examples include permanently wet areas characterized by waterlogged soils, scars in the landscape as a result of erosion ('blaiken' in Fig. 2.4), areas covered by a glacier with or without a snow cover, talus fans with exceedingly high infiltration, and lakes involving water storage and delayed discharge.

Runoff Generation Mechanisms and Modelling Perspectives

When focusing on the phenomena that can be observed during runoff events at the plot scale, various types of 'mechanism' can be identified (Fig. 2.5). These are intended to illustrate some of the ideas that lead to process-oriented conceptualizations in hydrology. The dynamic processes associated with different generation mechanisms differ widely. Briefly, Horton overland flow tends to occur on impervious soils and is associated with short response times, while Dunne subsurface flow tends to occur in pervious soils and is often associated with longer response times. Saturation excess flow generates rapid runoff responses from near-stream saturated areas, whereas return flow results in delayed runoff reactions and matrix throughflow is associated with very long timescales. Macropore flow (Beven and Germann, 1982) and other preferential flow phenomena may be associated with short or long timescales depending on flow path size and connectivity (e.g. Peters *et al.*, 1995; Sidle *et al.*, 2000). When also the geomorphology and the geologic structure of different catchments are considered, the complexity of the runoff process becomes even more stunning (e.g. Turner and Macpherson, 1990; Einsele and Lempp, 1992).

These and other properties of runoff generation mechanisms lead to the following implications for modelling:

- Different runoff generation mechanisms produce different 'typical' hydrographs (shape, steepness, non-monotonic behaviour, triggers).

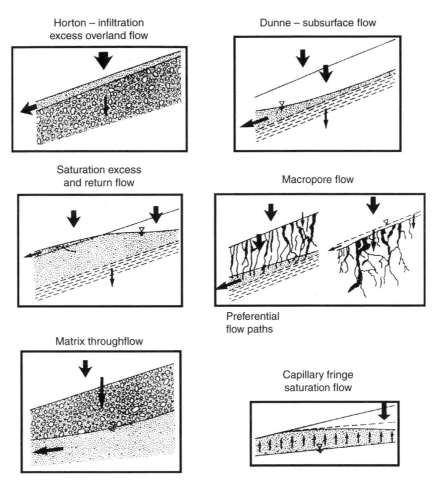

Fig. 2.5. Runoff generation mechanisms at the local scale.

- Capturing these features is essential because other processes – such as the transport of matter through the catchment – are linked to flow paths.
- A successful modelling approach should be based on the identification of hydrological catchment units (HCUs) on the basis of the predominant runoff generation mechanisms (process-based decomposition).
- Similarly, a phenomenon- and process-oriented approach should be applied during the acquisition of hydrologic data in the field (e.g. Kirnbauer *et al.*, 1996).
- Modelling should be preceded by an interdisciplinary drainage basin analysis.

River Processes and Patterns

Hydrological processes govern the input of water into the catchment and control biogeochemical reactions as well as the sediment and nutrient export (e.g. Soulsby, 1995; Pionke *et al.*, 1999). Water and sediment discharge have long been recognized as the two most important variables influencing channel form (Schumm, 1977). Consequently, they are also key factors in the development of flood plain ecosystems along a continuum from the sources to the river mouth. Sediment dynamics affects the development and the persistence of a variety of habitat patches. Sediment processes in the catchment system have been broadly subdivided into three zones (Schumm, 1977). The headwaters are the major sediment production area due to the erosion of channel banks and channel beds. It has been estimated that as much as 75% of the sediment produced in a river system has its source in the upper headwaters (Petts, 1984). The middle section of the river can be viewed as a major transfer zone with the lower reaches marking a depositional area where overbank deposition of fine sediment is commonplace (Schumm, 1977).

The development of catchment sediment systems has been mirrored in ecology by the development of models such as the River Continuum Concept (Vannote *et al.*, 1980), which explains how the structure and function of biotic communities, as well as the distribution of energy and matter, vary along the course of large-scale lotic systems. This influential concept did not consider the interactions between the river and its flood plain (Ward *et al.*, 2001), which were later addressed in the Flood Pulse Concept (Junk *et al.*, 1989) highlighting the ecological benefits due to regular, predictable flooding patterns. The Flood Pulse Concept illustrates the important role played by ecotones in lotic systems. Ecotones mark the transition zone between two adjacent ecosystems (Zalewski *et al.*, 2001; Ward *et al.*, 2001) and perform specific functions in retaining materials (sediments, nutrients, pollutants) and in keeping steep gradients between adjacent habitats (Naiman and Decamps, 1991). Within the category of land/water ecotones, flood plain forests and wetlands are of particular interest. From the hydrological point of view they have special impact on water stages (elevation of flood peaks and discharge velocities) altering the roughness of the greater flood channel. From an ecological point of view, ecotones are sites of extremely high assimilative (self-purification) capacity; however, this has not yet been appropriately quantified.

Previous catchment-scale theories have significantly enhanced our understanding of river systems although it is now recognized that a simple continuum from river sources to river mouth rarely exists due to complex internal processes (Petts and Amoros, 1996a). Theories such as the Fluvial Hydrosystems concept (Petts and Amoros, 1996b) and the River Styles approach (Brierley and Fryirs, 2000) suggest that catchments can be subdivided into a series of reaches and sub-reaches. Petts and Amoros (1996b) define these reaches in terms of functional sectors distinguished by contrasting process regimes, different habitat types and their stability over time. They further classify functional sectors at smaller scales such as functional sets, units and mesohabitats by virtue of their dimension and persistence. Brierley and Fryirs (2000) outline a similar geomorphologically based approach to river character and behaviour at four different scales. These range from the catchment/sub-catchment scale down to the assemblage of geomorphological units.

A further geomorphological process that is now being recognized as being important in ecohydrology is channel incision. Channel incision can be caused by a variety of factors but a key characteristic is that the river bed becomes depressed and degraded (*see* Simon and Darby, 1999 for further information on incised channels). Several Channel Evolution Models (CEMs) have been developed to explain this behaviour (Simon and Hupp, 1986; Watson *et al.*, 1986). The model by Simon and Hupp illustrates how channel incision is initiated through river canalization. Disturbance to the channel causes bed degradation, which in turn causes bank instability often leading to channel widening through bank failure. An aggradational phase ensues, followed by the formation of a new channel and a new flood plain at a significantly lower elevation. Channel incision can affect a whole river system as the degradation progresses upstream in the form of knick points until it becomes inhibited by hard control points. The ecological effects of channel incision can be extensive and can include disconnection between the river channel and the flood plain, the de-watering of flood plain aquifers, the loss of spawning gravels at riffles and point bars and increased in-stream velocities (Bravard *et al.*, 1999). The combination of these effects will have dramatic consequences for the riparian vegetation and for the fish and invertebrate life within river systems.

Catchment Functions in Ecohydrology

Any attempt to conceptualize complex ecohydrological processes can only be successful if the nature of the main factors influencing the development of patterns in the catchment at various temporal and spatial scales is understood. It is therefore important to emphasize the role of catchment features in transport, storage and buffering processes and to interpret their links to ecohydrological problem-solving. Process controls have been accordingly organized around the themes of transport, storage and buffering:

- *Transport* of water, sediment and substances (non-reactive/reactive) involving signals and properties, such as pressure and temperature. Here, important controls are the flow pathway, the 'route' of water and matter on its way to the outlet, as well as travel time.
- *Storage* in various zones and layers, interception, retention at the surface, within the upper/lower soil zone and within the groundwater zone. Storage zones can occur within the in-channel environment, the riparian corridor or in the wider catchment. Important properties are residence time, storage capacity and discharge patterns.
- *Buffering and filtering* including mechanical, chemical and biological effects.

All of these processes are related to mechanisms affecting the transport routes.

Catchment functions such as those described above can be moulded into a wider framework of processes characterized by the following properties (Gutknecht, 1996b):

- *Randomness* introduced by the more or less random occurrence of precipitation events and by the physical heterogeneity of natural catchments.
- *Disparity*, indicating the existence of distinct units or elements that display clearly distinguishable properties and response characteristics. Examples are

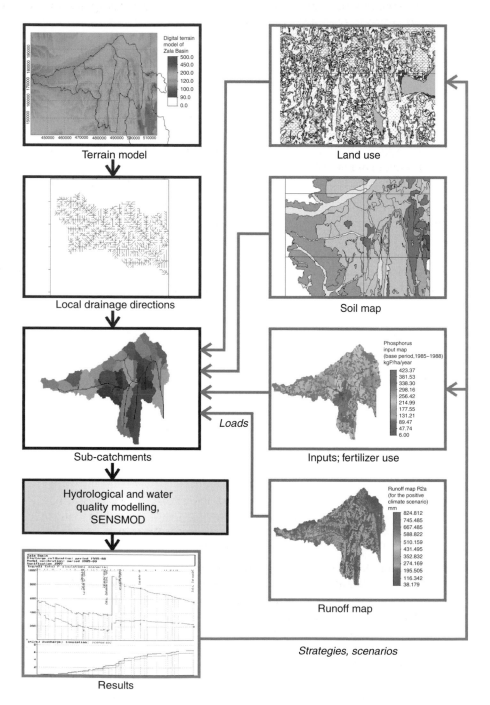

Fig. 2.6. Geographic Information Systems-based catchment modelling tools, Zala catchment (Hungary).

intermittent processes and systems showing clearly distinguishable patches. The various elements of which the system/process is composed are defined according to spatiotemporal boundaries.

- *Change over* occurring where thresholds become operative and capacities are exceeded. In some cases, a given process can undergo an abrupt change of the mode of operation. In other cases, the change may be a gradual transition to a new domain characterized by a different balance of the acting forces and parameters.

- *Organization*, addressing features such as systematic arrangements and sequences that establish connections between separate elements. Connections introduce 'structure'; for example, in the form of spatial dependence and preferential flow paths.

- *Storage*, which constitutes the most important element in persistent processes.

Examples taken from field observations or from hydrological data sets suggest that many aspects of flow variability can be attributed to the basic phenomena and process characteristics described above. Developing the ideas outlined above into a framework for a common methodology may link hydrological, geomorphological and ecological perspectives and methods. Efforts should be directed towards an in-depth integration of these basic concepts.

Quantification and Modelling of Catchment Processes

With the onset of the Water Framework Directive (WFD) of the European Union (EU) and the related River Basin Management Planning (RBMP), obligatory tasks for all EU Member States, the modelling of catchment processes gained or should have gained new impetus and importance, so as to provide the needed planning tools. Within Project 2.4 of Phase V of the UNESCO (United Nations Educational, Scientific and Cultural Organization) International Hydrological Programme, one of the original definitions of ecohydrology (and its main objective) is:

> To develop a methodological framework, through experimental research to describe and quantify flow paths of water, sediments, nutrients and pollutants through the surficial ecohydrological system of different temporal and spatial scales under different climatic and geographic conditions.

Considering this, it is easy to see that to reach the 'good status' of aquatic ecosystems, the ultimate objective of WFD, the planning tools of RBMP are just the tools of ecohydrological modelling in the sense that all hydrological, pollutant (nutrient) transport and ecological processes be modelled in their interaction (see also Fig. 2.3 above). Bearing this in mind, the authors and editors of this book decided to devote an entire chapter to this issue (Chapter 8), with ample examples taken from EU-supported integrated catchment, lake and wetland modelling projects. To illustrate the catchment modelling approach, Fig. 2.6 is a flow chart showing its application in the EU-supported INCAMOD Project (VITUKI, 1998; Jolánkai and Bíró, 2000), while more details of this project can be found in Chapter 8.

3 Nutrient Processes and Consequences

N. Pacini[1], D.M. Harper[2], V. Ittekkot[3], C. Humborg[4] AND L. Rahm[5]

[1]Department of Ecology, University of Calabria, Arcavacata di Rende (Cosenza), Italy; [2]Department of Biology, University of Leicester, Leicester, UK; [3]Centre for Marine Tropical Ecology – Bremen, Bremen, Germany; [4]Department of Systems Ecology, Stockholm University, Stockholm, Sweden; [5]Department of Water and Environmental Studies, Linköpings Universitet, Linköping, Sweden

Introduction

Living organisms require around 40 of the elements that naturally occur in the Earth's crust and atmosphere to sustain growth and reproduction. The most important – carbon – is usually considered separately from the others, because it is the energy locked into chemical bonds between carbon atoms and those with oxygen and hydrogen atoms which is the basis of the photosynthetic conversion of solar energy into living tissue. Oxygen and hydrogen are freely available in water under most circumstances. Other essential elements are usually considered in two groups: (i) the macronutrients or major elements, required in large quantities; and (ii) the micronutrients or trace elements, required in small quantities. Calcium, magnesium, potassium, nitrogen, phosphorus, sulfur and iron are the most important of the macronutrients, together with silicon (used in cell frustules by diatoms and a few other algal species), while copper, cobalt, molybdenum, manganese, zinc, boron, vanadium, chlorine, selenium and vitamin complexes are the most important of the micronutrients. Phosphorus and selenium are elements derived from the Earth's crust (lithosphere) essential to life, whose proportional abundance is lower in the lithosphere than in plant tissue. Phosphorus is thus often the limiting macronutrient for life. Selenium, followed by zinc, molybdenum and manganese, are potentially limiting micronutrients.

In a natural, undisturbed aquatic environment, the nutrient supply is derived from the drainage of the catchment together with direct rainfall and any internal recycling that may occur from the sediments. Studies that have been made of such catchments (and in the northern hemisphere, the more natural catchments are generally forested) have shown that nutrient runoff is very low because cycling within the vegetation of the terrestrial ecosystem is very tight. The same is true of tropical forests and savannahs. In the temperate zones, runoff from natural or secondary grassland is higher in nutrients than runoff from forested land, and runoff from arable land is higher still. Urban areas produce high nutrient loads.

The initial natural source of most material is the weathering of rocks. Using phosphorus as an example, igneous rocks contain apatite (complexes of phosphate with calcium) the weathering and subsequent marine sedimentation of which have led through geological history to phosphates being widely distributed in sedimentary rocks. The common weathering processes of such rocks lead to clays in which the phosphate is moved from apatite into the clay complex. It is both tightly bound into the clay lattice in place of hydroxyl ions and more reversibly bound by electrostatic attraction to aluminium or iron ions.

The atmosphere naturally contains few minerals of importance to aquatic systems other than those derived from nitrogen gas. The main source of nitrogen for all biological activity on the planet is the atmospheric reservoir of gaseous nitrogen, which is made available to organisms by fixation into a variety of oxides or reduction to ammonium. These events occur as a result of electrical or photochemical processes in the atmosphere but the major pathway is fixation by microorganisms in the soil, which is about seven times greater than the nitrogen from all atmospheric processes brought to Earth by rainfall.

Soil erosion processes, including material eroded from the river bank, from riparian areas, from agricultural soils and from deforested mountain slopes, provide the bulk of the suspended materials that accumulate in river systems. Of secondary importance are contributions provided by wash-off from urbanized areas, direct effluents produced by human activities other than farming (industrial, mining, transport) and autochthonous material formed within waterbodies (i.e. calcium carbonate precipitation, particulate organic matter formation). Anthropogenic particle sources that do not relate to agricultural practices provide the primary origin for most trace metals and persistent organic pollutants.

Under natural, undisturbed conditions, erosion processes would be concentrated in upland, higher-gradient areas. In addition to natural erosion forces associated with rivers and glacier movements, infrequent catastrophic events such as floods, avalanches and landslides cause the bulk of soil erosion. Much of this material becomes trapped in the higher parts of flood plains from where it is slowly mobilized by further infrequent high floods.

The largest portion of the particulate load, under the current conditions of anthropogenically accelerated erosion, comes from deforested and inappropriately farmed slopes, and cultivated flood plain soils. The flushing of the soil surface acts selectively and removes a disproportionate amount of fine fractions, rich in nutrients and organic matter. The enrichment ratio between the sediment eroded and the original soil is usually of the order of 1.2 to 2 times but may be as high as 12 times in fertile tropical soils.

The major contributions of material in early storm runoff are provided by rapid hydraulic transport pathways known as macropores – tiny (millimetre), vertical, preferential flow channels, through which surface soil solutions migrate rapidly, avoiding contact with the soil profile. Earthworms are primary natural causes of macropores, but many agricultural fields are often nowadays tile-drained, a practice that favours water percolation through soils leading to macropore development. Extensions of the hydrologic network through surface drains, pathways and cattle tracks throughout the catchment constitute additional sources. Thus, subsurface flow often transports significant fluxes. Unequivocal evidence for topsoil migration through macropores has been demonstrated by means of ^{137}Cs measurements. Soil type seems to be a determinant factor in this process, which still needs to be better understood. On the other hand, overland flow becomes relevant later in the course of a rain event, following soil saturation, which tends to be delayed especially in sandy soils. All materials enter water as it runs off or through rocks, vegetation and soils, either as soluble compounds (ions) or particulate material (usually eroded soil or rock particles).

Nutrient Ecohydrology

Phosphorus and nitrogen are the two most important nutrients biologically, providing contrasting examples of elemental cycles in the biosphere. Phosphorus is rare, likely limiting biological productivity and largely particle-bound in water; nitrogen is more abundant, rarely limiting biological productivity and largely soluble in water. Silicon is a third nutrient whose ecological importance is based upon the importance of diatoms to aquatic food webs and whose ecohydrological importance arises when the ratio of Si:N and Si:P changes in human-impacted systems.

Phosphorus

Phosphorus chemistry is complex; it is usually classified according to the procedure used for its analysis rather than its precise molecular form. The two most common analytical forms measured are total phosphorus (TP) and soluble reactive phosphorus (SRP). TP includes all forms of phosphorus, both dissolved (orthophosphate, inorganic polyphosphates, condensed phosphates and dissolved organic phosphorus) and particulate matter in suspension. The various forms of particulate phosphorus include organically bound phosphorus, phosphorus adsorbed on to clays/metal oxides, phosphorus occluded into the lattice structure of clay/metal oxides, phosphorus associated with amorphous metal oxides and the primary phosphorus found in particles of igneous rock such as apatite. TP is measured by incineration or acid digestion of a sample to convert all forms to orthophosphate, which is then measured as SRP.

SRP is a measure of orthophosphate (PO_4^{3-}) and the two terms are often used as synonyms. SRP may also include a small fraction of condensed phosphates and some organic phosphorus, however, both are hydrolysed during chemical determination. Total dissolved phosphorus (TDP), which includes both SRP and dissolved organic phosphorus, is determined by the same procedure as TP on a filtered sample. Particulate phosphorus can then be estimated as the difference between TP and TDP. Dissolved organic phosphorus can be estimated as the difference between TDP and SRP. Bioavailable phosphorus includes most of the SRP; it comprises orthophosphate, together with dissolved organic phosphorus and some of the phosphorus associated with suspended particles. Dissolved organic phosphorus from detrital matter is released through the action of enzymes produced by microorganisms, and the turnover of this form of phosphorus is more rapid than that of inorganic phosphorus forms.

Phosphorus is highly reactive, so the elemental form does not exist in natural conditions. Phosphorus exists in solution usually as phosphate, which may complex with metal ions and mineral particles. The chemical interactions governing the uptake and release of phosphate by mineral surfaces control the bioavailability of this nutrient. The sorption reactions that occur with sediment, whether benthic or suspended, are believed to buffer the dissolved reactive phosphate concentration in the water at some relatively constant value, despite short-term phosphorus inputs and biological removal (Froelich, 1988). Evidence for this comes from dilution of suspensions of riverine or estuarine sediments by varying quantities of distilled water or seawater and the constancy of dissolved phosphorus values observed by many authors. The term put forward to describe this is the 'phosphate buffer mechanism', analogous to pH buffering, which essentially suggests that there is a large pool of phosphate sorbed on to (or into) inorganic particles that is released to solution under conditions promoting desorption. The process is reversible, with an increase in phosphorus concentration in solution causing adsorption. The equilibrium phosphate concentration (EPC_0) is a distinctive property of soils and sediments, characterized by the greatest phosphate buffering capacity related to their composition. It is related to the general chemical properties of the ambient solution and to sediment past history. Behind the phosphate buffer mechanism, processes can be described with a two-step kinetics model implying a fast, non-specific, hydrophobic adsorption related to overall surface area and charge balance and a slow, specific, metal oxide-dependent adsorption which locks phosphate into a non-readily available pool.

The majority of phosphorus in rivers is transported bound to inorganic and organic particles. Dissolved phosphorus is estimated to account for only 5–10% of the total phosphorus transported to the oceans. Clay particles in soils play an important part in providing sites for ion exchange owing to their small size and relatively large surface area. They are chemically weathered minerals with a layering of geometrically arranged lattices. This large surface area of the particle allows for even greater chemical interaction. Clays are generically composed of silica sheets (an arrangement of silicon atoms and oxygen atoms/hydroxyl groups) and aluminium or magnesium silicate sheets (an arrangement of aluminium or magnesium atoms and oxygen atoms/hydroxyl groups) stacked alternately; different clay types having slight differences in ordering of these sheets and substitution of the aluminium, magnesium or silicon atoms for other mineral atoms.

Clay particles have an excess negative charge, which attracts cations, adsorbing them to the surface. Free phosphate can be adsorbed directly on to clay colloids where positive charge is found. This probably involves two mechanisms: (i) chemical bonding to positively charged Al^{3+} found at the edges of clay plates; and (ii) substitution of phosphate for silicate in the clay structure. Fe^{2+} and Al^{3+}

are the cations most commonly involved, particularly as their insoluble hydrated oxides, owing to their prominence in the Earth's crust. Other metal cations can act similarly. Hydroxides of iron and aluminium adsorbed on to clays act as a sponge for anions and cations and have a high affinity for phosphate (Viner, 1987). They can bind phosphate also as distinct complexes to form various metal-phosphate minerals. These complexes may be crystalline in structure or else amorphous and of no strict lattice structure.

The organic fraction of phosphorus is found mostly as particulate matter (dissolved organic compounds are less abundant). This particulate fraction consists chiefly of cellular phosphorus-based compounds within a variety of micro-organisms. During decomposition these organic compounds (a variety of phosphate esters and organic acids) are released and become part of the dissolved phosphorus fraction. Polyphosphates and metaphosphates introduced into watercourses are mostly man-made although a small proportion is generated by other organisms. Such complexes are unstable in water and are rapidly hydrolysed to orthophosphate.

Phosphate can also be indirectly adsorbed to organic particles as a result of binding to iron and aluminium. These metals are thought to become associated with organic particles in a similar way to the coating of metal hydroxides on clay particles. Such compounds are referred to as humic-iron-phosphate complexes and very little is known about how phosphate, once adsorbed, might be released. Probably organic particles transport a larger fraction of adsorbed phosphorus than inorganic particulates.

Sorption interactions between particles and solution control stream water phosphorus concentrations. Particles are either derived from the sediments and banks of the river in storm events or transported in from the catchment, settling on to sediments during low flow conditions. Interchange between sediment and water is thus important. The upper 4 cm of river sediment is that most actively involved in transport, mixing processes and storage. Sediment phosphorus capacity is related to the structure of the sediment particles, their degree of phosphorus saturation and sensitivity to environmental changes.

Sediment particle size, organic content and its iron and aluminium content are the main influencing factors. Fine-grained silty sediments were found to have a higher buffering capacity than coarse sediments in Bear Brook (Hubbard Brook Experimental Forest). A comparison of river sediments showed that the higher buffering capacity of one was linked to finer substrate particle size and organic content. Finer sediment is more likely to have a higher clay and metal oxide content, which would result in this enhanced ability to trap phosphorus.

Storage of phosphorus within the sediment potentially reduces eutrophication by reducing phosphorus bioavailability. Net retention depends on the settling flux of particulate phosphorus to the sediment and on the specific sediment retention capacity. Sediments do not provide a limitless sink for phosphorus storage and, in line with the buffering model, a reduction in ambient phosphorus concentrations in the water will promote net phosphorus release from the suspended and bottom sediments. This is evident from the instances where point sources of phosphorus have been reduced and a parallel reduction in stream water TP and SRP has not materialized.

Phosphorus release from lake sediment occurs in two stages. Phosphorus bound to particles or complexes is mobilized (desorption, dissolution of precipitates and complexes, ligand exchange and enzymatic hydrolysis of organic matter) and then transferred to the pool of dissolved phosphate in the pore water. Here the dissolved phosphorus may be transferred to another fraction before it has the opportunity of transport upwards from the sediment. For example, organic phosphates hydrolysed by microorganisms may quickly be absorbed by iron complexes; phosphate released from inorganic species may be taken up by microorganisms; and under pH >5.5 part of the phosphate is bound into calcium minerals.

Despite its predominant negative charge, phosphorus adsorbs readily on to clay and organic matter particles through the intermediation of their hydrated metal oxide coating. Freshly precipitated amorphous iron oxide achieves a surface area of between 200 and 500 m^2/g (Sigg, 1987) and represents the constituent with the greatest adsorption potential with an affinity for a wide variety of compounds. Under similar state conditions, the adsorption potential of soil or sediment is directly related to its iron content, although manganese and aluminium oxides, organic matter and other constituents may play a significant role. The specific phosphorus content of colloids can be high (>0.5%) due to their high specific surface and high iron hydroxide content; however, particulate phosphorus fluxes by this means do not appear to be highly significant. During base flow, natural dissolved phosphorus concentrations are mainly contributed by groundwater (and by treated effluents), with little or no phosphorus coming from surface runoff (or agricultural sources). Under these conditions in anthropogenically impacted watercourses, phosphate concentrations depend mainly on effluent discharges and other point sources since background phosphate concentrations are always very low. The free phosphate concentration can be high, and often higher than during rain events when phosphates in solution are partially buffered by higher suspended particle loads, and diluted by the contribution of rainfall. In nearly pristine catchments, in the absence of point-source pollution, free phosphate may come close to detection limits (1–10 $\mu P/l$) and base flow transport fluxes constitute a secondary component of the annual total phosphorus transport flux.

Nitrogen

The nitrogen cycle is complex, owing to the many chemical states in which nitrogen is found and the central role of bacteria in its transformation. The dissolved inorganic nitrogen (DIN) forms are ammonium (NH_4^+), nitrate (NO_3^-) and nitrite (NO_2^-). There are also gaseous forms such as N_2 and N_2O. Nitrate and ammonium are the biologically available forms, which can be assimilated directly. Excretion, decomposition and production of exudates are the principal pathways by which elements are recycled to an inorganic state. Dissolved organic nitrogen (DON), which includes urea, uric acid and amino acids as principal forms, and particulate organic nitrogen (PON) are thus important storage pools in aquatic ecosystems.

There are two main types of biological transformation in the nitrogen cycle: (i) those to obtain nitrogen for structural synthesis; and (ii) those which are energy-yielding reactions. In the former category are nitrogen fixation and assimilation of DIN. Nitrification and denitrification are energy-yielding processes. In nitrogen fixation cyanobacteria and bacteria convert elemental nitrogen to ammonium and incorporate it into biomass. In general, nitrogen fixation requires ATP, which is generated by photosynthesis, so this process is inefficient at night. However, cyanobacteria (primarily *Anabena*, *Aphanizomenon*, *Gloeotrichia*) can fix nitrogen directly, so do not have this diurnal limitation. Nitrogen fixation is important in eutrophic lakes with large algal populations depleting the nitrogen pool, leading to dominance by cyanobacteria. Ammonium-N is regenerated by excretion and decomposition and microbial reduction of nitrate. Microbial decomposition (ammonification) is oxygen-demanding and regenerates available nitrogen for re-assimilation by primary producers. It can result in rapid nitrogen cycling between the sediment and the water column. The rate of release of nitrogen from decomposing organic matter can be an important factor in determining nutrient limitation in freshwaters. Ammonia in aquatic systems can exist as the ammonium cation (NH_4^+) or as the unionized ammonium molecule (NH_3). High temperatures and high pH (>8) encourage the conversion of ammonium to ammonia, which is more toxic. This ammonium-N is used preferentially over nitrate, and nitrite for assimilation of nitrogen by autotrophs, bacteria and fungi. However, most bacteria can use a large variety of nitrogen compounds as sources of cellular nitrogen.

The nitrification–denitrification pathways result in loss of ammonium, which is first converted to nitrate, then to nitrogen gas. The nitrification process is carried out by two widespread genera, *Nitrosomonas* (NH_3 to NO_2^-) and *Nitrobacter* (NO_2^- to NO_3^-), and some closely related species. Nitrification is a two-stage oxidation process from ammonia to nitrite and subsequently to nitrate. The first step is usually rate-limiting, so nitrite is rarely present in appreciable concentrations in freshwaters. Nitrification is an oxygen-demanding process and requires minimum oxygen concentrations around 2 mg/l to function efficiently. The process requires an optimum pH of 8.4–8.6 and an optimum temperature above 15°C. Chemoautrophic nitrifying bacteria are usually dominant in freshwaters and their activity is generally highest at the sediment/water interface where ammonium-N generation is maximal.

Denitrification occurs largely in sediments and is controlled by both oxygen supply and available energy provided by organic matter. Most denitrifying bacteria are heterotrophic and able to utilize a wide range of carbon sources. The genus *Pseudomonas* includes the most commonly isolated bacteria. Other important groups are the *Alcaligenes* and *Flavobacterium*. Some bacteria (e.g. *Paracoccus*, *Thiobacillus*, *Thiosphera* spp.) can accomplish denitrification autrophically using hydrogen or various reduced sulfur compounds as energy sources.

Five factors control the rate of denitrification:

1.   Oxygen is an important competitive inhibitor. The gradual depletion of oxygen in semi-anaerobic conditions progressively favours denitrification.

2.   Organic carbon, which is required as an electron donor. A C:N ratio of 1 is required for 80–90% denitrification.

3.   pH, which has an optimum range of 7.0–8.0 (low pH favours N_2O production).

4. Temperature: denitrification decreases at low temperatures, although it is still measurable between 0 and 5°C. Generally, a doubling of the denitrification rate is possible with every 10°C increase in temperature.

5. Other compounds: a few inhibit denitrification, the most important being sulfur compounds. Sulfide depresses gaseous nitrogen production, but stimulates the reduction of nitrate to ammonium. Acetylene is also a well-known inhibitor preventing the reduction of N_2O to N_2. The most common analytical technique used for measurement of denitrification is based on this inhibition.

Denitrification occurs in lake sediments, where it is the most important reaction reducing nitrate concentrations of drinking water reservoirs to below World Health Organization standards; in riparian zones which can thus be used as water protection strips in areas of intensive agriculture (see below); and in river sediments (see below). Sediment/water interfaces, which are the primary sites for denitrification as for other nitrogen cycling processes, are unstable environments, highly dependent on the variability of hydraulic conditions. This characteristic represents a main difficulty in the monitoring of nitrogen forms for the assessment of nitrogen retention efficiency in different physical structures.

The main form of nitrogen available for transport from the soil is nitrate, due to microbial decomposition processes and direct fertilizer additions. In contrast to inorganic phosphorus forms, nitrate is not significantly retained by surface adsorption, is highly soluble and is characterized by high diffusion rates. The export of nitrogen from the catchment and its availability in waterbodies is linked to hydrological processes and represents a predominantly transport-limited system (i.e. not limited by the availability of nitrate in soils). The major routes for the transport of nitrogen are surface runoff, subsurface runoff, and deep and lateral groundwater flow. The relative importance of these pathways depends on the amount and pattern of precipitation, the surface water level, the specific retention time of groundwater systems, and the composition and slope of the soil. Agricultural drainage, in particular when using drains, favours transport in subsurface flow; this is recognized as the main pathway of nitrate transport from hillslopes to streams. Other forms of nitrogen such as particulate organic nitrogen and ammonia are also transported. These transport pathways may become important, depending on the nitrogen source. Organic and ammonium-N transport loss is high, particularly from heavily grazed grassland.

The availability of nitrogen in surface waters is thus critically dependent on the hydrology of the catchment and the hydraulic characteristics of the river channel. Absolute concentrations and their partition between different forms vary seasonally due to the changing efficiency of nitrogen cycling processes throughout the year. For example, ammonification of macrophytic organic detritus and organic sediments peaks in the autumn and spring periods. Under low flow conditions these can cause harmful effects due to the build up of high ammonia levels in surface waters. Meteorological conditions conducive to alternate wetting and drying of soils tend to increase nitrogen export by accelerating the decomposition and transport of organic nitrogen forms. Impacts due to human land use are superposed upon the natural seasonal succession of nitrogen cycling processes. Relevant factors are the quantity, frequency and timing of fertilizer applications in relation to transport events, the type of manure or inorganic fertilizer used, the time of ploughing, and related operations and activities relating to animal husbandry.

Eutrophication

Eutrophication is the general term given to the increase in concentration of plant nutrients in standing waterbodies and watercourses. The biological effects are manifest primarily in increased biomass of plants together with secondary responses of the primary producer community, such as shifts in the relative abundance of species and taxa through competition for nutrients or other limiting resources (e.g. light). This is followed by responses of the food chain to the direct (e.g. change of food organism abundance) and indirect (e.g. altered oxygen regime caused by decay of plant biomass) effects of the changed primary production base. In many waterbodies water quality changes follow the development of oxygen-deficient conditions (e.g. Jorgensen and Richardson, 1996), which promotes the production and emission to the atmosphere of climatically relevant gases such as nitrous oxide and methane. The new redox conditions further affect the mobility of metals and other organic pollutants with possible impacts on biogeochemical and ecological processes. Thus, the changes in land–water nutrient fluxes affect many socio-economic and regulatory functions of waterbodies.

Society's perspectives on eutrophication changed rapidly in both northern and southern hemispheres following long, hot, low-flow periods at the end of the 1980s, which led to pronounced blooms of cyanobacteria dense enough to cause toxic scums (Anon., 1992). One decade after Vollenweider and Kerekes produced a manual for the Organization for Economic Cooperation and Development clearly defining the problem, its causes and its management (Vollenweider and Kerekes, 1982), eutrophication was properly recognized as a widespread problem for all countries with developed agriculture and urbanization, and a growing problem in the developing world.

Eutrophication is caused by elevated nutrient input from a variety of sources in combination (natural background, atmospheric deposition, industry, agriculture, domestic); in a variety of manifestations (point, diffuse); leading to situations where different limiting factors act upon ecosystem productivity and stability (phosphorus, nitrogen, silicon, light). The most important single factor controlling the biological availability of nutrients in running or standing water systems is hydrology. This was recognized in the earliest models of lake eutrophication, which incorporated terms for water retention time, stratification/mixing and sedimentation rate. The subsequent recognition of eutrophication as a catchment problem rather than merely a lake problem led to the development of nutrient export models (Harper, 1992) that depended upon runoff and erosion components as well as land use. These have now been joined by sophisticated computer models, using satellite imagery to provide input data, integrated from single field to catchment, with the temporal frequency of the satellite pass.

The control strategies adopted therefore have to be holistic, structured in a hierarchical fashion from catchment level downwards, incorporate ecohydrological principles (Zalewski *et al.*, 1997), and link with other management strategies that initially might have had different objectives (e.g. erosion control, wetlands, buffer strips) (Straskraba, 1994) but are increasingly now seen as part of an holistic solution (Zalewski *et al.*, 2004).

River eutrophication is a more nebulous issue than in lakes, often because rivers are merely considered as transporters of nutrients to standing waters or as having problems manifest most obviously in the lower reaches where they become similar to long thin lakes (Descy, 1992). Almost nothing is known of the effects of nutrient enrichment separate from organic pollution on rivers (Sweeting, 1994) with the exception of specific works such as Holmes and Newbold (1984), Mainstone *et al.* (1994) and Woodrow *et al.* (1994) directed towards plant communities or plant indicators.

Marine and coastal eutrophication is a problem that has been widely addressed by states with major coastlines and enclosed waters such as The Netherlands, the UK and Denmark. In general, the causes of eutrophication are similar, although nitrogen limitation occurs more often in shallow marine waters than in freshwaters because of the recirculation of phosphorus coupled with the loss of nitrogen through denitrification. The increase in erosion rates on land is not always matched by increased sediment delivery to the coastal seas; there is an overall reduction in sediment delivery as a consequence of large-scale hydrological alterations on land such as river damming and river diversions. The impacts of these changes on coastal erosion and fisheries have been well documented (Milliman *et al.*, 1984; Halim, 1991). River transport of nutrients has changed as inputs from domestic, agricultural and industrial sources have increased. For dissolved nitrate and phosphate, an overall increase of more than twofold has occurred (Meybeck, 1998). The loss of particulate nitrogen from deforested soils further adds to the transport of nitrogen in rivers (Ittekkot and Zhang, 1989), especially so for the rivers of the Asian tropics which contribute substantially to sediment transfer from land to the sea globally (Milliman and Meade, 1983). The major fraction of this nitrogen is poor in protein and is mostly in the form of ammonium attached to minerals (Ittekkot and Zhang, 1989). Some of this nitrogen has the potential to be released via ion exchange reactions at the river/sea interface, adding to the nutrient loading of coastal seas.

The manifestations of coastal eutrophication are similar in some respects to those in lakes – excessive phytoplankton growth leading to toxic blooms in extreme events, deoxygenation through the decay of algal biomass leading to decline in biodiversity of food webs – and different in others. Major differences are the importance of benthic algal mats in shallow coastal waters, often mobile and thus accentuating problems of deoxygenation, together with the greater importance of damage to fisheries through loss of spawning and feeding grounds. Coastal eutrophication is potentially a more serious problem than lake eutrophication, since every river contributes to it; moreover, coastal towns and industries add wastes that are less well treated than those of equivalent-sized inland communities and hence with greater nutrient loadings.

It has been assumed for almost a century that the causes are nitrogen and phosphorus – nitrogen primarily derived from agricultural land and phosphorus from urban effluents, the origin of the latter approximately 50% human and 50%

detergent. Even rural catchments derive much of their phosphorus loads from a multitude of small domestic sewage effluents. Phosphorus loadings from diffuse agricultural sources are increasing, however, (McGarrigle, 1993) due to association with soil particles in erosion (Sharpley and Smith, 1990), increased stocking rates (Wilson *et al.*, 1993), runoff of applied animal-derived slurry, and saturation of the soil-binding capacities through continuous use of phosphate-based fertilizers (Sharpley *et al.*, 1994). Moreover, at times of the year when low flows persist, nitrogen loading from point sources can be considerable – over 50% in one study. In rivers, generally rural point sources (small village sewage works) may be important, and in estuaries and coastal waters effluent discharges may achieve a high importance because the absence or reduction of biological treatment stages results in higher loadings of nitrogen (as ammonia) as well as phosphorus. A new source of nitrogen, which grew in prominence towards the end of the 20th century, is forested land. It has been known for some time that afforestation may cause eutrophication through the disturbance of land by ploughing and the consequent erosion of phosphorus from superphosphate fertilizer (Harper, 1992), but in northern climates it has become apparent that mature forests are yielding higher loads of nitrogen in runoff streams, putting at risk strategies for eutrophication control which otherwise reduce nitrogen flow from agricultural land. This so-called 'nitrogen bomb' is believed to be linked with acid deposition and is potentially a problem for acid-sensitive areas in the UK (Fleischer and Stibe, 1989).

It is prudent to consider phosphorus and nitrogen both as limiting nutrients and silicon as a possible third at times. Phosphorus is the usual nutrient in shortest supply but some areas naturally rich in phosphorus show nitrogen limitation in summer; for example, the UK West Midlands meres (Moss *et al.*, 1994). Coastal waters are often nitrogen-limited due to phosphorus recycling from anoxic sediments combined with nitrogen removal by denitrification. Moreover, removal of phosphorus from point sources or recipient ecosystems will increase the probability that nitrogen becomes limiting and the two should, therefore, be considered together. Silicate plays a crucial role in growth and species composition of some algae (Officer and Ryther, 1980), see below.

In all environments the major manifestations of eutrophication are enhanced plant (usually algal) biomasses concentrated in a few 'nuisance' species, the worst being cyanobacteria. In lakes and coastal waters phytoplankton algae are the primary manifestations (Anon., 1992), with mats of benthic macroalgae on lake muds where light penetrates or coastal inter-tidal mudflats. In lakes nuisance macrophyte growth may be an intermediate stage in progressive eutrophication (or it may indicate recovery after successful management from a previous cyanobacterial-dominated state). The most widespread indication in rivers is enhanced growth and cover of the macroalgae *Cladophora* and *Enteromorpha* (Malati and Fox, 1985), with phytoplankton in the semi-impounded lower reaches.

Silicon

Silicon is a major element in the Earth's crust. Crustal erosion releases dissolved silicate to streams and rivers where, along with nitrogen and phosphorus, it is an essential nutrient for certain types for phytoplankton such as diatoms and silico-flagellates. They form frustules of silicon in the form of biogenic opal (Hecky *et al.*, 1973). Both are indicators of healthy waterbodies. Uptake by diatoms and their sedimentation represents the major sink for silicon in the sedimentary environment (Wollast and Mackenzie, 1983; Tréguer *et al.*, 1995).

In non-nutrient-limiting systems diatoms require nutrients such as nitrogen and silicon at a ratio of about 1:1 (Brzezinski, 1985). If the ratio falls below 1, then non-diatomaceous species take over. (The degree of recycling will of course play a role for the limiting nutrient but the slow dissolution of biogenic silica will not affect the diatom growth process.) In pristine waters these ratios are well above 1 and the systems are usually limited by nitrogen or phosphorus. Observations in recent years have shown that the amount of silicate in coastal waters has decreased (e.g. Conley *et al.*, 1993) and the changes in Si:N ratios have caused shifts from diatoms to non-siliceous phytoplankton in many waterbodies. Examples are the receiving waters of the Mississippi and Rhine rivers (Admiraal *et al.*, 1990; Turner and Rabalais, 1994). Such shifts have been suggested to be due to cultural eutrophication. When phosphate and nitrate inputs from anthropogenic sources increase, then the efficiency of silicate uptake by diatoms increases, causing a reduction in silicate concentrations and consequently the shifts in nutrients. Recent studies show that such changes can also be caused by a reduction in the initial inputs of dissolved silicate by rivers.

Changes in silicate transfer

There is some information on the reduction of dissolved silicate inputs to coastal waters by rivers that coincides with the construction of dams. The dissolved silicate (DSi) concentration in the Nile river has decreased by more than 200 μM after the construction of the Aswan High Dam (Wahby and Bishara, 1980). Dramatic changes have occurred in the inputs of dissolved silicate to the Black Sea from the River Danube (Humborg *et al.*, 1997). The River Danube, which contributes about 70% of the river inputs into the Black Sea, was dammed in 1970–1972 approximately 1000 km upstream at the Yugoslavian/Romanian border at the Iron Gates Dam, causing significant changes in the Danube's discharge pattern (Popa, 1993). The historical data sets of silicate DSi transport by the Danube showed that the dissolved silicate inputs have decreased from an average of 140 μM before construction of the dams in the 1950s to about 58 μM during 1979–1992. The observed decrease in silicate concentrations in Romanian coastal waters by about 60% coincides with the construction of the Iron Gates Dam in the 1970s (Fig. 3.1).

Dam constructions have an effect on the nutrient loads of rivers due to nutrient removal in reservoirs via primary production and sedimentation by what is known as the 'artificial-lake effect'. All nutrient elements – nitrogen, phosphorus and silicon – are retained by these processes. Whereas this removal might be overcompensated by anthropogenic nitrogen and phosphorus inputs downstream from the reservoir, no such compensation has been observed for silicate. This leads to dramatic changes in the nutrient mix entering the coastal seas.

In the Danube, there has been a strong decrease in DSi:N ratio from 42 to about 2.8 (Humborg *et al.*, 1997). The reduced silicate inputs from the Danube

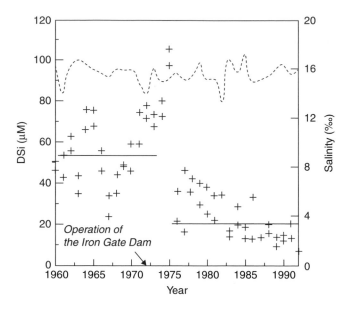

Fig. 3.1. Dissolved silicon (DSi) concentrations in the Black Sea. Mean winter silicate concentrations (+) at a coastal station (Constanta, Romania) about 60 nautical miles south of the Danube river mouth. The bold horizontal lines show overall medians from 1960 to 1972 and from 1973 to 1992. The salinity at these stations has remained at about $S = 15$ (dashed line), indicating that the effect of the Danube on salinity has not changed over time. (After Humborg *et al.*, 1997.)

appeared to also have an effect on the silicate concentrations in the biologically active surface layers of the open Black Sea. Comparison of silicate concentrations from 1992 with those collected during cruises in 1969 showed a decrease in silicate concentration in the upper mixed layer of the entire central Black Sea (Humborg *et al.*, 1997). Eighty per cent of this could be attributed to dam construction.

Regulation of rivers by damming and eutrophication seem to have reduced substantially the input of dissolved silicate also to the Baltic Sea, as evidenced from data compiled from Swedish rivers (Humborg *et al.*, 2000). The relationship between river DSi concentration and reservoir live storage of 20 Swedish and Finnish rivers is presented in Fig. 3.2. The reservoir live storage is expressed as the percentage of the virgin mean annual discharge (VMAD) of the river system that can be contained in reservoirs (Dynesius and Nilsson, 1994). An inverse relationship is apparent between the degree of damming and the DSi concentration. The damming effect can be seen in both more eutrophied as well as in very oligotrophic rivers, which are defined in Fig. 3.2 by their mean phosphate concentration.

Contrary to DIN and DIP, both the DSi concentration and the DSi:N ratio have been decreasing in the Baltic Sea since the end of the 1960s (Sandén *et al.*, 1991; Rahm *et al.*, 1996). The DSi:N ratio is approaching unity; that is, one that

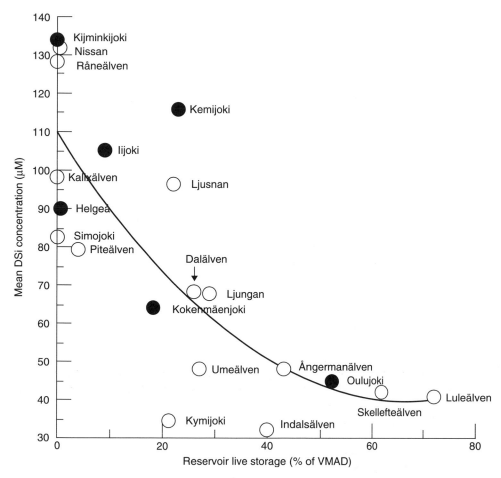

Fig. 3.2. Dissolved silicon (DSi) concentration versus reservoir live storage (as a percentage of virgin mean annual discharge, VMAD; see text) of 20 Swedish and Finnish rivers. The damming effect can be seen in both more eutrophied as well as in very oligotrophic rivers: O, <0.2 μM PO_4^{3-} ; •, >0.2 μM PO_4^{3-} . (After Humborg *et al.*, 2000.)

corresponds to the diatom demand. Thus, it was assumed that the decreasing trends in DSi concentrations were caused by the ongoing eutrophication of the Baltic Sea alone. However, major engineering works in the Baltic Sea catchment took place long before a monitoring programme started and occurred mainly in Scandinavia. Assuming a 'pristine' DSi concentration of about 110 μM (as in Fig. 3.2), the amount of DSi load reduced by dam construction in the Gulf of Bothnia amounts to 140,000 t per annum (Humborg *et al.*, 2000). These numbers agree surprisingly well with estimates of changes in the rate of DSi depletion in the Bothnian Sea and Bothnian Bay (Wulff and Rahm, 1988; Wulff and Stigebrandt, 1989; Sandén *et al.*, 1991) of about –4% per annum.

The falling trend in DSi concentrations is observed in the entire Baltic, but is most pronounced in the Gulf of Bothnia where almost all discharging rivers are heavily dammed and primary production is generally low. During the assessment period 1988–1993 in the HELCOM analysis (HELCOM, 1996) of the Baltic proper, it was reported that the proportion of diatoms in the spring bloom had decreased while flagellates had increased. Until that time it had only been attributed to mild winters (Wasmund *et al.*, 1996).

Ecological consequences

Information on the ecological consequences of a changing nutrient mix in the receiving waters of dammed rivers is scarce. The available information is not always consistent. In the Black Sea, the altered nutrient inputs from the Danube have resulted in a distinct increase in phytoplankton bloom frequency, in cell densities and in the number of bloom-forming species from the beginning of the 1970s (Bodeanu, 1988) (Table 3.1). Furthermore, dramatic shifts have occurred in phytoplankton species composition from diatoms to coccolithophores and flagellates. While diatom blooms increased by a factor of 2.5, blooms of non-diatoms such as dinoflagellates, the prymnesiophytes *Emiliania huxleyi* (cocco-lithophore) and the facultative toxic species *Chromulina* sp. as well as the Euglenophyte *Eutreptia lanowii* increased by a factor of 6 (Humborg *et al.*, 1997). Blooms of coccolithophores, which before the construction of the dam were only reported from far offshore regions and later in the season, were encountered near the Danube plume at salinities as low as 10‰.

In other systems such as the Mississippi plume and the northern Adriatic large blooms of diatoms continue to develop, suggesting that the impact is less severe (Turner and Rabalais, 1994; Justic *et al.*, 1995). It is conceivable that the effects of such changes will be severe in land-locked basins such as the Black Sea or the Baltic Sea. Furthermore, the existence of sources of dissolved silicates in the coastal seas other than river inputs, such as from groundwater seepage, sediment pore water as well as from offshore exchange, could make the impact less visible.

Environmental implications

Although the direct observational evidence is scarce, there are indications that environmental degradation of coastal waterbodies can be accelerated by changing

Table 3.1. Phytoplankton concentration in the north-western Black Sea. (From Humborg *et al.*, 1997.)

	1960–1970		1980–1990	
	Cell densities ($\times 10^6$/l)	Number of blooms	Cell densities ($\times 10^6$/l)	Number of blooms
Total diatoms	7–21	8	5–300	19
Total dinoflagellates	17–51	4	5–810	14
Total Euglenophytes	–	–	5–108	6
Total prymnesiophytes	–	–	220–1000	3
Total blooms		12		42

nutrient inputs from rivers due to river damming. The large number of dams in operation around the world today can thus affect the biogeochemical and ecological characteristics of the receiving waters. Since many of them are already affected by population pressure, habitat modification, aquaculture development and pollution, the impact of the changing nutrient mix will be an additional stress on an already overstressed environment.

Diatoms play a major role in the marine food chain and contribute substantially to primary productivity in the sea (Dugdale *et al.*, 1995). In near-shore waters, diatom populations are sustained by silicate inputs from rivers. Any change in the river nutrient mix resulting from silicate reduction can thus have a significant impact on diatom-based food web structures and will affect the biogeochemical cycles of carbon in coastal marine regions (e.g. Ittekkot *et al.*, 2000).

In many marine regions, diatoms are being replaced by non-siliceous organisms, some of which are toxic and adversely affect the food web, especially fish (Smayda, 1990). Although over-fishing and potential changes in climate are factors that have caused enormous problems in the sustainable use of fisheries resources, in many parts of the world a reduction in fish catch appears to have occurred due to toxic algal blooms. A possible example is that of the fish kills in the eastern Mediterranean, which appeared to coincide with the construction of the Aswan High Dam in Egypt (Milliman, 1997).

The impact of changing nutrient concentrations, ratios and silicate inputs from river damming is an example of the concept ecohydrology at work at the land/ocean interface (Zalewski *et al.*, 1997). Similar ecohydrological investigations will help us to better understand the functioning of aquatic systems, to better appreciate the influence of sectorial activities on them and to implement a more comprehensive approach to managing coastal waters.

4 Lotic Vegetation Processes

G.A. JANAUER[1] AND G. JOLÁNKAI[2]

[1]Department für Limnologie und Hydrobotanik, Faculty of Life Sciences, University of Vienna, Vienna, Austria; [2]Environmental Protection and Water Management Research Institute (VITUKI), Budapest, Hungary

Introduction

Hydrology has a rich tradition in dealing with channel vegetation, but it is the generalizations and assumptions necessary for making a mathematical model work in the practical world that let it fall short of an ecohydrological view of aquatic vegetation. This chapter points out some aspects of the aquatic ecosystem that are related to the 'macrophyte' compartment.

Wetlands and the littoral reaches of rivers, ponds and lakes have one important feature in common: in general, life-supporting water is always abundant. Yet, the peculiar conditions in these habitats require special adaptations for survival (den Hartog, 1970; Wetzel, 1975). Only about 700 aquatic plant species are known. Among them some 20 species severely impact upon human interests (Spencer and Bowes, 1993). Despite intensive investigation since the late 1970s knowledge of the biology and ecology of aquatic plants is still limited (Pieterse and Murphy, 1993), as is that of their impact upon hydrology.

Aquatic Macrophytes

The term 'macrophytes' describes aquatic and semi-aquatic plants that can be determined to the species level by eye (Westlake, 1974; Wetzel, 1975). This is a definition based mainly on practical aspects, and comprises a set of taxonomically mixed plants – macroalgae, mosses, ferns and seed-producing plants (Wetzel, 1975; Casper and Krausch, 1980, 1981; Frahm and Frey, 1992). Submersed macrophytes live inside the waterbody. They are found in fewer than 20 plant families, most of them monocotyledons (Wetzel, 1975). Floating-leaf rooted species (e.g. water lilies) extend their leaves on the water surface, some non-rooted plants float with their entire body on top of the water (pleustophytes *sensu*; Luther, 1949) (e.g. duck weed, water hyacinth). Emergent macrophytes (helophytes) are tall erect plants living mainly on river banks and lake shores (e.g. common reed). Amphibious plants can grow under both wet and fully aquatic conditions (e.g. arrowhead).

Functional Adaptations to the Aquatic Environment

The growth forms of macrophytes are of importance owing to the stress that flow exerts on the plants (Hejny, 1960; den Hartog and Segal, 1964; Hutchinson, 1975b) (Tables 4.1 and 4.2). In some cases the growth form is modified by environmental conditions like salinity, water depth (Table 4.3), flow or wave action (Pieterse and Murphy, 1993), and is often flexible within individual species to suit the flow environment of an individual plant. Growth in water is limited by effects which relate to the higher density of water compared with air: reduced gas diffusion rates, which influence the flow of oxygen and carbon dioxide, light attenuation, hydraulic pressure, and the effects of water flow and wave action.

Typical submersed plants are characterized by special anatomical and morphological features different from those in terrestrial plants (Sculthorpe, 1967). Submersed leaves are thin, often divided or thread-like, with a thin cuticle.

Table 4.1. Species associated with different flow types. (Adapted from Haslam, 1978 based on surveys in the UK.)

Species	Flow type					
	nof	mtf	slf	mof	faf	naf
Ceratophyllum demersum	▓					
Glyceria maxima	▓					
Phragmites australis	▓					
Elodea canadensis		▓				
Enteromorpha intestinalis		▓				
Lemna minor		▓				
Sagittaria sagittifolia		▓				
Nuphar lutea			▓			
Potamogeton pectinatus			▓			
Schoenoplectus lacustris			▓			
Mosses (all spp.)					▓	▓
Callitriche spp.						▓
Phalaris arundinacea						▓
Polygonum amphibium						▓
Potamogeton natans						▓
Potamogeton crispus						▓
Potamogeton perfoliatus						▓
Rorippa nasturtium						▓
Zannichellia palustris						▓

nof, no flow; mtf, more than no flow; slf, slow flow; mof, moderate flow; faf, fast flow; naf, not associated with flow type.

Table 4.2. Species with different hydraulic resistances. (Adapted from Haslam, 1978 based on surveys in the UK.)

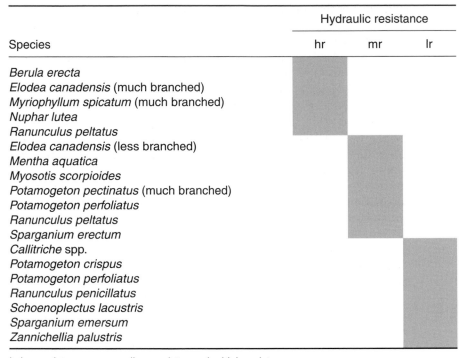

Species	Hydraulic resistance		
	hr	mr	lr
Berula erecta	■		
Elodea canadensis (much branched)	■		
Myriophyllum spicatum (much branched)	■		
Nuphar lutea	■		
Ranunculus peltatus	■		
Elodea canadensis (less branched)		■	
Mentha aquatica		■	
Myosotis scorpioides		■	
Potamogeton pectinatus (much branched)		■	
Potamogeton perfoliatus		■	
Ranunculus peltatus		■	
Sparganium erectum		■	
Callitriche spp.			■
Potamogeton crispus			■
Potamogeton perfoliatus			■
Ranunculus penicillatus			■
Schoenoplectus lacustris			■
Sparganium emersum			■
Zannichellia palustris			■

lr, low resistance; mr; medium resistance; hr, high resistance.

Table 4.3. Species associated with different water depths. (Adapted from Haslam, 1978 based on surveys in the UK.)

Species	Water depth		
	sm	fd	in
Callitriche spp.	■		
Ranunculus spp.	■		
Lemna minor		■	
Nuphar lutea		■	
Sagittaria sagittifolia		■	
Schoenoplectus lacustris		■	
Sparganium emersum		■	
Ceratophyllum demersum			■
Potamogeton crispus			■

sm, shallow and moderately deep water; fd, fairly deep and deep water; in, indifferent to water depth.

The epidermal cells contain chloroplasts; stomatal openings are absent or non-functional (Wetzel, 1975). Water is taken up through either special cells in the epidermis or through the whole plant surface. Mechanical and lignified tissue is lacking and water-conducting xylema are reduced. Plants are held upright in the water by buoyancy-providing air spaces (lacunae) in stems and leaves.

Floating leaves – and the leaves of pleustophytes – are built of more cell layers and all surfaces exposed to the air are protected against water loss by a thicker cuticle, giving the leaf a leathery appearance. Petioles of floating-leaved plants are longer than the distance between the bottom and the water surface to avoid submersion in case of water-level changes. Through stomatal openings in the upper epidermis of the floating leaves, these plants make use of the aerial carbon dioxide pool. Large air spaces in leaves, petioles and stems (aerenchyma) contribute to buoyancy and gas exchange within the plant (Wetzel, 1975). Lateral overlap of leaves on the water surface can lead to strong competition for light and nutrients in tightly packed stands (Hutchinson, 1975b).

Helophytes extend most of their body into the air. They are built much like terrestrial plants – sensitive stomatal regulation, cuticles and wax covers protect them against enhanced water loss. Oxygen transport to the rhizomes and roots, which are embedded in oxygen-deficient or anoxic sediments, is mediated by extensive lacunar systems. During extended anoxic periods the metabolism of rhizomes may switch from respiration to fermentation (Hutchinson, 1975a).

Light attenuation in the water and self-shading within the plant stands (Westlake, 1974) force many aquatic plants to adapt to low light conditions (Fair and Meeke, 1983). Red and ultraviolet wavelengths are attenuated more, thus green and blue shades dominate the underwater light field (Gessner, 1955). Only a few species can sustain deep shade (e.g. *Berula erecta* (Hudson) Coville; Janauer, 1981). Light mediates the growth and development of aquatic plants as it does in the terrestrial environment (Chang *et al.*, 1998).

The temperature niche of aquatic plants ranges from just above freezing to about 40°C (Bowes *et al.*, 1979). Most water plants are tolerant of pH variations (Hutchinson, 1975b): during periods of high photosynthetic activity, highly basic conditions (pH 11) may be reached (Brown *et al.*, 1974; Wetzel, 1975; Jorga and Weise, 1977; Spencer and Bowes, 1993).

In natural freshwater environments phosphorus often is present in suboptimal concentration; therefore a small, often man-made enrichment in this critical element causes eutrophication, which may be followed by mass growth of phytoplankton and macrophytes. A few aquatic plants (*Utricularia* spp., *Aldrovanda* spp.) access additional nitrogen sources by catching and digesting small animals (Hutchinson, 1975b; Pieterse and Murphy, 1993).

Sexual propagation is of little importance in many aquatic plant species. Following their terrestrial ancestors, most macrophytes flower above the water surface. Pollination in the water is rare (e.g. *Zannichellia*, *Ceratophyllum*; Casper and Krausch, 1980, 1981). Asexual reproduction is the main path of propagation: fragmentation, the dispersion of tubers, turions and winter buds (Pieterse and Murphy, 1993) and the growth of rhizome systems are competitive strategies (Spencer and Bowes, 1985). This results in plant stands often composed of one species only, which is typical for aquatic vegetation.

The Role of Macrophytes in the Aquatic Environment

Macrophytes as primary producers

Macrophytes are photo-autotrophic plants: light is used as energy source to produce starch, other photosynthates and oxygen. In turn they are used by herbivores and detritivores in a food web. During periods of low light intensity and at night, the plants' respiration draws on the oxygen pool of the aquatic system, which, in polluted waters, can limit the oxygen available for fish (Hough, 1974; Hutchinson, 1975b; Jorga and Weise, 1977). Biomass production as a means of carbon input to the system is highly dependent on the growth form and the species. Only reeds and tropical pleustophytes (e.g. water hyacinth, water lettuce) are among top producers, owing to unlimited water supply and access to the aerial pool of carbon dioxide. All other aquatic plants are in general less productive than terrestrial plants (Pieterse and Murphy, 1993) (Fig. 4.1).

In addition to carbon dioxide, some water plants can utilize bicarbonate as an additional carbon source (Sand-Jensen and Gordon, 1984; Bowes and Reiskind, 1987), some rely on carbon dioxide taken up through the roots from the sediment (Sondergaard and Sand-Jensen, 1979) and a few species of *Isoetes* show crassulacean acid metabolism. By successfully coping with such situations aquatic plants are considered as S-strategists (Hutchinson, 1975b; Phillips *et al.*, 1978).

Macrophytes and their relationship to other organisms

Structural elements
The stems and leaves of macrophyte stands provide surfaces and structural elements in a waterbody, which by itself would be without any definite structure.

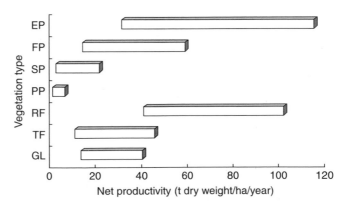

Fig. 4.1. Net productivity of vegetation types (GL, grassland; TF, temperate forest; RF, tropical rainforest; PP, phytoplankton; SP, submersed plants; FP, floating-leaf plants; EP, emergent plants). (Adapted from Pieterse and Murphy, 1993.)

The composition of growth forms in special habitats critically influences the spatial distribution of abiotic and biotic environmental factors (Orloci, 1966; Williams and Lambert, 1966), such as light, temperature (Carpenter and Lodge, 1986; Kornijow and Kairesalo, 1994), water flow (Hutchinson, 1975a), oxygen and carbon dioxide. Biotic environmental factors influenced by plant structure include phyto- and zooplankton, sessile invertebrates, young and adult fish (Fig. 4.2).

Phytoplankton and phytobenthos

The algal communities in the free water and on the sediment surface are primary producers like the macrophytes, which makes them direct competitors for light and plant nutrients (Dokulil and Janauer, 1989; Sondergaard and Moss, 1998; Janauer and Wychera, 2000). Macrophytes can reduce the trophic basis of algae due to their higher ability to store nutrients (Balls *et al.*, 1989). The reduction of water movement within – and around – their stands enhances sedimentation processes, which can reduce the concentration of planktonic algae in the water column and thereby enable the zooplankton to intensify its grazing on the phytoplankton. However, these effects become evident usually only in still or slow-flowing environments, and if a certain threshold of macrophyte biomass is surpassed (Sondergaard and Moss, 1998).

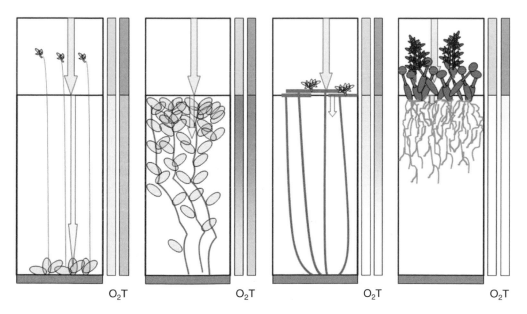

O₂T   O₂T   O₂T   O₂T

Fig. 4.2. Growth forms of macrophytes and their influence on environmental parameters (schematic). From left to right: submerged rosette plants (e.g. lobelia); submerged tall macrophytes (e.g. pond weed); floating-leaf plants (e.g. water lily); (acro-) pleustophytes (e.g. water hyacinth). Wide arrow above water line indicates incident light; narrow arrow in the water indicates approximate depth to which light will support photosynthesis. Oxygen depth distribution in the water is represented by O_2 bar; temperature distribution in the water by T bar.

'Aufwuchs'

The surface area of macrophytes can be much larger than the area on which the plants grow (Wetzel, 1975; Carpenter and Lodge, 1986; Park-Lee, 1986; Kornijow and Kairesalo, 1994). It is the prime habitat for numerous groups of organisms. Bacteria, fungi and algae (Park-Lee, 1986; Schwencke-Hofmann, 1987) cover the aquatic plants with a 'biofilm', which in turn is a nutrient source for many animals like invertebrates, fish and birds. In general the taxonomic composition of the algal community in the aufwuchs is not too unique to the macrophyte species it grows on (Schwencke-Hofmann, 1987; Jeppesen et al., 1998a). Yet, the species diversity and spatial density of invertebrates correlate positively with the diversity of growth forms and the expanse of the macrophyte stands (Lillie and Budd, 1992; Wollheim and Lovvorn, 1996).

Zooplankton and fish

Zooplankton and juvenile stages of fish use macrophyte stands as an important shelter and refuge against predators (Jeppesen et al., 1998b; Persson and Eklöv, 1995), which in turn may only use these food sources during certain stages of development (Persson, 1988). The predators, limited by their sensory capacity, suffer reduced efficiency with increasing density of the aquatic vegetation (Jeppesen et al., 1998b), with growth form, morphology and size of the macrophyte stands as additional functional parameters (Diehl, 1988).

Water birds

On dense vegetation birds find support to move around, feed and breed. The size of bird populations is often directly correlated with the biomass and density of the aquatic plants (Mitchell and Perrow, 1998) and the distribution of macrophyte propagules may be critically controlled by birds (Pieterse and Murphy, 1993).

Herbivory

Relatively few animals use macrophytes directly as a food source (Lodge et al., 1998); the best known are crayfish, herbivorous fish, turtles, water fowl and some mammals (beaver, muskrat, nutria, hippopotamus). The protein-rich aufwuchs enhances the nutritive value of the aquatic plants. Grazing by herbivorous fish introduced to foreign regions as a weed control agent can lead to the total extinction of water plants in slow-flowing rivers and irrigation channels, ponds and small lakes. Nutrient leaching from excreta of these fish then usually leads to algal blooms, high turbidity and oxygen depletion. In still waters clear states dominated by macrophytes have been reported to alternate with turbid states dominated by phytoplankton (van Donk, 1998; Donabaum et al., 2004; Irfanullah et al., 2004).

Macrophytes and Man

Macrophytes can develop high biomass, if temperature and nutrient conditions are favourable. In the temperate climate, usually submersed forms like *Elodea canadensis, Elodea nuttallii, Ranunculus* spp. and *Potamogeton* spp., and some rather recent neophytes like *Crassula helmsii* (Western Europe) and *Myriophyllum spicatum* (North America), are known for their vigorous growth. In tropical and subtropical regions more species show weedy character (Coordinatiecommissie Onkruidonderzoek, 1984; Pieterse and Murphy, 1993; Gopal, 1998): submersed macrophytes like *Hydrilla verticillata* (Haller, 1976), pleustophytes like *Azolla* spp., *Salvinia molesta*, *Eichhornia crassipes* and *Pistia natans*, floating mats of *Alternanthera philoxeroides* (Pieterse, 1993b) and even helophytes like *Typha australis* (Hellsten *et al.*, 1999) cause serious problems when humans use rivers and still waters as a resource (fishing, communication, navigation).

The invasive behaviour of water plants is promoted by animals, which can transport diaspores of macrophytes over large distances (Pieterse and Murphy, 1993), but not least by man himself: using weedy species as ornamentals (e.g. *E. crassipes*) or chicken food (e.g. *Salvinia natans*) is the most common way by which aggressive propagation is started (Murphy and Pieterse, 1993). Lack of information, too little knowledge about the potential danger associated with these plants, and sometimes plain ignorance, are a starter for future environmental disasters, most commonly occurring in Africa and Asia today.

Negative impacts of the mass growth of aquatic plants include restrictions to boat and ship traffic, drastic decrease in subsistence fisheries, and risks to health and life by predatory fish, electric eels, snakes, crocodiles, etc., which find habitats in such plant stands, mainly pleustophytes. Such conditions also lead to closer contact with vectors of infectious diseases like malaria or bilharzia. Safe body care and laundry is often impossible under these conditions. Water movement in irrigation and drainage channels may be obstructed and channels leading to the water inlet of hydropower plants can be blocked. In the tropics water loss from drinking water reservoirs and hydropower reservoirs is largely enhanced by the transpiration of pleustophytes (e.g. *Salvinia*) and helophytes (e.g. *Typha*). These problems generally lead to control measures to keep the weeds on a tolerable level (Clark, 1982; Murphy and Pieterse, 1993). Integrated control (van den Bosch *et al.*, 1971; Murphy and Pieterse, 1993) combines the application of easily degradable herbicides, manual/mechanical methods and biological control like herbivorous insects (*Agasicles hygrophila* against alligator weed: Foret, 1974; *Neochetina eichhorniae* against water hyacinth: Haag, 1984; *Neohydronomus pulchellus* against water lettuce: Harley *et al.*, 1984; review: Pieterse, 1993a,b) and herbivorous fish. In many cases permanent control was achieved only when biological agents were supported by herbicide application (Murphy and Pieterse, 1993).

Making use of macrophytes as cattle feed and for the preparation of compost is common in warm climates around the world. Pollen, seeds, flower buds, leaves and stems, rhizomes and tubers are used for human nutrition and therapeutic applications in alternative medical practices (Shrestha, 1997a,b; Shrestha and Janauer, 2001). Macrophyte biomass is used for roofs and paper pulp production. Natural or artificial wetlands serve in sewage treatment units (Joyce, 1993). Water plants with attractive flowers are used as ornamentals, and some are bioindicators in toxicological tests (Schwertner, 1995). Bio-gas production has not surpassed the experimental or pilot level to a great extent due to numerous difficulties associated with the plant material (Polprasert *et al.*, 1986; Chynoweth, 1987).

In the member states of the European Communities the aquatic macrophyte vegetation is one of the essential biological elements for assessing the ecological status of waterbodies (Directive 2000/60/EC).

Macrophytes and Water Flow

Water flow probably is the most dominant factor determining the species composition and spatial distribution of plant stands in the running water environment. Yet, aquatic vegetation itself modifies the flow pattern in a channel and may even block water movement; for example, in irrigation systems. Thus many studies on channel hydraulics have devoted special interest to the subject of water plants, which are a prime feature regarding hydraulic resistance and interfere with calculated channel properties (Kaenel and Uehlinger, 1998; Kaehnel *et al.*, 1998).

A study of the channel conditions of small Hungarian streams measured the 'roughness' of nine channels. The values of Manning–Strickler's *k* (see below; VITUKI, 1993) described the following situation:

Description of channel	k (m$^{1/3}$/s)
Channel lined with concrete and stones	60–70
Gravel channel without vegetation	50
Sandy/fine gravel channel without vegetation	40
Channel with deposited sediment and light vegetation	35
Channel heavily overgrown with aquatic plants	28
Rocky channel with large boulders	25

The study revealed that channel roughness increased in importance for defining the flow-carrying capacity as flow depth decreased. However, it noted that 'the roughness due to the vegetation of the channel depends on a high number of factors and this field is less investigated in the relevant literature' (VITUKI, 1993).

Pitlo and Dawson (1993) showed that the presence of vegetation in a channel can affect flow by reducing water velocity, which in turn raises the water level in the channel and on the adjacent land by increasing flooding and overland spill, enhances the deposition of suspended sediment, alters the magnitude and direction of currents – which may enhance or reduce bank erosion in a given location – and interferes with water uses, such as navigation and recreation. They give a short review on the mathematical background used to describe flow phenomena in channels (Pitlo and Dawson, 1993), referring to the work of Prandtl (1904; see Pitlo and Dawson, 1993, 75) and Chow (1981) for calculation of the vertical distribution of velocity and to Manning's equation (Manning, 1891) which relates velocity, roughness, hydraulic gradient and depth. These approaches describe flow conditions in channels with cross-sections free of vegetation, and different attempts have been made to account for the change in channel roughness due to aquatic plant stands (Dawson, 1978; Kouwen and Li, 1980; Pitlo, 1986; Kaenel and Uehlinger, 1998). Additional complication arises from the fact that aquatic vegetation is flexible and therefore the hydraulic resistance of aquatic plant stands is not constant. This variation occurs both during the seasonal growth cycle and with different flow velocities or with the effect of 'pseudo-braiding' (Pitlo, 1982; Pitlo and Dawson, 1993).

Very little information is available on the direct relationship between aquatic plants and water flow. Macrophyte surveys along the course of a stream or river reveal changes in species composition, dominance and abundance, some of which are related to the change in flow conditions from source to mouth. Starting with field investigations in the late 1960s, total length surveys were performed in many rivers in Germany (Kohler *et al.*, 1971), triggering investigations in other countries (e.g. Austria: Janauer, 1981; Sweden: Sonntag *et al.*, 1999). At about the same time detailed studies started in the UK (Holmes and Whitton, 1975, 1977). These investigations just noted the change in species distribution along the river course, but did not report on flow conditions in detail. For river systems in the UK, Haslam's remarkable work (Haslam, 1978) was the first to describe macrophyte distribution in relation to flow, as well as spates and storm flows and substrate type. Hydraulic resistance was linked with morphology and growth form (Table 4.2), anchoring strength, susceptibility to turbulence, rooting depth (Table 4.3) and susceptibility to erosion and turbidity (Table 4.4). The distribution of plants was interpreted in relation to basin-scale properties such as drainage order, man-made constructions, water depth and storm damage, which links this

Table 4.4. Species with different turbidity tolerances. (Adapted from Haslam, 1978 based on surveys in the UK.)

Species	Turbidity tolerance		
	tol	int	let
Ceratophyllum demersum	■		
Lemna minor	■		
Nuphar lutea	■		
Polygonum amphibium	■		
Schoenoplectus lacustris	■		
Callitriche spp.		■	
Myriophyllum spicatum		■	
Potamogeton natans		■	
Potamogeton pectinatus		■	
Sparganium emersum		■	
Sparganium erectum		■	
Elodea canadensis			■
Potamogeton perfoliatus			■
Ranunculus spp.			■
Mosses			■

tol, tolerant; int, intermediately tolerant; let, least tolerant.

work to the more recent river habitat surveys conducted by the UK's Environment Agency (Environment Agency, 1997).

In other member countries of the European Union, first publications were summarized in Whitton (1984). Some contributors to this book paid attention to the longitudinal distribution of river vegetation, but reports on flow conditions were never directly linked with macrophyte data. A little later Haslam (1987) studied numerous rivers in the European Union, as she had done in the UK a decade earlier, and added to the general picture of macrophyte habitat conditions in continental Europe. A new level of sophistication was reached by the long-term monitoring of natural and man-induced changes in the macrophyte composition in rivers by Würzbach *et al.* (1997). These earlier works followed the distribution of species along the river course: some species are more confined to the upper reaches than others, which in turn may dominate the lower stretch. In a very broad sense this pattern is associated with an increase in current velocity – and water depth and turbidity – in the central parts of the downriver reaches. Yet, many aquatic plants susceptible to fast flow avoid the centre of the channel by growing closer to the banks, and occur in the lower reaches, too. A survey without measurement of flow conditions right next to the individual weed bed will reveal little about the effects of current velocity on species distribution.

Studies in the impoundments of hydroelectric power plants situated on the River Danube in Germany and in Austria have recently added some insight into the flow-dependent distribution of water plants. In the German reach a cascade of five power plants was surveyed (Pall and Janauer, 1995). Progressing down river (river-km 2552 to river-km 2511.8; in the Danube river-kilometres run from mouth to source), species numbers increased with the length of the impoundment. This is due to the slower flow conditions near the dam in longer impoundments (the longest one is 70% longer than the shortest, see Table 4.5). Rheophilic species like *Fontinalis antipyretica* and *M. spicatum* developed more biomass in the upper reaches, whereas some *Potamogeton* species tended to have a higher abundance closer to the hydropower station. Several species were more evenly distributed over the whole length of the impoundments, as there are lentic microhabitats near the banks everywhere.

In Austria, ten reaches with hydropower plants and two free-flowing stretches (Wachau Valley, 52.3 km long; Vienna to Slovak border, 55.8 km long) were

Table 4.5. Number of submersed and floating-leaf macrophyte species in hydropower impoundments of the River Danube in Germany. (Adapted from Pall and Janauer, 1995.)

Power plant	Left bank	Right bank	Approximate length (km)
Faihingen	16	14	6.3
Dillingen	21	16	6.7
Höchstädt	22	19	8.2
Schwenningen	23	20	8.5
Donauwörth	23	22	10.5

studied (Pall and Janauer, 1998). The impoundments are much larger than those in Germany (16–40 km). Constrained reaches, free-flowing as well as impounded, were dominated by mosses (Table 4.6). Impoundments situated in wide geological basins showed a rich development of higher vegetation (e.g. Ottensheim, Abwinden, Wallsee) closely associated with reaches near the barrage. Stretches with higher bank development also favoured higher macrophyte abundance. Bryophytes preferred two disjunct flow classes: high abundance was recorded either below 0.3 m/s or above 0.7 m/s. Only half the number of moss species was found in the intermediate flow class (0.3–0.7 m/s) for reasons that are still unclear. In contrast to the German stretch, higher plants occurred along the banks of the parallel levees and in harbours, side channels and the mouth sections of tributaries, always in rather stagnant conditions (flow below 0.3 m/s).

The connectivity of the main river stem with its side channels and oxbows is another important determinant of the relationship between macrophyte distribution and the flow effect. High hydrological dynamics associated with frequent changes in flow direction, which are typical for mouth sections of side channels, lead to drastically reduced macrophyte growth (Janauer, 1997; Bornette et al., 1998). In side channels and oxbows with negligible direct flood impact in the Danube river corridor in eastern Austria, 18 species were found and plant mass estimates reached highest levels. In waterbodies located close by but flooded

Table 4.6. The macrophyte vegetation in the Austrian reach of the River Danube. (Adapted from Pall and Janauer, 1998.)

Power plant/Free-flowing reach (FFR)	C/B	H	G	L	BD	M	S	F
Jochenstein	C	15.3	0.7	23.0	1.000	5	4	0.13
Aschach	C	15.3	0.4	40.0	1.041	5	4	0.25
Ottensheim-Wilhering	B	10.5	0.7	16.0	1.069	5	5	0.69
Abwinden-Asten	B	9.3	0.3	27.0	1.304	5	4	0.44
Wallsee-Mitterkirchen	B	10.8	0.4	25.0	1.198	4	5	0.56
Ybbs-Persenbeug	C	10.9	0.3	34.0	1.203	5	3	0.12
Melk	C	9.6	0.4	22.5	1.044	3	3	0.27
Wachau Valley (FFR)	C	–	–	26.2	–	5	0	0.00
Altenwörth	B	15.0	0.5	30.0	1.100	5	3	0.40
Greifenstein	B	12.6	0.4	31.0	1.077	5	3/5[a]	0.23
Freudenau	B	8.5	0.3	28.0	1.029	5	0[b]	0[b]
Reach east of Vienna (FFR)	B	–	–	55.8	–	5	4	0.07

C, constrained stretch; B, wide basin; H, water level difference at the dam (m); G, mean gradient (‰); L, length (km); BD, bank development (similar to shore development in lakes); M, occurrence of mosses; S, occurrence of higher plants and mosses, e.g. M5, S3 = the highest value reached on the five-level cumulative Kohler index scale, a measure for the total mass of macrophytes (here, mosses and higher plants are separated; for details see Pall and Janauer, 1998); F, frequency of the occurrence of higher plants with respect to the total length of the impoundment.
[a]The higher value is relevant for the harbours.
[b]This number refers to status only 1 year after completion of the power plant impoundment.

more frequently, only 12 species occurred and the plant mass estimates were considerably lower (Janauer and Kum, 1996).

Pall *et al.* (1996) found 29 species and a high ratio of pleustophytes (e.g. *Lemna minor*, *Ceratophyllum demersum*) in the active Danube flood plain in the Szigetköz region (Hungary). Many microhabitats with rather stagnant conditions were detected, but rheophilic species (e.g. *Potamogeton*) were abundant as well. Waterbodies sheltered by the flood protection levee were characterized by 25 species and a large ratio of submersed macrophytes and high abundance (Janauer *et al.*, 1993; Kohler and Janauer, 1995). Well-structured near-natural river banks like in the Szigetköz can support aquatic species despite the high connectivity (see also Janauer, 1997; Janauer and Schmidt, 2004; Goldschmidt *et al.*, 2006; Janauer, 2006; Janauer *et al.*, 2006; Schmidt *et al.*, 2006). However, deeper knowledge of this subject is still missing (Puckridge *et al.*, 1998).

Little information is available about flow within macrophyte stands (Madsen and Warncke, 1983; Marshall and Westlake, 1990; Minarik, 1990; Machata-Wenninger and Janauer, 1991; Stephan and Wychera, 1996). More information will be needed to better describe the habitat conditions of organisms living within the structures provided by macrophyte stands. Although the lack of detailed information is evident for many aspects of the aquatic vegetation, ecohydrological strategies can be formulated today to propagate, as well as to eradicate, aquatic vegetation in the process of managing running waters (Janauer, 2006; Janauer and Schmidt, 2006; Janauer *et al.*, 2006).

Flood Plain Vegetation

In the larger rivers of Europe, and elsewhere with similar flood protection measures, it is the width, man-made obstacles and the vegetation of the flood plain (within the levees, also called the flood berm) that substantially influence the flood-flow-carrying capacity of the river, thus the flood water stages, flood levels and, consequently, flood safety. During floods the flow velocity can be substantially reduced over the flood berm by its vegetation cover. This affects the flow-carrying capacity of the larger channel and thus increases flood levels. This effect can lead to highly dangerous flood situations when not controlled, particularly if man-made objects or intrusions also confine the flood channel cross-section. On the other hand, such flood plain ecosystems (e.g. in Eastern Europe and very few places elsewhere in the developed world) are highly valuable natural resources, last remnants of the once vast flood plain ecosystems of lower reaches of rivers. Therefore a careful balancing of actions is needed: keeping sufficient flow-carrying capacity yet maintaining the flood plain ecosystem's biodiversity. The two objectives are conflicting in many aspects. The hydrological–hydraulic approach to resolving this dilemma deals with the roughness or smoothness of the channel as a function of the vegetation present in the channel and/or in the flood plain.

Nagy (1997) studied the effects of flood plain forests on the flood load on levees. The importance of protection forests in reducing wave impacts on the levees was emphasized. The levee protection forests were planted along the levees of the River Tisza (Hungary) in 1930–1940. A combination of wind-reducing poplar trees with wave-reducing low willow trees seems to give the highest protection. The lower branches of the canopy should not be higher than the maximum flood level, so as to achieve the optimum wave attenuation effect. Studies on the effect of flood plain vegetation on the flood capacity of the full flood plain show that it has a dramatic effect through its impact on the smoothness of the channel (VITUKI, 1991). The effect may be expressed as:

$$C_{tree} = f(h, d, S_{t,\%})$$ (4.1)

where

C_{tree} = Chezy roughness coefficient for forested flood plain ($m^{1/2}$/s),
h = water depth over the flood plain at flooding (m),
d = diameter of the tree trunk (m),
$S_{t,\%}$ = percentage of the total area covered by tree trunks.

The Chezy equation (a basic hydraulic formula) relates flow velocity to channel slope and the hydraulic radius as:

$$v = C(RS)^{1/2}$$ (4.2)

where

v = flow velocity (m/s),
C = roughness coefficient,
R = hydraulic radius (wetted cross-section area divided by the wetted perimeter; in larger flood plains this can be well approximated by the average water depth),
S = channel slope (dimensionless, e.g. m/m).

An additional value C was defined for the undergrowth (brush) on the basis of the Manning tables (Chow, 1981). The resultant effect was C varying in the range 9–17 $m^{1/2}$/s for different Danube forested flood plains (particularly the Gemenc flood plain, where detailed vegetation and ecological mapping is also available).

Since flow is linearly proportional to velocity v ($Q = Av$), for a given cross-section, this means that the flow-carrying capacity of forested Danube flood plains can be doubled or halved, depending on the vegetation (and the season). A similarly quantitative (numerical) conclusion was obtained in another study (Kozák and Rátky, 1999), showing the smoothness factor k of the Manning–Strickler equation to vary from 5 $m^{1/3}$/s (for dense, scarcely penetrable forests with heavy brush undergrowth) to 30 $m^{1/3}$/s (for pasture with short cut grass). The equation is:

$$Q = Av = AkR^{2/3}S^{1/2}$$ (4.3)

where

Q = flow (m^3/s),
A = cross-sectional area (m^2),

v = flow velocity (m/s),
k = smoothness factor ($m^{1/3}$/s),
R, S = as above.

The conversion between the two factors C (Chezy) and k (Manning–Strickler) is:

$$C = kR^{1/6} \tag{4.4}$$

These calculations show that the flow velocity, and thus the flow-carrying capacity, can be six times higher in 'clean' flood channels than in heavily overgrown ones. On the basis of applying a non-steady-state hydraulic model the authors concluded that, for a characteristic Hungarian Tisza river flood plain at a flood of 3000 m^3/s, the increase of flood level would be 1.8 m higher in a wooded section than in a pasture one. The difference would be approximately 2.6 m at a flood flow of 4000 m^3/s. While the water-carrying capacity of a flood plain of high hydraulic resistance (k = 5 $m^{1/3}$/s) is only about 0.5–25% of the main channel (depending on the depth and the width of the flood berm), this might increase to 150% at flood plains of high flow-carrying capacity (k = 30 $m^{1/3}$/s).

Kozák and Rátky (1999) concluded that, when the space between tree trunks is less than ten times the trunk diameter, the roughness due to the trees will be dominant while the bottom roughness (the undergrowth) will be less important, and the roughness varies only slightly with the water depth. When the canopy of the trees is higher than the design flood stage H_{max} (e.g. is always above the water level) and there is no undergrowth, then the hydraulic resistance of the flood plain forest is very low.

Overall, our state of knowledge about the hydraulic consequences of aquatic vegetation needs to be improved so that the roughness and smoothness of a wider range of examples of flood plain and channel types can be defined, to enable a compromise to be made between flood protection and biodiversity maintenance. In river management, compromises need to be found between channel clearance and the channel macrophyte diversity, and in flood plain management the protection of levees against wave erosion adds a component that makes this possible. Vegetation strips for levee protection can be turned into more linear ecotone elements, which maintain biodiversity as well as offering nutrient and pollutant filtering effects.

Conclusion

Macrophytes are one of the key ecological elements in many aquatic ecosystems. Planning and management strategies for waterbodies must consider them as an indisputable aspect in flowing and still waters, no matter if artificial or natural. Macrophytes provide habitat for numerous other groups of organisms, making them an ecosystem structure of extremely high importance. Considering their value as the starting element for many food webs, it is evident that they also contribute positively to human uses, even when seen in an economic perspective. It is thus necessary to balance the ecological benefits with the real and potential negative impacts for man. In heavily populated areas and intensively cultivated landscapes the aquatic macrophyte vegetation of flood plain water-bodies is a rare and vanishing ecosystem element under present land-use conditions, suffering from regulation regimes in streams and rivers. Macrophyte stands are often composed of rare and/or endangered species, deserve protection where needed and are a substantial element in assessing the ecological status of waterbodies according to the European Water Framework Directive.

5 Processes Influencing Aquatic Fauna

J.A. Gore[1], J. Mead[2], T. Penczak[3], L. Higler[4] and J. Kemp[5]

[1]Program in Environmental Science, Policy and Geography, University of South Florida, St Petersburg, Florida, USA; [2]Division of Water Resources, North Carolina Department of Environment and Natural Resources, Raleigh, North Carolina, USA; [3]Department of Ecology and Vertebrate Zoology, University of Lodz, Lodz, Poland; [4]ALTERRA, Wageningen, The Netherlands; [5]SEPA Aberdeen, Aberdeen, UK

Longitudinal Patterns

As water flows down a river basin from headwaters to mouth, changes in slope and the input of suspended load from tributary streams have the effect of altering the manner in which the body of water dissipates energy. Flow patterns, the distribution and the frequency of occurrence of various depth, velocity and substrate combinations can be expected to change with distance from the headwaters (Vannote *et al.*, 1980). In general, headwater streams have shallow slopes and gentle flows but, at a distance relatively close to the source, undergo a shift to high-velocity flows as the slope increases dramatically. As the river nears the foothills or lowland areas, another transitional zone is encountered in which flows are decelerated and the river begins to braid and deposit suspended loads. Further downstream, the river begins to meander and undercut bank material. This can be a zone of unpredictable flows as the river begins to cut backwater areas and form oxbow lakes (Table 5.1).

Statzner and Higler (1986) proposed that an index of hydraulic stress, laminar sublayer thickness and slope could be used to describe a 'typical' pristine flow, for the purposes of comparison. The source and headwaters, then, are characterized by low hydraulic stress while the first transitional zone (approaching the high-slope zone) is characterized by elevated hydraulic stress. At the zone of transition to lowland streams, high hydraulic heterogeneity exists as the river begins to braid but is characterized by lower hydraulic stress. As the river meanders and forms backwaters and oxbow lakes, numerous hydraulic discontinuities occur. In an examination of 14 stream systems around the world, Statzner and Higler (1986) found a significant correlation between high benthic invertebrate diversity and the location of the transitional zones between low and high hydraulic stress (Fig. 5.1). This is not a surprising result as the availability of substantially

Table 5.1. Ecological characteristics along the course of running waters in Europe. (From Higler and Statzner, 1988).

Source	Glacier	temperature $(T) < 7°C$; high turbidity in summer; no canopy; very few species
	Hot spring	$T > 30°C$; high turbidity (often by sulfur); no canopy; very few species
	Lake	T depends on origin; so does turbidity; number of species and organisms in accordance with trophic status and origin
	Limnocrene	groundwater of low T follows annual T; clear water; in natural conditions canopy comprises lentic organisms
	Helocrene	as limnocrene, but T tends to be lower and with lotic organisms in addition
	Rheocrene	$T = 5–12°C$; clear water; in natural conditions under timberline canopy comprises lotic organisms and organisms specialized to living in vertical thin water layer
	Seepage area	groundwater of low T follows annual T with smaller amplitude; clear water; generally canopy species composition as in helocrene, but more adapted to semi-permanent conditions (The last four types are dependent on the slope of terrain)
Upper course		T and turbidity depend on the origin of the source; $P/R < 1$ (production is generally lower than respiration); allochthonous production from the bank vegetation is not considered here but may play an important role; sometimes these courses are temporary; order 1–3; dimensions small (standard type) or large (lake outlet: Karstic water); slope related to source is an important ecological factor (C_t); canopy results in allochthonous organic matter and then shredders dominate; if there is no canopy, grazers and predators dominate at high current velocities, but non-filtering collectors, grazers and predators dominate at low ones; fish species like trout and bullhead
Middle course		extreme temperatures of glacier streams and hot springs tend to stabilize, the others have larger amplitudes than upper courses; $P/R \cong 1$ or $P/R > 1$; order 4–7; dimensions and slope are important ecological factors (R and C_t); high numbers of species; primary producers are algae on stones and beginning production of plankton; macrophytes in soft-bottom stretches; invertebrates are filter feeders, collectors, grazers and predators; fish in temperate streams are grayling/barbell; higher nutrient content than in upper course by input from the upstream part of drainage area
Lower course		daily temperatures have been stabilized; annual T follows air T; $P/R > 1$ or in disturbed systems $P/R < 1$; under natural conditions many stagnant and periodically overflown waters in the flood plain can add to high production; depth is the most important dimension variable (R); sedimentation of sand and silt in the lower reaches; generally low numbers of species in modern streams by pollution and weirs; highest number of species in natural conditions by flood plain waters; in stream predominantly collectors; under natural conditions all types of functional groups according to local variation; high nutrient content; plankton and macrophytes; barbel/bream in temperate streams

different hydraulic microhabitats would increase in these transitional zones and a diverse fauna, existing at the limits of their hydraulic tolerances, could be expected to be collected. Indeed, Statzner and Higler suggested that, in the zones between the transitions, communities are more stable in their structure; their resilience to disturbance might be different from those communities at the transition zones. This general conclusion was echoed by other studies outside North America; that zonational patterns could not be attributed equally to patterns of particulate organic carbon use, as in the River Continuum Concept (Vannote *et al.*, 1980). In Australian streams, for example, temporal and spatial flexibility of the fauna is more attributable to variability in streamflow (Lake *et al.*, 1986), while King (1981) found in South Africa that changes in physico-chemical conditions (including water velocity) correlated well with changes in faunal distributions along the length of a short coastal river. Brussock and Brown (1991) also found a stronger longitudinal distribution based on changes in riffle pool geomorphology and concluded that this pattern generally held true for alluvial gravel streams.

The relationship between changes in hydraulic condition and benthic invertebrates extends to the interstitial meiofauna as well. Ward and Voelz (1990) found that substrate condition and localized hydraulic conditions were more important than elevational factors in determining meiofaunal distributions. The areas of transition between meiofaunal communities coincided with the zones of hydraulic transition, as predicted by Statzner and Higler (1986).

Longitudinal transition in fish communities has been well documented (Sheldon, 1968; Jenkins and Freeman, 1972; Evans and Noble, 1979; Platania, 1991). Most of the patterns of species replacement or change in community structure are related to identifiable physical habitats associated with site-specific hydraulic conditions. Chisholm and Hubert (1986) found that brook trout densities varied as a function of gradient, as well as stream width-to-depth ratios. There were three identifiable zones of population density, varying from highest densities in low-gradient streams to lowest densities in highest-gradient areas. Platts (1979) also identified an abrupt transition in salmonid-dominated communities at the first zone of transition to increased hydraulic stress in low-order streams of Idaho.

The importance of hydrologic and hydraulic conditions in determining the preferred microhabitat of aquatic organisms extends to the longitudinal distribution and zonation of lotic biota as well. Indeed, Statzner (1987a,b) had predicted that zonational changes occur at major hydrologic breaks (usually at geomorphic changes in slope). In these areas where a high diversity of hydraulic conditions occurs, diversity of biota has been found at its highest value. These points of hydrologic change usually coincide with optimum points of placement for most hydraulic control structures, particularly impoundments. Thus, these ecotonal areas, which are reasonably fragile in structure (Naiman *et al.*, 1988), are often the first communities to be altered by anthropogenic changes that influence discharge regimes. The settlement pattern in North America shows this. European settler communities were often established near hydrologic breaks that afforded a source of initially hydromechanical, later hydroelectric, power.

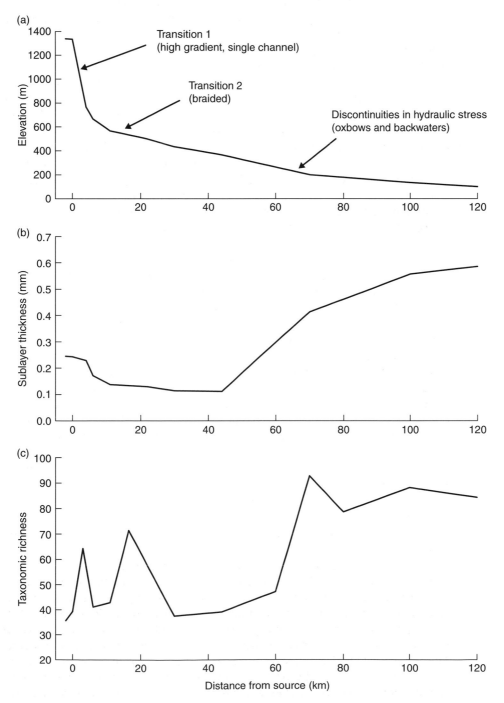

Fig. 5.1. The suggested relationship between hydraulic heterogeneity (a), complex hydraulics at the substrate (as sublayer thickness) (b) and diversity of aquatic organisms (as taxonomic richness) (c) over the length of a typical river. (Adapted from Statzner and Higler, 1986.)

Not only does the physico-chemical nature of release water from impounded waters in these areas influence the structure and function of downstream communities, but alterations in flow patterns also influence the distribution of substrate particles and the pattern and distribution of flow which, in turn, creates different patterns of microhabitat availability. Furthermore, impoundments serve as a sediment trap for all of the upstream catchment area. These discontinuities can have the effect of 'resetting' community structure to mimic those of areas further upstream where physico-chemical conditions are similar (Ward and Stanford, 1983). The ability of the lotic biota to recover from these changes in physical environment is an index of the ultimate severity of the disturbance.

Superimposed upon these longitudinal and within-reach changes in hydraulic conditions are changes in the seasonal hydrograph, short-term events such as floods, potentially long-term events such as floods, and the effects of human intervention on the flow regime (i.e. afforestation, impoundments, diversions and abstractions), which have the potential to alter the temperature regime, flow regime and the availability of particulate organic food to lotic biota.

Within-channel Distribution Patterns

In order to translate the hydrological changes described above into some sort of response variable by riverine biota, it is necessary to understand the interactions of within-channel hydraulics and the distribution of biota among various habitat types in the channel. Statzner (1981) used Manning's formula to create an important index, called hydraulic stress. It is the impact caused by the powers of running water on animals that live in conditions of flow. These animals have the advantage of a permanent supply of oxygen and food. Their position is perilous because they are in constant danger of being swept away by the current. In the course of evolution, many species have adapted in one way or another to life under swiftly changing conditions of less and more current velocity, of shifting sediments and, sometimes, drying out of the environment. These organisms are so well-adapted to the circumstances in running waters that they survive only in streams and cannot compete with organisms in standing waters.

While Statzner was the first aquatic ecologist to break into the field of hydrology, a Canadian hydrologist, Newbury (1984), was experimenting with small-scale hydraulics to find out what forces were working on small objects such as water insects. In general, there are four major hydraulic conditions that most affect the distribution and ecological success of lotic biota. These are: (i) suspended load; (ii) bedload movement; (iii) water column effects (such as turbulence, velocity profile); and (iv) near-bed hydraulics (substratum interactions). Singly, or in combination, the changes in these in-stream conditions can alter the distribution of biota and disrupt community structure. Within a stream reach the interactions of these hydraulic conditions with the morphology and behaviour of the individual organisms govern the distribution of aquatic biota (Table 5.2).

Table 5.2. In-stream hydraulic characteristics which have been shown to influence the distribution of riverine biota.

Simple hydraulic characteristics:

D	mean depth (cm)
g	acceleration due to gravity (cm/s^2)
k	substrate characteristic (cm)
Q	discharge (m^3/s)
S	slope of water surface
U	mean current velocity (cm/s)
w	stream width (cm)
v	kinematic viscosity (cm^2/s)
ρ	density of water (g/cm^3)

Complex hydraulic characteristics relevant to describe flow for an area within a stream reach (e.g. IFG4 cell, benthic sampler, etc.) or for an entire stream reach:

Fr	Froude number (dimensionless); $Fr > 1$ (shooting), $Fr < 1$ (tranquil)
Re	Reynolds number (dimensionless); drag characteristics
Re^*	boundary Reynolds number (dimensionless)
U^*	shear velocity (cm/s)
δ'	thickness of viscous sublayer (cm)
τ	shear stress (dyne/cm^2)

Formulae:

$$Fr = U/(gD)^{0.5}$$

$$Re = UD/v$$

$$Re_1^* = U_1^* k/v$$

$$Re_2^* = U_2^* k/v$$

$$U_1^* = (\tau/\rho)^{0.5}$$

$$U_2^* = U/5.75\log_{10}(12D/k)$$

$$\delta_1' = 11.5v/U_1^*$$

$$\delta_2' = 11.5v/U_2^*$$

$$\tau = gSD\rho$$

Suspended load

Transported sediment has been observed in rivers to increase the velocity and consequently the fluid discharge (Karim and Kennedy, 1986). The velocity augmentation is greater at increasing distance from the substrate and with higher concentrations of suspended sediment (Coleman, 1981). These changes follow the logarithmic velocity distribution found in most open channels, with the maximum velocity increase roughly equivalent to the boundary shear velocity, U^* (Coles, 1956) (Table 5.2). Thus, any hydrological change (either in-stream, such as channelization; or in the catchment area, such as altered logging or agricultural practices), which increases sediment–discharge relationships, might be expected to alter the water column velocity profile and, through increased shear and scour, the pattern of flow and depth combinations in downstream areas. This can have a significant effect on the distribution of the majority of biota occupying reaches downstream from the point of increased sediment loading. Of course, the concentration of suspended particles may also have an effect on scour of the substrate and mechanical damage to biota in the water column.

Bedload movement

In general, suspension of particles in the water column varies as a function of shear velocities. Suspension of substrate particles occurs when the shear velocities exceed the fall velocity of the particle, while deposition of those same particles is also a function of the vertical distribution of sediment and turbulent dispersion upwards in the water column (Engelund, 1970). This means that the transport of particles past a given point is a function of conditions upstream from the observer. Bed material is not completely mobilized until the stress is great enough to move the material that forms the matrix or framework of the bed deposit (Milner *et al.*, 1981). At any given time then, bedload may not be characteristic of the bed material (Everts, 1973). Indeed, the composition of substrate at any one time is characteristic of the frequency of high-discharge events. Bedload mobilization and movements are more frequent in areas with high flood frequency. Anthropogenic changes in flood frequency (impoundment or hydropower operation) can have a significant effect on bedload movement and downstream patterns of shear stress and other near-substrate complex hydraulics.

Water column effects

Turbulence within the water column is a rather simple function directly correlated with stream velocity and the inverse of depth (see Froude number, Table 5.2). In an ideal channel, maximum water velocity is centred below the surface. Just outwards from the centre of flow, velocity is reduced but turbulence is highest. In natural systems channels are asymmetrical and the zone of greatest flow is shifted towards the deepest side, with areas of highest turbulence on the shallow side of the zone of maximum velocity (Leighley, 1934). The transition from tranquil to shooting flow (at $Fr = 1$; the hydraulic jump) is likened to the transition from potential to kinetic energy and is the point at which the river is able to move large particles of substrate.

Deposition occurs, then, where minimum turbulence is common or where turbulence is high above the bottom. On the other hand, helical flow, at the bends in rivers, leads to turbulence with a downwards, converging movement of water on the outside of the bend where erosive forces are strong. On the inner side of the bend, the helical flow is upwards and deposition occurs as the formation of a point bar (Leopold *et al.*, 1964). Thus, the frequency of turbulent flows combined with sinuosity of the river can determine the distribution of substrate particles and, therefore, a significant component of habitat for lotic biota.

Any channelization or alteration (or creation) of hydraulic jumps can cause a significant change in substrate distributions and the distribution of biota linked to distinct substrate sizes. Indeed, the hydraulic fitness of swimming organisms, and by implication fish community structure, is also affected by the distribution of turbulent zones (Scarnecchia, 1988).

Near-bed hydraulics

For any organisms living on or near the substrate of a river or stream, the interactions of depth and velocity with the profile of the substrate particles is of critical importance to the range of potential microhabitats available (Statzner *et al.*, 1988). Table 5.2 summarizes some of the critical hydraulic conditions that a benthic organism must encounter. Although studies of flow in flumes have indicated the existence of a non-moving boundary layer (up to 2 mm in thickness) (Nowell and Jumars, 1984) and lotic scientists have indicated that morphological and behavioural adaptations are directed towards existence within this boundary layer (Hynes, 1970), recent investigations have determined that very few benthic organisms are sufficiently streamlined to take advantage of this boundary layer (see discussions below and Statzner, 1987a). Indeed, a true boundary layer rarely exists under natural flow conditions because of the complex profile of the substrate. Thus, aquatic organisms are restricted to those combinations of velocity, depth and substrate that allow morphological and/or behavioural resistance to flow to be exceeded by energetic gains from foraging in these areas. That is, conditions such as shear stress and the thickness of the slower-moving laminations of water that constitute the viscous sublayer are the major determinants of the distribution of most benthic organisms within a given stream reach.

Davis and Barmuta (1989) classified near-bed flows as hydraulically smooth or hydraulically rough (chaotic, wake interference, isolated roughness, or skimming flow). Each of these flow types is associated with recognizable assemblages of benthic biota. Any hydrologic change that leads to increases in shear velocities or shear stresses or reduction in the thickness of the viscous sublayer will reduce the availability of adequate microhabitats to some species while increasing it for others and, presumably, alter the abundances of individuals in the community. For example, Ciborowski and Craig (1989) found that increases in velocity resulted in increased drift and significant changes in aggregations of larval blackflies. Indeed, the position of the nearest upstream organisms influenced the hydraulic conditions of the downstream individual and its ability to feed. In turn, community functioning and food web dynamics may be changed to the extent that some species will be locally eliminated from the community.

Near-bed conditions are further altered with changes in substrate composition or distribution since increased hydraulic roughness increases the rate of sediment deposition. Increased sedimentation rates have been shown to significantly alter the composition of benthic communities as interstitial microhabitats are eliminated (increased predation success; Brusven and Rose, 1981) and scour or deposition eliminates primary production (Chutter, 1969; Brusven and Prather, 1974).

Life Cycles

The life cycles of lotic biota appear to be regulated by two major physical factors: temperature and hydraulic conditions. In many circumstances, hydrological conditions influence temperature distribution along the length of the river and certainly influence the zonation of lotic biota (Gordon *et al.*, 1992).

The impacts of temperature change on aquatic insects are well documented (Hynes, 1970; Allan, 1995). Temperature and day length seem to be the regulating features that synchronize hatching, maturation of larvae, emergence and mating of adults and, in some cases, egg-laying behaviour. Not only are there certain temperature optima for the hatching of various eggs but a diel cycle also appears to enhance development (Humpesch, 1978). Indeed, it appears that greater daily temperature changes reduce the degree-hours required for hatching of some species of aquatic insects (Sweeney and Schnank, 1977). Although there is some debate about the importance of this diel temperature cycle, alteration of thermal regimes by change in hydrologic condition could result in reduced success of many macroinvertebrate species.

Thermal cues also regulate the ability of aquatic invertebrates to enter a period of quiescence or rapid development prior to emergence as adults. In general, rising temperatures induce diapause and declining temperatures terminate that period of dormancy (Ward, 1992). These responses pre-adapt a species to survive drought periods (or periods of artificially low flow) and to colonize temporary waters. In temperate zone rivers, rising temperatures in the spring cue growth and maturation among larval and nymphal aquatic insects that have overwintered. In combination with day length, the summation of degree-days synchronizes the emergence of adults. Declining autumnal temperatures may also cue the emergence of the more rapidly growing summer communities of benthic invertebrates (Ward, 1992). Artificial disruption of thermal patterns and/ or day-length cues has been shown to disrupt emergence patterns and reduce population success (Ward and Stanford, 1982). Higher than normal discharge during snowmelt periods may depress downstream water temperatures and delay emergence (Canton and Ward, 1981).

The combined influence of temperature and hydraulic preferences has also been demonstrated for lotic fish species, particularly benthic species, where distinct thermal tolerances correlate well to preferences for pool (higher temperature) or riffle (lower temperature) habitats (Hill and Matthews, 1980; Ingersoll and Claussen, 1984). Any shift in hydrologic pattern that leads to an alteration of the established thermal regime of a lotic ecosystem will ultimately lead to a dramatic change in the composition and survival of lotic biota. Statzner *et al.* (1988) summarized the results of many examinations of the relationship between the distribution of lotic biota and the heterogeneity of flows within a stream reach.

There is little doubt that a strong interaction between body morphology and hydrodynamic stresses controls the ability of lotic species to forage and maintain position within a preferred habitat. The relationship between flow conditions and life cycle requirements is not as clear. However, there is evidence to indicate that flow patterns also control life cycle success as well. Dudley and Anderson (1987) demonstrated that receding water level cued pupation for the tipulid, *Lipsothrix nigrilinea*. In this case, discharge was the governing factor rather than the hydraulic conditions. Pupation by the larvae could be delayed for a year or longer if water levels did not decline. As growth leads to an alteration of both morphology and cross-sectional area of the organisms exposed to various hydrodynamic conditions, it could be expected that the organisms must migrate to areas of suitable hydraulic fitness in order to continue to forage and mature. Gersabeck and Merritt (1979) calculated distinct velocity preferences for different instars of larval blackflies, while Gore et al. (1986) demonstrated significantly different depth distributions for the final instars of the caddisfly, *Hydropsyche angustipennis*. Although insufficient physiological data existed to verify, Gore and Bryant (1990) speculated that the apparent short-term preference of egg-bearing female crayfish for high-velocity riffles is a result of aeration requirements for successful incubation. Statzner et al. (1988) concluded that the distribution of the various life stages of the hemipteran, *Aphelocheirus aestivalis*, was controlled by the ability of each instar to respond to changes in the distribution of hydraulic stresses across the substrate. Thus, early instars required low-stress areas (pool and moderate runs) while adults lived almost exclusively in high-velocity riffles. In stream reaches where a high diversity of depth and velocity combinations existed, there also existed a high diversity and density of instars and adults. Diversity of velocities and depths in stream reaches is directly related to discharge and to diversity of fish and macroinvertebrate species (Gore et al., 1992, 1998). Artificial discharge patterns that promote more homogeneous flow patterns might be expected to sustain lower diversities of lotic species, and reduce or eliminate some species that are cued to certain hydraulic conditions. This can be the result of extending the duration of low flows (diversion and abstraction projects), increasing the frequency of high flows (hydropeaking), or reducing the frequency of flushing flows and flooding (regulation).

Organism Morphology

Ambühl (1959) indicated that most benthic invertebrates were sufficiently streamlined to avoid the shear conditions along the substrate and could, therefore, forage across the substratum with relative impunity. This notion was generally accepted by most lotic ecologists and contributed to the search for other land–water interactions that might serve as a major template to the distribution of aquatic biota (both within reaches and along the length of river ecosystems). In the past 2 decades a number of scientists have examined the interactions of complex hydraulics and the responses of biota to changes in these conditions. Gore (1983) demonstrated a significant correlation between shell length (as an indicator of exposure to shear velocities) and velocity distributions among pleurocerid snails. Using laser Doppler anemometry, Statzner and Holm (1982, 1989) and Statzner (1987a, 1988) demonstrated that benthic invertebrates do not exhibit the streamlining conditions which Ambühl reported. Indeed, Reynolds numbers (the interaction between frontal area exposed to the current, the length of the body and the velocity and viscosity of the water) were sufficiently high to predict that, in order to compensate for the forces of pressure drag and friction drag and maintain an energetic gain from foraging activity along the substrate, most benthic organisms were quite limited in the range of conditions in which they could exist.

Although high drag coefficients may be of some advantage to organisms swimming in still waters where the edges of the prothorax or elytra may aid predators in making fast turns and stops (Nachtigall and Bilo, 1956), resistance to flow in lotic ecosystems implies that body shape (a compromise between hemispherical and ovoid) and behavioural responses to changes in flow restrict the ability of lotic organisms to move freely within stream reaches and between stream reaches by limiting energy gains to occupancy of a narrow range of flow microhabitats. Indeed, because of the changes in surface-area-to-volume ratio during the growth of an individual, many species (mayflies, for example) must move to higher velocity as they grow larger in order to satisfy oxygen require-ments (Kovalak, 1978). Resistance to entry into the drift increases as much as 1000-fold during the growth of benthic invertebrates (Waringer, 1989) and is a function of increase in size and alteration in cross-sectional profile presented to the direction of flow (Dussart, 1987). Gore (1983) suggested that exposure of shells to higher-velocity lamina might result in the erosion or decollation of the top spires of lotic pleurocerid snails. This decollation can result in further bacte-rial and fungal deterioration of the shells (Burch, 1982). Thus, increases in aver-age velocity in a channel might be expected to differentially affect cohorts of the same benthic population. Statzner *et al.* (1988) used these hydraulic parameters to demonstrate that the density and distribution of specific cohorts of benthic organisms could be predicted if complex hydraulic conditions such as Reynolds velocity, laminar sublayer thickness and even Froude number were known at the point of sampling.

The effect of flow is not limited to the interactions of complex hydraulics at the substrate level. Brewer and Parker (1990) determined that the break force on the stems determined the upslope distribution of macrophyte species. The tensile strength was different for all species and the force required for stem breakage varied as a power function of the cross-sectional area. Thus, thick-stemmed spe-cies are the only species occurring in high-gradient streams while lowland rivers and large-order rivers tend to present more flow microhabitats for a variety of macrophyte species. Similarly, Scarnecchia (1988) used the fineness ratio (FR; ratio of length to maximum diameter) of Webb (1975) to demonstrate that homogeneity of flows in a channelized river tended to reduce the diversity of fish communities. Where an FR value of 4.5 gives minimum drag for fish, stream reaches of high uniform flows were characterized by fish with this more highly streamlined characteristic while species with FR values less than 4 were virtually eliminated from these areas.

Behaviour and Physiology

As hydraulic conditions change within a stream channel, the effects on stream biota are either direct mechanical damage and/or removal of the organisms from the preferred microhabitat, or are exhibited by changes in the behaviour and/or physiological conditions of the biota. These changes in behaviour and physiology may mean a net loss or reduction in energy for growth and reproduction, or the inability to breed or forage in appropriate habitats.

Although Pfeifer and McDiffett (1975) indicated the potential for large increases in nutrient uptake with increased velocity, a minimum current of at least 5 cm/s across the surface of periphyton beds will stimulate increased primary production in lotic ecosystems (Lock and John, 1979). The influence of changes in current velocity extends to composition of periphyton communities themselves. Peterson (1986, 1987) found that reduced mean daily flows supported higher diatom biomass in partially exposed areas with a higher diversity of taxa including unattached colonial species. Those species inhabiting slow current areas were less resistant to desiccation, probably due to slow nutrient renewal.

Among fish species, changes in hydraulic conditions may differentially affect certain cohorts or functional groups within the community. These changes are related to energetic balances tuned to the body morphology discussed above and physiological abilities. Among fish, morphological adaptations that minimize exposure to shear in a microhabitat or preference for low velocities (and low oxygen consumption) can be predicted by an energy-cost hypothesis (Facey and Grossman, 1992). Although it is well known that fish can rapidly adjust their swimming speeds and positions in streams with changes in current velocity (Jones, 1968; Hanson and Li, 1978), Trump and Leggett (1980) were able to demonstrate that the most efficient swimming conditions (particularly during migration) were under conditions in which current velocities can be rapidly detected and are not rapidly varying. Under conditions other than these, energetic losses may lead to decreased spawning success or condition factor. In the same manner, Gibson (1983) has noted that the change from aggressive to schooling behaviour for migration is triggered not only by changes in temperature in the stream ecosystem, but is also tied to changes in buoyancy in association with changes in current velocity.

Shifts in behaviour and/or physiological demand according to changes in near-bed flows are also common among benthic species. At a minimum, many species of lotic insects have been shown to increase respiratory activity with decreases in current velocity (Feldmeth, 1970). Kovalak (1976) found larvae of the trichopteran, *Glossosoma*, to occupy upstream faces of rocks during times of higher temperatures and lower dissolved oxygen concentrations. Similarly, mayfly nymphs change position on the surface of substrate particles in relation to current velocity and dissolved oxygen concentrations. As dissolved oxygen decreases, nymphs take positions in higher current velocities, usually exposing themselves to greater shear and mechanical removal (as drift) or greater predation (Wiley and Kohler, 1980). At least some species of crayfish exhibit the same apparent high-risk activity (exposure to predation) in order to oxygenate egg clusters (Gore and Bryant, 1990). With increases in stream velocity, many crayfish species can alter body posture to counteract the effects of drag, maintain position and take advantage of increased oxygen in these areas (Maude and Williams, 1983).

The phenomenon of downstream drift of benthic species has been attributed to a variety of factors including negative phototactic responses, escape from introduced toxic substances or increased sediment load, escape from predation, interference competition as density increases, and as a response to a deteriorating microhabitat. Brittain and Eikeland (1988) have provided an excellent review of the phenomenon of drift. Although some of the factors are quite controversial in their interpreted influence, there is little doubt that changes in velocity pattern result in increased drift from areas in which the hydraulic habitat is no longer suitable. Varying discharges, particularly as a result of hydropower operation, are known to initiate catastrophic drift among stream invertebrates and larval fish (Cushman, 1985). However, it is not certain if the rate of change of discharge is the primary influencing factor. That is, releases that simulate large floods do not seem to initiate greater drift densities than gradual increases in discharge (Irvine and Henriques, 1984). However, it does appear that initial changes in flow initiate more drift than subsequent similar flow changes (Irvine, 1985), which seems to indicate that areas with unpredictable flow patterns are continually occupied by invertebrates at lower densities than the hydraulic habitat might support under more predictable or stable flow conditions (Ciborowski and Clifford, 1983). The addition of sediment to increased flows further increases drift density, as scour tends to remove some individuals maintaining position in adequate flow habitats (Ciborowski *et al.*, 1977). The importance of the hydraulic habitat to the success of benthic species is underlined by the observations that drift also increases during conditions of decreasing flow. Catastrophic drift is usually initiated at some discharge greater than conditions in which the majority of the substrate is de-watered, indicating that lower velocity tolerances or preferences can also be exceeded (Gore, 1977; Corrarino and Brusven, 1983). In either case, increased or decreased discharges, which promote hydraulic conditions that are outside the bounds of preferences of the benthic species (and/or increase sediment loads to the flow; Culp *et al.*, 1986), can result in a net loss of benthic species from a given stream reach. Indeed, the redistribution process of macroinvertebrates along the length of a stream is largely dependent on channel hydraulics (Lancaster *et al.*, 1996). Thus, changes in channel hydraulics may have a significant impact on the community composition of macroinvertebrates along the continuum from headwater to mouth. Tachet *et al.* (1992) found that only those species of net-building caddisflies capable of adjusting their filtering activity in the face of decreased velocities were able to survive in rivers with lowered discharge.

Bioenergetics

The effects of hydrological processes on ecology are more usually expressed as distribution patterns and abundances, as in the examples given above. The response of single species, at population level, is usually confined to larger animals that are easier to record and to conduct experiments on. Fish are ideal for this, in terms of both their size and their diversity in river basins, as well as their economic value to man for food and recreation. Many studies have been made on fish distribution, providing parallel information to that of macroinvertebrates for classification and prediction. Fish bioenergetics, however, provides an insight into how individuals respond to the constraints of their environment and the anthropogenic changes imposed upon it.

Energy budgets have a long history in aquatic ecology, stretching back to the middle of the 20th century (Winberg, 1956). Warren and Davis (1967) used an expanded form of Winberg's model for fish (with some symbols changed):

$$C = (R_s + R_d + R_a) + (P_s + P_g) + (F + U + Mu) \qquad (5.1)$$

where

C = rate of energy consumption,
R_s = standard metabolic rate,
R_d = metabolic rate increase due to specific dynamic action,
R_a = active metabolism rate, which is recorded in animals undergoing sustained activity,
P_s = somatic growth rate,
P_g = gonads' growth rate,
F = faeces rate,
U = excretory products rate, e.g. urea and ammonia,
Mu = mucus secretion rate.

Data derived from such a model have been used to estimate how much energy a given population transforms during an arbitrarily established time (e.g. instantaneous energy budget) or during all ontogeny (cumulative energy budget) (Klekowski and Duncan, 1975).

The basic energy budget parameters, such as C, P, R and $F + U$, can be related to one another in a non-dimensional form (or percentage) expressed as coefficients of ecological efficiencies to become simple and useful tools in understanding and explaining the importance of energy and nutrients allocation in investigated populations (Klekowski and Duncan, 1975). The three below were developed theoretically by Ivlev (1939) and have subsequently been extensively applied, both for energy and nutrients and either instantaneously or cumulatively (Winberg, 1965; Naiman, 1976; Mann, 1978; Penczak *et al.*, 1982, 1984, 1999, 2001; Knight, 1985; Penczak, 1985, 1992, 1999; Ney, 1990; Kamler, 1992; Hanson *et al.*, 1997):

$U^{-1} = A/C$ is the coefficient of assimilation (A) efficiency,
$K_1 = P/C$ is the gross ecological efficiency coefficient (one of the most commonly used),
$K_2 = P/A$ is the coefficient of energy assimilated for growth.

Energy budgets have now been calculated for many populations and in different ecosystems, such that it is possible to analyse changes in the food base utilization in a given year (or ontogeny) by comparing populations influenced by human impacts or climatic abnormalities. Water velocity patterns and organisms' energy expenditure are strongly interdependent. In natural aquatic ecosystems there is a mosaic of velocities, resulting in areas where animals can rest and feed with low energy expenditure. The lack of such areas is very detrimental (Calow, 1985), especially for juveniles, resulting in dramatic mortality (Schlosser, 1985; Schiemer *et al.*, 1991) and leads to an understanding of why, for example, are fish becoming extinct in channelized, regulated rivers (Zalewski *et al.*, 1998).

Flow changes

Riverine fish normally hold position in a current where the velocity does not exceed a critical value for a given species or its given ontogenetic stage. At higher velocities an individual must actively swim to maintain its position, but then its oxygen consumption (metabolic cost) increases rapidly and the energy available for growth or reproduction is lost. This was well documented for male and female longnose dace (*Rhinichthys cataractae*) in spring (Facey and Grossman, 1990). Velocities up to the increase in oxygen consumption were 7.3 L/s for males and 8.6 L/s for females (Fig. 5.2), where L is the fish length. The importance of higher current velocities for larvae (5 mg dry weight) and early juveniles (50 mg dry weight) of *Chondrostoma nasus* was assessed experimentally at constant temperature (Schiemer *et al.*, 2001). Faster growth during critical larval periods gave an increased probability of survival. The bioenergetic performances were already significantly higher in juvenile fish than in larvae because their energy gain by drift feeding and the cost of swimming (respiration) increased by an order of magnitude. This was shown by velocity thresholds, i.e. current speeds at which assimilated energy was equal to respiratory cost (Fig. 5.3). Schiemer *et al.* (2001) showed that the experimental results were in full agreement with the observed distribution pattern of 0+ fish in the field.

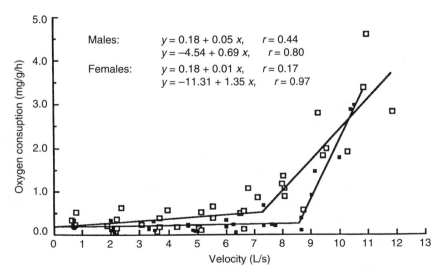

Fig. 5.2. Relationship of oxygen consumption to relative flow for male (□) and female (■) longnose dace (*Rhinichthys cataractae*); $n = 5$ males, $n = 4$ females, one fish not sexed; L is the fish length. (After Facey and Grossman, 1990.)

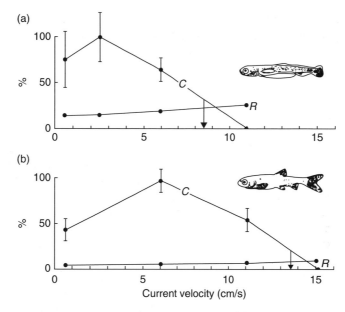

Fig. 5.3. Dependence of bioenergetic performance on current velocity in larval (a) and juvenile (b) *Chodrostoma nasus*. In order to allow for an immediate comparison, the rates of food consumption (*C*) and respiration (*R*) are given as percentages of maximal *C* values. The arrows indicate velocity thresholds, i.e. current speeds at which the assimilated energy is equal to *R* cost. (After Schiemer *et al.*, 2001.)

Adams *et al.* (1982) used laboratory and field data to compile a seasonal energy budget for largemouth bass, which showed that active metabolism ranged from 18 to 144% of standard metabolism. Moderate swimming for benthic taxa can increase the standard metabolic rate 1.1 times, but up to 3.8 times this value has been recorded for reophilic *Salvelinus fontinalis* kept in field enclosures (Boisclair and Sirois, 1993), while Brett and Groves (1979) recorded the swimming metabolism of sockeye salmon during migration as 8.5 times standard metabolism. Such high energetic costs recorded in swimming respirometers have been confirmed by acoustic telemetry, recording the heart rate of fish in natural, diversified habitats (Lucas, 1994), as well as by a stereo-video method (Krohn and Boisclair, 1994). All of these different approaches to measuring swimming cost show that preserving rivers with natural, diverse flow patterns is essential for maintaining high fish diversity (Boisclair and Sirois, 1993) and profitable harvests through continued recruitment of riverine fish species (Schlosser, 1985; Schiemer *et al.*, 1991).

Impoundments

Impoundment and flow changes are strictly connected with each other in disrupting the natural disturbance regimes that maintain a dynamic complex of riverine ecosystems (Ward and Stanford, 1995). A dam interferes with organic matter transport and drift, and may lead to unsuitable habitat conditions for the eggs and larvae of obligatory riverine fish species and for hyporheic invertebrates. Also, as a result of the dam's impact the flood plain may become isolated from the river channel, which leads to a reduction of habitat heterogeneity and biodiversity (Ward and Wiens, 2001). Monitoring studies in the dammed Warta river (Penczak, 1999) showed that gross assimilation efficiencies (K_1) were 24.1 and 25.8 on a backwater site and a tailwater site respectively in the pre-impoundment year, while for 9 years after impoundment research gave corresponding values of 16.6 ± 2.3 and 17.4 ± 2.1. The biggest differences in K_1 values occurred in the site downstream of the dam (tailwater), where construction of the hydropower plant resulted in 'pulse releases' that exerted a negative impact on fish and their food base (Moog, 1993; Penczak, 1999). Changes in fish food base after impoundment were especially conspicuous in the tailwater site (Grzybkowska *et al.*, 1993). Fish tried to adapt to the new diet spectrum after impoundment, but this was connected with higher maintenance costs (Penczak, 1999). These costs were particularly high where associated with manipulation of the sluice gates' operation. Each time the sluices were opened fish that had remained there drifted downstream, while their return involved high energy expenditure.

Ecological Structure of Rivers seen by Faunal Response to Hydraulics

The responses of animals to the hydrological parameters shown above are expressed as the distribution and abundance of species within a watercourse. This pattern can be used practically for verifying on a large scale (e.g. between basin) conclusions derived from investigations at small scale (e.g. mesohabitat). It can also be used for a classification of natural aquatic systems. This also verifies the species–environment conclusions and enables predictions to be made about the effects of anthropogenic changes.

Figure 5.4 shows a hierarchical scheme for the effect of hydrological processes on fauna in running waters. The example is taken from The Netherlands, but it can be applied in other countries with different values for the parameters. The highest level, climate and geomorphology, gives the natural conditions least altered by human influence (other than by global climate change). The parameters used are those from Manning's formula (such as J for slope of the terrain). The characteristics of the discharge area, the next level, are heavily influenced by man in the densely populated area of north-western Europe. Especially smaller discharge areas of tens of square kilometres are modified to agricultural needs by, for example, drainage, water distraction and damming. In the lower level, all types of human impacts are manifest and a natural situation is hard to be found. How does the ecosystem react to these alterations and what is the natural situation?

Most running waters start as small trickles, gradually collecting more water from tributaries and seepage until small streams have become large rivers. Small mountain streams have a swift flow while the current velocity in small lowland streams seldom exceeds 30–50 cm/s. The slope is an important determinant for

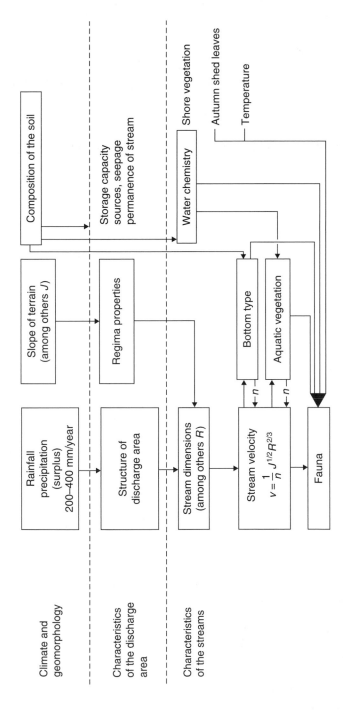

Fig. 5.4.   Factors controlling the conditions for aquatic fauna in running waters.

current velocity. Manning combined dimensions and slope together with resistance or roughness in his well-known law:

$$v = \frac{1}{n}J^{1/2}R^{2/3} \qquad (5.2)$$

where

v = mean current velocity (m/s),
n = roughness,
J = slope of the energy line, taken here as the slope of the terrain,
R = hydraulic radius, i.e. wetted perimeter of the cross-section (m).

Ground slope and roughness are local conditions, whereas the hydraulic radius is the result of factors concerning the total discharge area and the amount of precipitation. Higler and Mol (1984) used the elements of Manning's law to describe different types of running water as predicted by Statzner (1987a). The ground slope and roughness were taken together as a single local factor, called the terrain factor C_t:

$$C_t = \frac{1}{n}J^{1/2} \qquad (5.3)$$

From equations (5.2) and (5.3) it follows that:

$$v = C_t R^{2/3} \qquad (5.4)$$

which can also be written as:

$$\log R = \frac{3}{2}\log\ v - \frac{3}{2}\log C_t \qquad (5.5)$$

In a logarithmic plot of C_t versus R (Fig. 5.5), the current velocity is represented by straight lines with a slope of –3/2. The two components of C_t can be easily obtained. Ground slope is measured as the difference in height between two stream localities, divided by the distance along the stream course. Roughness is experimentally determined as varying between 0.01 and 0.05 (Nortier and van der Velde, 1968). An extreme case is a straight concrete gutter with little resistance

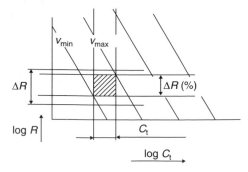

Fig. 5.5. The terrain factor (C_t)– hydraulic radius (R) diagram, in its original form (explanation in text).

on the one hand; the opposite a natural, meandering stream with stones and vegetation. In natural streams, n may vary from 0.03 to 0.05. Therefore, there is a variation in C_t described by:

$$\Delta C_t = \log C_{t(max)} - \log C_{t(min)} = \log\frac{C_{t(max)}}{C_{t(min)}} = \log\frac{1/n_{(max)} \times J_1^{1/2}}{1/n_{(min)} \times J_2^{1/2}} \qquad (5.6)$$

At some localities the ground slope is a constant. The natural variation of C_t may therefore be described by:

$$\Delta C_t = \log\frac{n_{(min)}}{n_{(max)}} = \log\frac{0.050}{0.035} = 0.155 \qquad (5.7)$$

This is illustrated in Fig. 5.6, where it follows that all localities with the same ground slope are found in a vertical belt with a width of 0.155 times the scale unit. The hydraulic radius (R) is composed of the wet cross-sectional area of the stream (A), divided by the wet outline of this cross-section (O).

R is not a constant but varies with the water level, as variation of the water depth affects both A and O in different ways. The natural variation of R, ΔR, of a locality can be reduced by excluding the extremes to $\Delta R_{x\%}$. In this way, each stream locality can be represented by a limited area, including the percentage of all natural conditions occurring on that locality (hatched area in Fig. 5.5). The maximum and minimum current velocity on that locality can then be read off.

Several hundred measurements and data from the literature were plotted to make the C_t–R diagram in Fig 5.6. Watercourses with approximately the same features are found together in certain parts of the diagram. Measurements in a stream system 'move' from source and headwaters to lower course, upwards through the diagram. These different parts of stream systems may be named, such as lowland stream or large river, without defining them in concrete and measurable terms. In this C_t–R diagram, blocks have been drawn to indicate stream types together with their hydraulic characteristics. The vertical borders of blocks separate types of landscape whereas the horizontal borders separate watercourses of different dimensions as directed by the hydraulic radius.

The distribution of 34 Trichoptera species in running waters of the province of Overijssel, The Netherlands, confirmed that the habitats of lotic species are found in defined places of the C_t–R diagram (Higler and Verdonschot, 1992). From the measured abiotic features at the sampling stations, the measurements of C_t and R were selected. Each of the species was plotted on the C_t–R diagram with the help of the figures for C_t and R. Clusters were found for each; the smaller the cluster, the more defined were the hydraulic characteristics for those species. Species with a similar distribution pattern were grouped and in this way six groups were found that partly overlapped, forming a series of clusters from springs to large rivers (Fig. 5.7). The overlap seems to confirm the River Continuum Concept (Vannote et al., 1980), which hypothesizes that a gradual replacement of species groups forms a continuum from source to mouth of a riverine system.

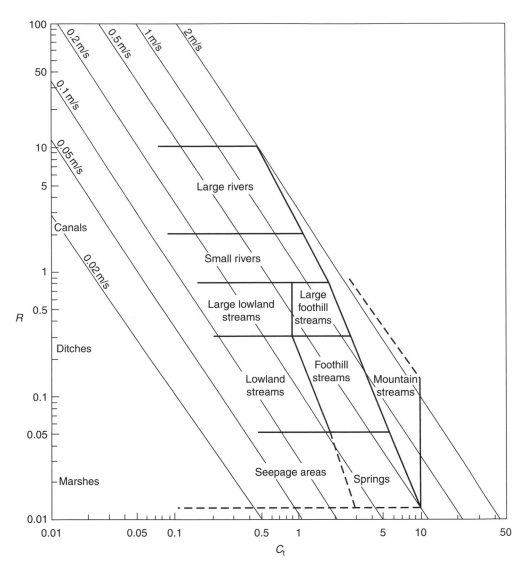

Fig. 5.6. The terrain factor (C_t)–hydraulic radius (R) diagram defining types of running water based on measurements of the slope of the water surface and the hydraulic radius. The roughness is incorporated in C_t. The mean current velocity follows from the diagonal lines.

The same 34 species together with 87 abiotic variables that were measured at the sampling stations were analysed by clustering and ordination in several steps (Higler and Verdonschot, 1992), culminating in de-trended canonical correspondence analysis in which the axes are chosen in the light of the environmental variables. The method indicates which variables are responsible for the grouping of the organisms and which ones are of minor importance. Figure 5.7 shows only the most distinctive variables, together with the same six groups of

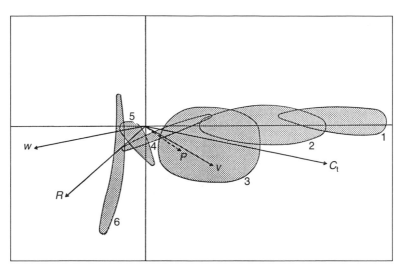

Fig. 5.7. Ordination (CANOCO) diagram for axes 1 and 2 with groups of caddisfly larvae and the main environmental variables: width (*w*), permanence (*P*), current velocity (*v*), hydraulic radius (*R*) and combination of slope of the water surface and roughness (C_t).

Trichoptera from the first method. The arrow for each environmental variable points in the direction of the steepest increase in the weighted average number of species for that variable and the rate of change in that direction is equal to the length of the arrow. This means that the relationship of a set of species (or one species) with one or more important variables is visualized by its perpendicular projection on the environmental arrow or the imaginary extension in both directions of the arrow. Within the diagram the species and arrows can be seen as relative projections upon each other. Species–environment correlations of the axes indicate the degree of explanation, through the variables, of the variation within the data set.

The most important variables out of 87 are terrain factor C_t, current velocity *v*, width *w*, hydraulic radius *R* and permanence *P*. Groups 1–4 are beautifully arranged along the horizontal axis, indicating that *v*, C_t and *w* are the dominant steering variables. Group 5 and 6 have an orientation along the vertical axis with a high relationship to *R*. Groups 5 and 6, and to a lesser extent group 1, show only a slight overlap with the other groups. This evidences the zonation concept of Illies and Botosaneanu (1963) in which changing current velocity and water temperature from source to mouth of a stream form the basis of discrete biocommunities in different zones. Statzner and Higler (1986) showed that changes in water temperature must be of lesser importance and stream hydraulics greater, in a hierarchical pattern of abiotic environmental variables. In reality, the stream is neither a fluent continuum of overlapping species distributions nor a strict zonation of distinct communities, but a combination of both concepts in which hydraulic factors are the major determinants of benthic invertebrate zonation patterns.

Ecological Structure of Rivers seen by Faunal Response to Geomorphology

It has been shown above that individual species respond to hydraulics in a predictable fashion and that benthic faunal assemblages can be recognized in different flow types (Davis and Barmuta, 1989). This link has also been taken into the relationship between fauna and substrate types, indeed fauna and all physical habitat types, since physical habitat is a result of flow patterns. Brooker (1982), working on the River Tefi (UK), quantified the invertebrate assemblages found in each of ten subjectively named habitats (riffle, fast run, slow run, pool, slack, backwater, tree roots, grass roots, *Ranunculus penicillatus* and *Potamogeton natans*). In Australia, Barmuta (1989) also investigated the distribution of invertebrates according to visually distinguishable habitats (riffles, pebble beds, cobble pool or run, silted pebbles, exposed pebbles, silted sands, clean sands and mixed sands). Sixty three of the 133 taxa investigated showed little habitat preference, but there were enough 'high fidelity' species to distinguish two major habitat groups: erosional and depositional. Erosional habitats were riffles in summer and winter; exposed pebbles in summer (absent in winter); pebble beds in winter; cobble pool or run in winter. Intermediate habitats were pebble beds in summer; cobble pool or run in winter. Depositional habitats were silted pebbles in summer (absent in winter); silted sands in summer (absent in winter); clean sands in summer and winter; mixed sands in summer and winter). Species found in erosional habitats required fast water for feeding (e.g. *Asmicridea edwarsi*) or for physiological reasons (e.g. Hydrobiosidae) while those in depositional habitats were intolerant of turbulent flow (e.g. *Triplectides proximus*) or adapted for burrowing in silty substrata. Velocity was found to be the most influential environmental variable. Bournaud and Cogerino (1986) identified 12 'prospective' habitats on the submerged banks of a canalized reach of the Rhone. These were then validated by the actual species distribution of invertebrates, except for some overlap occurring between habitats within each of the larger-scale classifications of erosional, depositional and vegetative habitats.

The idea that macroinvertebrate habitat units exist, each containing a characteristic assemblage of macroinvertebrates, was developed in lowland UK by Harper and co-workers (Harper *et al.*, 1992, 1995, 1998; Harper and Smith, 1995) and by Armitage and co-workers (Armitage *et al.*, 1995; Pardo and Armitage, 1997). Harper called them 'functional habitats' because they were identified as 'building blocks' for conservation assessment (Harper *et al.*, 1992), while Armitage called the same entities 'mesohabitats'. Armitage *et al.* (1995) identified eight mesohabitats, according to their macroinvertebrate assemblages, in a lowland chalk stream (Mill Stream, Dorset). These were '*Ranunculus* fast', '*Ranunculus* slow', 'silt', '*Nasturtium*', '*Phragmites*', 'sand', 'gravel fast' and 'gravel slow'. These habitats were examined in spring, summer and autumn. Even though community changes were observed between seasons, the different habitat units tended to remain distinct. Discontinuities between the habitat types were greatest in the summer, during the period of lowest discharge. In the spring, with highest discharge, the habitats were least distinct. They concluded that depth and velocity were the main factors determining the distinctiveness of the mesohabitats.

Harper *et al.* (1995) added information from the literature to build a working list of 16 functional habitats (Table 5.3) which were then validated by local flow and depth conditions associated with each of them (Kemp *et al.*, 2000, 2002). The habitat diversity that a river contained was shown to fall when the river was either over-widened or over-deepened by engineering works (Kemp *et al.*, 1999), and this information was used to guide river rehabilitation (Kemp and Harper, 1997) as well to inform the choice of parameters to be recorded by the UK Environment Agency in its River Habitat Survey methodology (Raven *et al.*, 1998).

Studies of the habitats of two Italian rivers, which were of very different character to the English river sites, confirmed the generality of the physical habitat concept. The habitat units identified were similar to those of lowland England (allowing for investigator naming differences) but with different frequencies. In the River Ticino, a large, braided, low-gradient piedmont river, five habitats were identified. These were named 'river margins with macrophytes', 'margins without macrophytes', 'backwater', 'run-riffle' and 'macrophytes in flow'. The scale at which this habitat study was operating was appreciably larger than that of the English studies as a result of the size of the river, and the resulting functional/ meso habitats are larger-scale features than those identified in lowland English rivers. None the less, clear parallels were found in the results, related to the importance of macrophytes, flow and organic matter (Buffagni *et al.*, 2000). In a typical north Italian mountain stream, the Pioverna, only four habitats were found. These were 'riffle', 'pool', 'transition' and 'bedrock' habitats (Crosa and Buffagni, 1996). These were identifiable more by their flow type than by their substrate type. The Pioverna study demonstrated that the functional/meso habitat concept can be transferred from English lowlands to dramatically different systems.

Table 5.3. Functional habitats in running waters of the UK. (From Harper *et al.*, 1995.)

Rock (bedrock, boulders and rocks)
Cobbles
Gravel
Sand
Silt
Marginal plants
Emergent macrophytes
Floating-leaved macrophytes
Submerged, fine-leaved macrophytes
Submerged, broad-leaved macrophytes
Moss
Macroalgae
Submerged roots
Trailing vegetation
Leaf litter
Woody debris

Functional/meso habitats differ significantly from habitat definitions used in other studies (e.g. Scarsbrook and Townsend, 1993; Cellot *et al.*, 1994) as they are biologically validated instead of based on experience and intuition. They have the advantage of 'taking the organism's eye view' as recommended by Hildrew and Giller (1994), but at a scale usable for pragmatic river management rather than the individual organism scale. Each habitat supports a characteristic assemblage of macroinvertebrates, because each provides a unique physical and biological environment. In terms of physical characteristics, these include factors such as typical particle size and interstitial space provision for the substrate habitats, and structure of the stem and leaf matrix for the macrophyte habitats. Each of these structures channels water in a distinctive manner and is associated with a typical range of velocities and shear stresses. Armitage *et al.* (1995) identified the characteristic depth and velocities of mesohabitats as being important; Kemp *et al.* (2000) showed the importance of hydraulics by correlating Froude number with habitat.

Ecohydrology and Water Management

Water managers have to deal with a variety of problems concerning running waters, such as flooding, water shortage, pollution, lack of fish passages and many others. Often, they want to rehabilitate conditions towards more natural system to increase biodiversity or promote a better fish population.

One tool for management, developed in The Netherlands, has been derived from elements in Fig. 5.4. Verdonschot *et al.* (1998) formalized this scheme into five compartments, called the '5-S Model':

- 'System conditions' is the first S. It is climate and geomorphology in Fig. 5.4 and management is not able to influence these conditions.
- 'Stream hydrology', 'Structures' and 'Substances' are at second level, the second, third and fourth Ss. They comprise hydrological, morphological and hydrochemical conditions, which are subject to water management.
- 'Species' is the final S, as a response to the functioning of all above-mentioned groups of controlling factors; in Fig. 5.4 represented by fauna and aquatic vegetation. Because of the dependency of all abiotic factors, species form a reflection of each situation in which hydraulics is shown to be the master factor.

In an example of application of the 5-S model, the suitability of certain well-defined streams for the noble crayfish, *Astacus astacus*, was analysed by Verdonschot *et al.* (1998). For each of the five Ss, a table was constructed with appropriate variables aggregated to the life conditions of the crayfish and the streams where the species has been found. Subsequent comparison between ideal and present conditions indicated the suitability of the streams for a healthy population and, at the same time, ways to improve conditions for such a goal. Table 5.4 generalizes this comparison.

The conclusion for this particular case is that the stream is too deep, the current velocity too high and the number of ponds not enough to meet the optimal

Table 5.4. Suitability of an example situation for a specified goal with respect to stream hydrology. (Adapted from Verdonschot *et al.*, 1998.)

Controlling factor	Parameter	Habitat conditions	Actual stream	Suitability
Groundwater	Supply	Continuous	Continuous	Suited
Hydrology	Maximum depth	<x cm	x+30	Insufficient
	Surface (ha)	<y ha	Ponds <50	Suited
Hydraulics	Current (cm/s)	<·>	>·	Insufficient
	Running	+++	+++	Suited
	Stagnant	+++ (ponds)	+	Insufficient

conditions for the hypothetical needs. The water manager may draw the conclusion that his/her goal with respect to hydrological conditions is not met and that he/she has to take measures to change this if possible or to accept an alternative goal. In this particular imaginary case, perhaps it is not possible to make the stream shallower, but it is easy to reduce the current velocity and to create more ponds. As a result, the depth of the stream may change and optimal conditions are within reach.

This is just one example of how to deal with some hydrological variables, but practice has shown that it is an applicable tool for managers to handle ecological problems concerning hydrology. However, the system conditions are the principal steering processes and they have to be considered in the first instance.

Conclusions

Ecohydrology is a science that integrates the hydrology and the ecology of running and still waters. In this way, the freshwater parts of the hydrological cycle are connected with different ecosystems like source areas and upper, middle and lower sections of river systems. Hydrology explains the geomorphological and ecological differences of these ecosystems. Hydraulics, as a major determinant of benthic invertebrate distribution in running waters, explains the conditions for lotic biocommunities on both small and larger scales.

6 Ecohydrological Modelling for Managing Scarce Water Resources in a Groundwater-dominated Temperate System

J.-P.M. WITTE[1], J. RUNHAAR[2] AND R. VAN EK[3]

[1]Kiwa Water Research and Vrije Universiteit, Institute of Ecological Science, Amsterdam, The Netherlands; [2]Kiwa Water Research, Nieuwegein, The Netherlands; [3]Eindhoven, The Netherlands

Introduction

The need for ecohydrological models

Everywhere in the world man's demand for water is growing, both as a result of an increasing human population and an intensified use per person of water for agriculture, industry, domestic purposes and recreation. In many places in the world, surface water from rivers and lakes is transferred to areas that suffer from water deficits. Often this surface water is polluted, or it has a chemical composition that strongly differs from the original water. Ecosystems that depend on a high groundwater or surface water level have suffered from falling groundwater tables and the inlet of 'foreign' water.

This is particularly the case in The Netherlands, a country of 35,000 km^2 that is almost entirely situated within the estuarine area of the rivers Rhine, Meuse and Scheldt. Due to the low level of the land, groundwater tables are very shallow in The Netherlands: in 90% of the country the water table is less than 1 m below the surface in winter and less than 2.5 m below the surface in summer (Colenbrander et al., 1989). As a consequence, a considerable part of the indigenous plant life is characteristic of wet to moist conditions. Therefore, many conservation values in The Netherlands are associated with wet and moist ecosystems, such as fens, bogs, dune slacks, wet heathlands, swamp woodlands and wet meadows.

The Netherlands is heavily industrialized and densely populated (462 inhabitants/km^2), which inevitably results in a strong claim for land. About 90% of the surface area is used for economic or urban purposes so that no more than about 10% of 'waste land' is left for nature and forestry. Many animal and plant species are to a considerable extent restricted to small reserves (often measuring less than 1 km^2) or spots within an agricultural landscape. As a result, they are strongly dependent on the water management of the surroundings.

The water regime of most areas has changed drastically as a result of human interference in the second half of the 20th century. Since 1950 several hundred land improvement plans have been implemented, which have made a radical revision of their water management. Moreover, annual groundwater extraction for industrial and drinking water purposes increased from 0.3×10^9 m^3 in 1955 up to 1.1×10^9 m^3 in 1990 (Beugelink *et al.*, 1992). As a result, groundwater levels have dropped, especially in the eastern and southern part of the country, where drainage is almost entirely by gravity. Feddes *et al.* (1997) estimated that since 1950 the groundwater table in this area has dropped 60 cm on average. To compensate for the negative effects of low groundwater levels on agriculture, polluted surface water from the rivers Rhine and Meuse is distributed over the land.

Ecosystems have suffered severely as a result of this man-induced 'drought'. A nationwide investigation into the loss of conservation value since 1950 showed moderate to severe damage to ecosystems in 50–60% of the groundwater-dependent nature reserves (Runhaar *et al.*, 1996a). Ecosystems of nutrient-poor sites were especially affected (Fig. 6.1).

Water management in The Netherlands used to be attuned mainly to economic interests, such as shipping and agriculture, as well as to the safety of the human population against flooding. However, biodiversity value is now recognized as a valuable commodity, so water management also includes the conservation, restoration and creation of both nature reserves and other

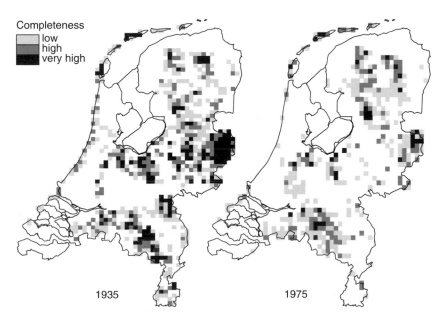

Fig. 6.1. Changes in the completeness (relative species richness) of ecosystems of wet, nutrient-poor and neutral soils over the period from 1935 to 1975, approximately. (From Witte and Van der Meijden, 1990.)

natural elements in the rural environment. The new interdisciplinary field of eco-hydrology now gives this water management a scientific basis.

In The Netherlands, ecohydrology is nearly always associated with the study of the hydrological characteristics and requirements of ecosystems of land and small stagnant surface waters (terrestrial and semi-terrestrial ecosystems), where only the vegetation is considered to be representative of the biotic part of the ecosystem. There are two reasons for this approach: (i) the relationship between the abiotic environment and plant life is quite direct, whereas the relationship between the abiotic environment and fauna is more indirect; and (ii) the value of an area for biodiversity conservation is usually deduced from the vegetation (Van Wirdum, 1986).

The Dutch Government (at regional, provincial and national level) uses computer models for directing policies for quantitative and qualitative water management of The Netherlands. Examples range from those for the calculation of groundwater levels and the quality of surface waters to the water depth of the main shipping routes, and now include several ecohydrological ones.

The following example shows the kind of problems that ecohydrological modelling in The Netherlands deals with. A protected area, in a valley, has vegetation with rare plant species such as *Carex pulicaris* and *Cirsium dissectum* and its maintenance is dependent on upward seepage of lithotrophic groundwater. Water, infiltrating in an adjacent hill, flows through the subsoil to the direction of the valley, where it will eventually exfiltrate. During its transport the groundwater becomes enriched with calcium and bicarbonate ions, and its pH rises. The upward seepage creates the wet, nutrient-poor and pH-neutral site that is typical for the vegetation in this area. A water supply company proposes a plan to extract groundwater from the hill. The effects of that measure on the vegetation may be predicted and judged by a general scheme (Fig. 6.2) based on the way ecohydrological modelling has been carried out in practice.

First, the water management measures (scenarios, interventions) that are to be analysed are formulated. In the example, the measure is the intended groundwater extraction, although it might also be drainage of arable land, felling of coniferous trees in order to reduce evapotranspiration or the distribution of Rhine waters to new areas for the benefit of agriculture, for example. Second, the effects of these measures are calculated with hydrological models for the saturated zone (often a steady-state computation) and for surface waters. These models compute water flows in the horizontal plane, so may be referred to as spatial hydrological models. Present hydrological models are capable of computing a spatial picture of hydrological variables, such as the phreatic groundwater level, the seepage intensity and the discharge of rivers, rivulets and channels. A steady-state, (quasi-) three-dimensional groundwater model is used for this example. The prediction for our protected area would probably be a fall in the groundwater level and a decrease in the intensity of upward seepage.

Models that are capable of computing the chemical composition of the groundwater have hardly been used, mainly because they require data – particularly data about the chemical characteristics of the subsoil – that are rarely available.

The model prediction is divided into 'modules' (Fig. 6.2). Changes in the hydrological variables form the input to *site modules*, which compute how site

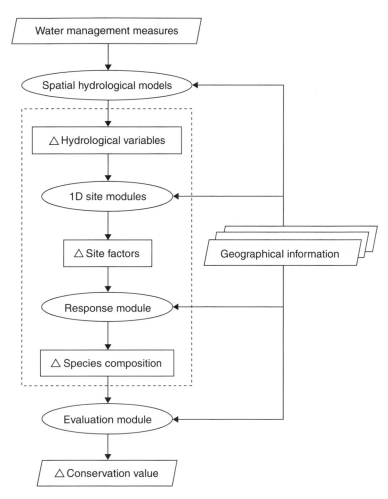

Fig. 6.2. General process chain for an ecohydrological prediction. Input and output of data and variables are indicated by a rhombus, interim variables by a box, models and modules by an ellipse, direction of data flow by an arrow and change in a variable by a triangle (1D, one-dimensional). This chapter focuses on the elements framed by the dotted rectangle. (From Witte, 1998.)

factors react on the hydrological changes. These site modules do not contain spatial relationships, in contrast to the hydrological models for the saturated zone and the surface waters: calculations are carried out for *plots*, i.e. one-dimensional and vertical representations of reality. Examples of site factors are the depth of the groundwater table in spring, the soil moisture supply and the availability of nutrients. The predicted effects of the intended groundwater extraction on the site factors in this example would probably be: (i) a lower availability of soil moisture; (ii) a higher nutrient richness; and (iii) a lower pH (Fig. 6.3).

How the changes in these site factors, in their turn, influence the species composition of the vegetation is predicted with a *response module*. This module

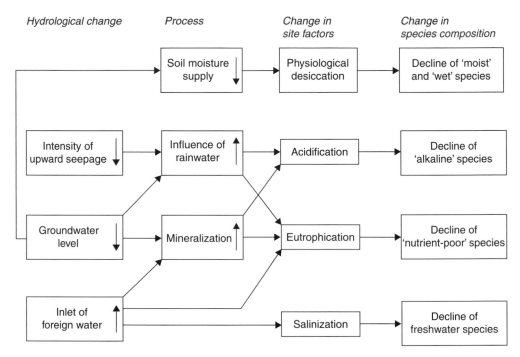

Fig. 6.3. Main negative effects of a groundwater fall on important operational factors of wet and moist sites. Direction of change indicated by arrows: ↑, increase; ↓, decrease. A falling groundwater level may lead to shortages in the water supply to the vegetation and, as a result, to a *physiological desiccation* of the vegetation: species that are adapted to wet and moist environments will disappear. A groundwater fall may also cause increased aeration which, in turn, promotes mineralization and, consequently, *eutrophication*. Hence, species characteristic for nutrient-poor sites will disappear. When organic matter is mineralized, protons are released and acidification of the soil takes place, causing species of neutral and alkaline sites to vanish. The availability of phosphorus is largely regulated by adsorption on calcium hydroxides (at pH > 6.5) and iron hydroxides (pH < 6.5) (Stumm and Morgan, 1981). In this adsorbed form it is not available for plants. Hence, an influx of calcium- or iron-rich water by upward seepage may lead to the development of mesotrophic and oligotrophic sites. Moreover, calcium and bicarbonate in upward seepage water form an important buffer against acidification by percolating rainwater. Hence, when the root zone is originally influenced by lithotrophic upward seepage, a decreasing groundwater level may enhance both eutrophication and acidification. Of course, both effects may also take place when the intensity of upward seepage diminishes. The inlet of foreign surface water may lead to eutrophication, especially of aquatic ecosystems, since this water – in many cases from the rivers Rhine and Meuse – is often rich in phosphorus and nitrogen. Even when nutrient concentrations are low, system-alien water may stimulate mineralization, leading to 'internal eutrophication'. An explanation for this phenomenon is that the inlet water often has a higher alkalinity than the original water. Moreover, the high sulfate concentrations may contribute to internal eutrophication, since sulfur is capable of forming a complex with iron, leaving less iron for the fixation of phosphorus (Caraco *et al.*, 1989). Finally, a higher salinity of the inlet water will have marked effects on the vegetation: freshwater species will disappear.

consists of empirical or expert rules about the way species react to site factors. Here, *C. pulicaris* and *C. dissectum* are predicted to disappear and new species, such as common grasses, predicted to take their place.

In most ecohydrological models the effect on the vegetation is calculated in these two steps, by means of site modules and a response module respectively. In some ecohydrological models however, a direct link is made between hydrology and vegetation.

Sometimes the effect on the species composition is also evaluated in terms of nature conservation value. In Fig. 6.2 this activity is indicated with the *evaluation module*. Evaluation allows the outcomes for different species or, if these are the classification units applied, for different vegetation types/ecosystem types, to be combined. To evaluate on the basis of the criterion of 'national rarity' would leave no doubt that the final output of the ecohydrological prediction would be a lower conservation value than its current value.

This example gives an overall picture of the way ecological effects of water management measures can be modelled. The chapter now focuses on the ecological part of the prediction: the effects of hydrological changes on site factors and the resulting species composition. Three major approaches in ecohydrological modelling, their benefits and their drawbacks are considered.

Description of Dutch Ecohydrological Models

We may classify ecohydrological models into three types, according to the causality of the modelled relationships (Fig. 6.4): (i) *correlative* models; (ii) *mechanistic* models, with causal relationships; and (iii) *semi-mechanistic* models, which

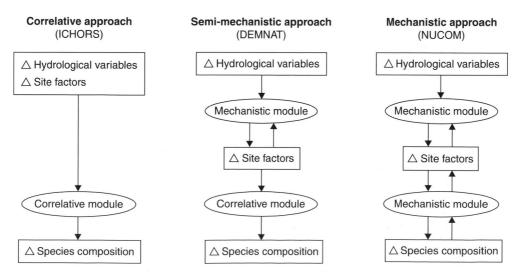

Fig. 6.4. Three types of ecohydrological models, differing in the causality of the modelled relationships. Examples of models are given in parentheses; for an explanation please see text. (Adapted from Runhaar, 1999.)

contain both correlative and causal relationships. From each of these types we describe one example and refer to similar ones.

The correlative approach: ICHORS

The basic feature of a correlative approach is that it considers the ecosystem as a black box: the occurrence of species is statistically correlated with a broad spectrum of measured site factors, irrespective of their supposed ecological importance. Completely correlative approaches do not exist however, as one will always try to select site factors that are expected to have at least some ecological meaning. So the adjective 'correlative' has a relative meaning, to distinguish between 'more correlative' and 'more mechanistic'.

A good example of a (relatively) correlative approach is ICHORS (Influence of Chemical and Hydrological factors On the Response of Species), which is a model for species of small fresh surface waters in the western, Holocene part of the Netherlands (Barendregt, 1993). ICHORS comprises the response module of Fig. 6.2 only. In fact, the model consists of a set of probability functions for species. Each of these functions describes the species occurrence probability in relation to abiotic variables, like surface water depth, magnesium ion content and pH. To illustrate this, Fig. 6.5 shows the computed occurrence probability of a particular plant species (*Achillea ptarmica*) as a function of both surface water level and chloride concentration.

The probability functions are obtained as follows. Data are collected in the study area only once: plants are recorded and abiotic variables are gathered, such as width and depth of the surface water, pH, thickness of the sapropel layer and turbidity of the water, from many locations. Also water samples are taken, which are submitted to a thorough chemical analysis (including for instance silicon

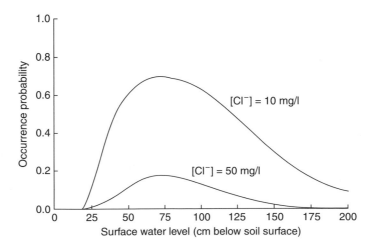

Fig. 6.5. Occurrence probability of *Achillea ptarmica* as a function of surface water level at two different chloride concentrations. (From Barendregt, 1993.)

and iron) in the laboratory. Then, logistic multiple regression (Hosmer and Leme-show, 1989) is carried out – for instance, with GENSTAT – fitting a function through the frequency occurrences in the data set (Fig. 6.6) for each species. About 25 variables are involved in the existing ICHORS versions, but only a limited number (seven at most) of variables is selected for the function, namely those that significantly improve the goodness-of-fit.

Because of the black-box character, for each region a special version of the model is made, with unique occurrence probability functions.

In scenario analysis, ICHORS is applied by first estimating the new abiotic variables and, subsequently, by computing the occurrence probability for each species by means of the probability functions (Table 6.1).

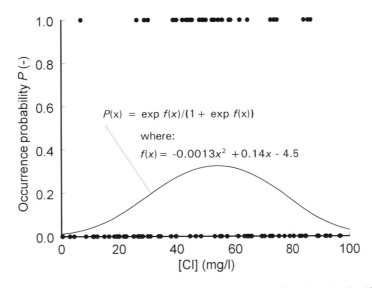

$$P(x) = \exp f(x)/(1 + \exp f(x))$$

where:

$$f(x) = -0.0013x^2 + 0.14x - 4.5$$

Fig. 6.6. Imaginary example of how an occurrence probability function for ICHORS is obtained. Out in the field, many sites are visited where the chloride concentration is measured and the occurrences of plant species are recorded. After collecting these field data, logistic regression is applied to fit an occurance probability function $P([Cl])$ through the presence–absence data for each species.

Table 6.1. Response of four species to two chemical types of surface water: clean litotrophic water (rich in calcium and bicarbonate ions) and polluted River Vecht water (rich in, among other nutrients, sodium, chloride and sulfate ions). Presented are species name (with number of observations used for logistic regression) and occurrence probability P (including 95% confidence interval). (From Barendregt *et al.*, 1986.)

Plant species	Lithotrophic water	River Vecht water
Phragmites australis (483)	0.51 (0.44, 0.58)	0.58 (0.50, 0.66)
Lemna minor (408)	0.51 (0.39, 0.63)	0.44 (0.33, 0.56)
Equisetum fluviatile (223)	0.91 (0.83, 0.96)	0.03 (0.01, 0.07)
Hottonia palustris (48)	0.05 (0.00, 0.39)	0.00 (0.00, 0.00)

Comparable correlative models are HYVEG (HYdrology–VEGetation; Noest, 1994) – meant for species of young dune slacks – and ITORS (Influence of Terrestrial site conditions On the Response of Species; Ertsen, 1998) for species of terrestrial ecosystems in the province of Noord-Holland.

The mechanistic approach: NUCOM

A mechanistic model contains causal relationships that are well known from experimental studies or that have been derived theoretically. Like 'correlative', the adjective 'mechanistic' also has a relative meaning, since any model will always contain processes that to a certain degree have been lumped.

Ecohydrological models that are mechanistic do not exist. Nevertheless, NUCOM is a good example of a mechanistic model, which in the future might be adapted to be used for ecohydrological purposes.

NUCOM (NUtrient cycling and COMpetition; Van Oene et al., 1999) was developed to analyse the effects of environmental changes on terrestrial ecosystems, in particular the effects on plant species composition. This model is of mixed species type, linking vegetation and soil processes. NUCOM describes an ecosystem including the complete cycling of nutrients, carbon and water and a variable number of plant species that compete for light, nitrogen and water. External driving forces are climate data, atmospheric carbon dioxide concentrations, and atmospheric loads for nitrogen, sulfur and other macronutrients. Each plant species produces a certain kind of litter that – through decomposition and mineralization – influences soil development. As succession proceeds the organic matter content of the soils increases and, with that, the internal cycling of nitrogen and the availability of soil moisture grows. This favours certain species, which take over from others. Outputs of the model are time series of biomass per species, soil and plant organic carbon, soil and plant organic nitrogen, mineralized nitrogen, soil moisture content and soil pH.

Typical for this mechanistic approach is that it contains feedback mechanisms between vegetation and its site (Fig. 6.4): soil characteristics determine the growth of plant species; these plant species then determine soil development as well as species' growth (through the competition for light, nutrients and water).

Van Oene et al. (1999) used NUCOM for a relatively simple ecosystem, containing seven plant species and a free-draining sandy soil profile. Their model successfully simulated the observed succession from pioneer vegetation dominated by *Corynephorus canescens* to woodland with *Pinus sylvestris*, *Betula pendula*, *Quercus robur*, *Vaccinium myrtillus* and *Deschampsia flexuosa*.

NUCOM thus seems a powerful tool for understanding ecosystem processes and for predicting the effects on ecosystems of climate change and the atmospheric deposition of nutrients. In its present state, however, it is not suitable for analysing the effects of a groundwater regime, be it natural or driven by water management. Groundwater is not an external diving force of NUCOM and, consequently, the model was used for a free-draining profile with a deep groundwater table. All in all, NUCOM is not an ecohydrological model; it contains much 'eco' but not enough 'hydro' to deserve this classification.

Hydrology may be involved in a model like NUCOM by using the techniques that have been developed in agrohydrology; for instance, by linking NUCOM to SWAP (Soil–Water–Atmosphere–Plant; Van Dam *et al.*, 1997), a model that describes the transport of soil water as dependent upon climate, vegetation characteristics, soil characteristics and groundwater regime. In the past, SWAP has already been linked to models for crop growth and soil chemistry. It contains feedback mechanisms between vegetation and soil (e.g. vegetation extracts water from the soil for transpiration; transpiration and vegetation cover are reduced when the soil dries up). In contrast to NUCOM however, the soil characteristics themselves do not change, as soil development is not a part of the present SWAP version. Therefore, technical adaptations are needed.

The semi-mechanistic approach: DEMNAT

Most ecohydrological models are of a semi-mechanistic approach. In the semi-mechanistic approach, the species composition of the vegetation is regarded as a function of a limited number of site factors. Those site factors used are expected or have been proved to have the largest influence on the species composition of the vegetation. This approach is partly mechanistic and partly correlative (Fig. 6.4). How environmental changes influence site factors such as moisture regime and nutrient availability is – as far as possible and practical – modelled in a mechanistic way, on the basis of the present knowledge of the processes that take place in soil and groundwater. The relationship between site factors and species composition, however, is determined in a correlative way.

A good example of this approach is DEMNAT (Dose-Effect Model for terrestrial NATure), a national prediction model meant for analysing the effects of water management on ecosystems (Witte *et al.*, 1993; Witte, 1998; Van Ek *et al.*, 2000). Except for the hydrological input, in its standard configuration it contains all of the elements of Fig. 6.2 that are needed by an ecohydrological prediction: (i) a geographical schematization of The Netherlands, containing information on hydrology, soil and vegetation (element 1); (ii) a module to translate hydrological changes into changes in species composition (element 2; the dotted frame in Fig. 6.2); and (iii) a module to evaluate these changes (element 3). This chapter focuses on element 2, but to be able to understand this element sufficiently, we first have to give information about the geographical schematization (element 1).

In contrast to most ecohydrological models, DEMNAT uses ecosystem types for its computations instead of individual plant species. These ecosystems are defined on the basis of five factors that explain important differences in the species composition of the plant cover of The Netherlands: (i) salinity; (ii) moisture regime; (iii) nutrient availability; (iv) acidity; and (v) vegetation structure (Runhaar *et al.*, 1987; Runhaar and Udo de Haes, 1994; Witte, 2002). The first four (abiotic) factors may be directly or indirectly affected by hydrological changes (see e.g. Fig. 6.3) and this is exactly the reason why this ecosystem classification is very suitable for impact assessment studies concerning water management. For ecohydrological modelling on a national scale, the ecosystem classification by Runhaar *et al.* (1987) was modified, which resulted in the distinction of 18 ecosystem types (Table 6.2).

The 18 ecosystem types have been mapped nationwide on a 1 km × 1 km national grid (Witte and Van der Meijden, 1995, 2000; Witte, 1998). These maps were derived from the national flora database FLORBASE (Van der Meijden *et al.*, 1995) in combination with indicator values of separate plant species deduced from Runhaar *et al.* (1987). By way of example, Fig. 6.7 shows part of the ecosystem map of one of the ecosystem types (type K27: herbaceous vegetation on a wet, moderately nutrient-rich soil) in the province of Drenthe. For each km^2 on the grid the completeness (relative species richness) of the ecosystem type is given in four classes: 'noise', 'low', 'high' and 'very high'.

The thin lines in Fig. 6.7 indicate areas with a high groundwater table. Most of these are brook valleys. According to DEMNAT, the infiltration areas of the province of Drenthe are dominated by herbaceous ecosystems of wet and moist, nutrient-poor, acid soils, such as heathlands (i.e. K21 and K41). Ecosystems typical of wet and moist, nutrient-poor, neutral soils (H22, K22) and of wet, moderately nutrient-rich soils (H27, K27; Fig. 6.7) are abundantly present in the brook valleys, where upward seepage occurs.

DEMNAT also needs geographical information on soil characteristics (e.g. organic matter content and soil texture) because the reaction of a vegetation type to hydrological changes is largely dependent on controlling soil factors. With the aid of a 1:50,000 national ecological soil map (Klijn *et al.*, 1996), the km^2 information on the ecosystem map is ascribed to those soils types which the ecosystem type in question prefers by nature ('down scaling' or 'degregation' of the ecosystem maps).

Table 6.2. Description of the 18 ecosystem types used in DEMNAT.

Code	Description
K21	Herbaceous vegetation on wet, nutrient-poor, acid soil
K22	Herbaceous vegetation on wet, nutrient-poor, neutral soil
K23	Herbaceous vegetation on wet, nutrient-poor, alkaline soil
K27	Herbaceous vegetation on wet, moderately nutrient-rich soil
K28	Herbaceous vegetation on wet, very nutrient-rich soil
K41	Herbaceous vegetation on moist, nutrient-poor, acid soil
K42	Herbaceous vegetation on moist, nutrient-poor, neutral soil
H22	Woods and shrubs on wet, nutrient-poor, neutral soil
H27	Woods and shrubs on wet, moderately nutrient-rich soil
H28	Woods and shrubs on wet, very nutrient-rich soil
H42	Woods and shrubs on moist, nutrient-poor, neutral soil
H47	Woods and shrubs on wet, moderately nutrient-rich soil
A12	Vegetation in stagnant, nutrient-poor, neutral/alkaline water
A17	Vegetation in stagnant, moderately nutrient-rich water
A18	Vegetation in stagnant, very nutrient-rich water
bK20	Herbaceous vegetation on wet, very nutrient-rich, brackish soil
bK40	Herbaceous vegetation on moist, very nutrient-rich, brackish soil
bA10	Vegetation in stagnant, very nutrient-rich, brackish water

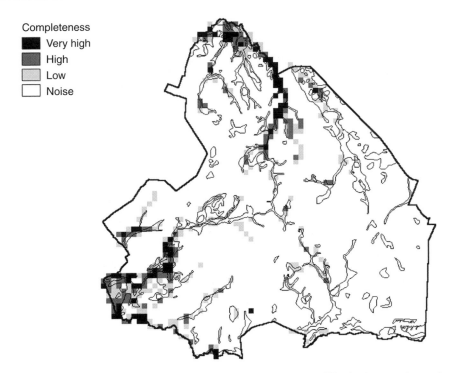

Fig. 6.7. Distribution and completeness of ecosystem type K27 in the province of Drenthe, The Netherlands, based on the national flora databank FLORBASE-2d. (From Witte and Van der Meijden, 2000.)

Input to DEMNAT are changes in hydrological variables which can be predicted with existing hydrological models and which are decisive for the plant species composition of the vegetation. These are changes in: (i) spring groundwater level (SGL); (ii) surface water level; (iii) seepage intensity; and (iv) percentage of 'foreign' water in summer. Examples of hydrological models that in the past have been linked to DEMNAT are NAGROM (National GROundwater Model; De Lange, 1996), MOZART (Vermulst *et al.*, 1996) and LGM (Landelijk Grondwater Model; Pastoors *et al.*, 1993).

The core of the model, the way these hydrological changes are translated to changes in plant species composition, is executed in two steps (Fig. 6.8). In the first step (the 'site module' of Figs 6.2 and 6.4) it is determined which changes in the four operational site factors (salinity, moisture regime, nutrient availability, acidity) are expected to occur, given a certain hydrological input. These changes in site factors are derived from computer models, empirical data and expert judgement. The changes depend on soil characteristics, which can be deduced from the ecological soil type. In the second step (response module of Figs 6.2 and 6.4), correlative relationships between ecological species groups and the four operational site factors are used to predict how these abiotic changes will affect the completeness of the ecosystem type. By way of example, Fig. 6.3 shows which changes in the four operational site factors can be expected as a

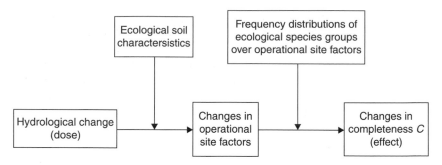

Fig. 6.8.   Calculation of the effect of a hydrological change (dose) on the completeness of the vegetation (effect). (Adapted from Runhaar *et al.*, 1996b.)

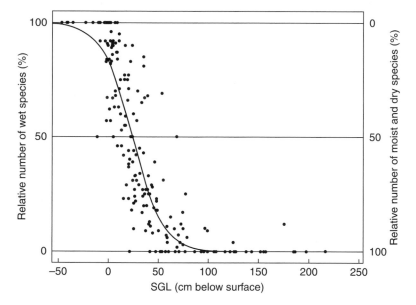

Fig. 6.9.   Relative number of wet species (hygrophytes) in the vegetation as a function of the mean spring groundwater level (SGL) on all soil types. Each point represents a vegetation sample (relevé). (From Runhaar *et al.*, 1997a.)

result of harmful hydrological changes (step 1) and which groups of plant species are likely to be affected (step 2).

The site factor 'moisture regime' is a measure of the availability of both oxygen and moisture (Runhaar *et al.*, 1997a). In wet circumstances, oxygen is decisive for plants whereas in dry circumstances the availability of moisture is what matters. In step 1, changes in the site factor moisture regime are directly related to the computed SGL. Figure 6.9 shows the correlative relationship that is used in step 2 to calculate the reaction of 'wet' species (hygrophytes) to the SGL. This figure clearly illustrates how strongly hygrophytes react to groundwater lowering.

These species are physiologically adapted to survive in wet anaerobic soils, for instance because they have air-space tissues transporting oxygen to their roots. For the reaction of 'moist' and 'dry' species (mesophytes and xerophytes) to the SGL, soil texture and organic matter content are also taken into account, since these factors determine (together with the groundwater level) the amount of moisture that is available for plants. To this end, the ecological soil types have been divided into a number of hydrological relevant classes. As an example, Fig. 6.10 shows the reaction of dry species to SGL on soil types that are classified as 'non-loamy sand'. The response of moist species is deduced from the relative numbers of wet and dry species. On non-loamy sand they have an optimum at an SGL of 40 cm below the soil surface. At higher levels they are being replaced by wet species (Fig. 6.9) and at lower levels by dry species (Fig. 6.10).

We expect in the near future that not only the SGL but also the computed annual moisture deficit of the soil will be used as a measure of the moisture regime site factor. Currently, research is being carried out to relate the response of moist and dry species to this deficit (Runhaar *et al.*, 1997b).

The site factor 'nutrient availability' is related to the nitrogen mineralization rate. On the basis of model calculations by Kemmers (1993), the relationship between SGL and mineralization rate is derived in step 1 for various kinds of soils, differing in the type and content of organic matter. For the reaction of the vegetation (step 2), correlative relationships are used between mineralization rate on the one hand and 'nutrient-poor' species and 'moderately nutrient-rich' species on the other. In a similar way, the process chain 'SGL–soil pH–fraction of 'acid', 'neutral' and 'alkaline' species' is established.

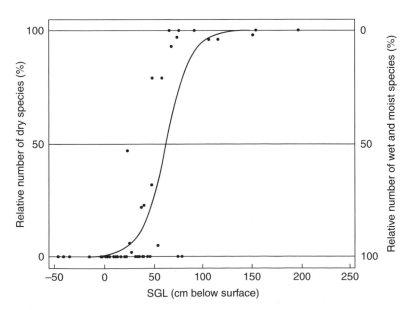

Fig. 6.10. Relative number of dry species (xerophytes) in the vegetation as a function of the mean spring groundwater level (SGL) on non-loamy sand. Each point represents a vegetation sample (relevé). (From Runhaar *et al.*, 1997a.)

By combining the separate reactions of a hydrological dose on, respectively, the four site factors and the ecological species groups, the final response of the ecosystem type is calculated (Runhaar *et al.*, 1996b, 1997b). As an example, Fig. 6.11 shows dose–effect functions of ecosystem type K21 (herbaceous vegetation on wet, nutrient-poor, acid soil) for a fall in the groundwater table on four ecological soil types. Similar functions have been established for other ecological soil types and the other three hydrological doses.

From Fig. 6.11 it appears that ecosystem type K21 is the most susceptible to a groundwater lowering on meso-eutrophic peats (V04). This is due to the fact that the effects of physiological desiccation, mineralization as well as of acidification are strong on this soil type. On old strongly weathered clays (K12), the susceptibility is limited because no mineralization effect is expected.

To demonstrate DEMNAT's performance, we show the results of an imaginary scenario for a certain part of The Netherlands, the province of Drenthe. In this scenario the groundwater extraction for drinking water and industrial water supply is to cease completely. Figure 6.12 shows the location of the pumping wells, as well as their current extraction.

The computed effect of the extraction stop on the average SGL is shown in Fig. 6.13. A rise in SGL of more than 30 cm is expected to occur near pumping wells, whereas a smaller rise is computed for a much wider area, also outside the provincial border. The computed intensity of upward seepage has a more or less similar pattern, although the effects are somewhat more concentrated in the brook valleys. The intensity of upward seepage in such areas might increase by 0.5–2 mm/day.

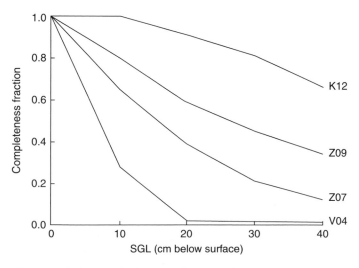

Fig. 6.11. Predicted changes in the completeness fraction of ecosystem type K21 as a result of a lowering of the mean spring groundwater level (SGL) for various ecological soil types: K12, old strongly weathered clays; Z09, loamy non-calcareous sands with thin topsoil; Z07, non-calcareous sands with thin topsoil (not loamy); V04, primary meso-eutrophic peat. (From Runhaar *et al.*, 1997b.)

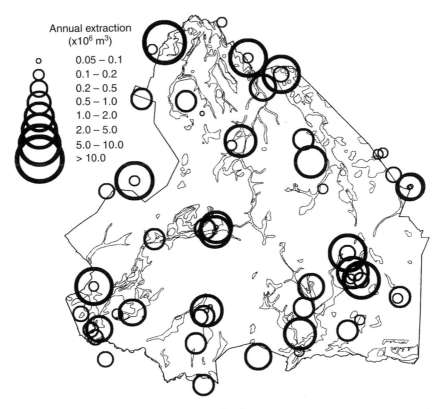

Annual extraction
(×10^6 m^3)

- 0.05 – 0.1
- 0.1 – 0.2
- 0.2 – 0.5
- 0.5 – 1.0
- 1.0 – 2.0
- 2.0 – 5.0
- 5.0 – 10.0
- > 10.0

Fig. 6.12. Location and magnitude (× 10^6 m^3) of annual groundwater extraction in the province of Drenthe, The Netherlands, in 1990.

The ecological effects of these hydrological changes were predicted with DEMNAT. Aquatic ecosystem types (A12, A17 and A18) have been omitted, since effects of this scenario on the surface water level and the supply of foreign surface water were not computed. Figure 6.14 gives a geographical picture of the predicted gains in conservation values. Especially for the brook valleys and for areas near the pumping wells, a substantial increase in conservation value is predicted. Figure 6.15 shows which groundwater-dependent ecosystem types will profit most from the extraction stop.

The extraction cessation results in a substantial increase in the completeness and abundance of ecosystems typical of brook valleys with upward seepage (Fig. 6.15: H22, H27, K22, H27). The recovery is particularly large in areas where remnants of these ecosystem types are present. Ecosystems that are typical of an infiltration regime (K21, K41), however, do not profit very much from this scenario. An explanation is that the groundwater rise in infiltration areas is very small (Fig. 6.13) because of the presence of a thick layer of bolder clay in Drenthe. Another explanation is that these ecosystems occur largely on soils with a perched water table. Water management cannot influence such soils.

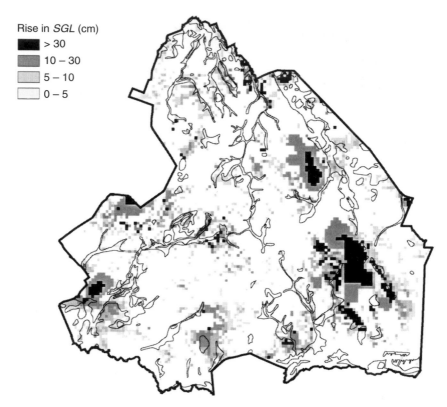

Rise in *SGL* (cm)
- ■ > 30
- ▨ 10 – 30
- ▨ 5 – 10
- ☐ 0 – 5

Fig. 6.13.   Predicted rise in mean spring groundwater level (SGL) in Drenthe, The Netherlands, due to a 100% reduction of groundwater extraction.

The semi-mechanistic approach: other models

There are also some other models that may be classified as 'semi-mechanistic'.

WAFLO (Water–FLOra; Gremmen, 1990) was the first Dutch ecohydrological model. This model is meant for the evaluation of the increase in groundwater extraction in the Pleistocene parts of The Netherlands. It comprises both the response module and the evaluation module of Fig. 6.2. It contains 'if–then' expert rules applied to the indicator values of Ellenberg (1979); for example, 'if the final SGL exceeds or equals 100 cm below the soil surface, then any species with an Ellenberg moisture indicator value of 6 or 7 will disappear'.

Like DEMNAT, NTM (Nature Technical Model) also makes predictions for ecosystem types. The applied ecosystem classification is based on Ellenberg's (1979) indicator scales for moisture, acidity and nutrient availability, which all have been reduced to three classes. The moisture scale, for instance, has the classes 'wet' (Ellenberg's indicator values 8–10), 'moist' (5–7) and 'dry' (1–4). A combination of the classes results in a matrix of 3^3; which is 27 elements, each of which represents a certain site type, such as 'wet, nutrient-rich, acid'. On the basis of Ellenberg's indicator values, ecological species groups are assigned to each site type. Furthermore, each site type is given a potential conservation

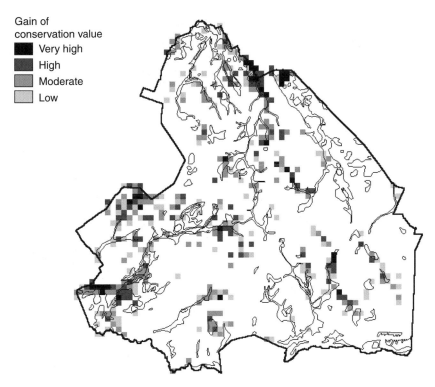

Fig. 6.14.   Predicted increase in conservation value in Drenthe, The Netherlands, due to a 100% reduction of groundwater extraction.

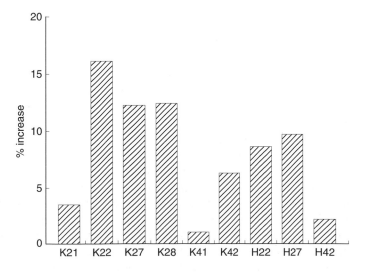

Fig. 6.15.   Predicted increase in the most common terrestrial ecosystem types (see Table 6.2 for a description) in Drenthe, The Netherlands, as a result of a complete stop of groundwater extraction.

value, calculated only once from the number of highly valued species in the corresponding ecological group. To facilitate predictions, the class boundaries have also been defined in physical terms. The boundary between 'wet' and 'moist', for example, corresponds to an SGL of 20 cm below the soil surface. Computed changes in the site factors may bring about the crossing of class boundaries and, as a result, a new site type with its associated new potential conservation value.

A model still in full development is MOVE (MOdel for the VEgetation; Latour *et al.*, 1993). For the response module of MOVE, a method of Ter Braak and Gremmen (1987) is applied. This method combines the statistical approach of ICHORS with the indicator values of Ellenberg (1991). Instead of abiotic field data, MOVE uses a large database of vegetation samples (relevés) to obtain information about the habitat of species. For each relevé, average Ellenberg indicator values for moisture regime, nutrient availability, acidity and salinity are calculated. These averages are then processed with GENSTAT as if they were measured abiotic factors. The result is a set of equations describing the occurrence probability of species as a function of Ellenberg's indicator values (Wiertz *et al.*, 1992). Currently, the Ellenberg scales are calibrated to physical site factors, like soil pH and nitrogen mineralization (Alkemade *et al.*, 1996). These site factors can be modelled by the site module SMART (Kros *et al.*, 1995).

Finally, a model similar to DEMNAT, but intended for local applications, is NICHE (Nature Impact assessment of Changes in Hydro Ecological systems; Meuleman *et al.*, 1996). It uses more detailed geographical information and makes predictions for phyto-sociological vegetation types (of Braun–Blanquet) instead of ecosystem types.

Discussion

One general assumption made is that the response modules of the correlative and the semi-mechanistic approach assume that vegetation is in *equilibrium* with its site, and that the predicted site conditions correspond to a clear-cut, new species composition. Succession from the old to the new equilibrium is not modelled, nor is the interaction between plants of different species incorporated in the prediction. Such simplifications of reality are necessary for practical applications. They arise from a lack of both ecological knowledge and data. In the mechanistic approach of NUCOM, interactions between plants do exist, but for a relatively simple ecosystem which is not influenced by groundwater.

A second assumption for all three approaches is that the site of a vegetation type is *homogeneous*, i.e. that all the plants have their roots in the same environment. This may be true for many ecosystems but – as demonstrated in detail by Van Wirdum (1991), for instance – it is definitely not the case for many, such as floating rich fens and hayfields on wet, nutrient-poor, pH-neutral soils. This simplification of reality is made, however, in order to get a model that is workable in practice.

A third assumption is that all models cover different fields of applications: DEMNAT, for example, is meant for the evaluation of various kinds of water management measures in the whole of The Netherlands, whereas ICHORS is intended to be used for small surface waters of a particular region.

Finally, the models differ considerably in their practical applicability. For example, WAFLO was constructed to be practical, and indeed this model has been applied in several cases for the evaluation of a proposed groundwater extraction. However, the utility of ICHORS is very limited because this model requires input data that are difficult, if not impossible, to obtain without the help of expert judgement (like the future concentrations of bicarbonate, silicon and iron).

Fundamental benefits and drawbacks of the model approaches

An advantage of a correlative approach is that – presuming enough relevant factors are included – the model successfully describes the variation in the given study area, and that no presuppositions about the relative importance of ecological factors are required. Correlative studies may reveal so far unknown relationships and a correlative approach may be very useful as a first approximation, especially when the available knowledge about underlying processes is limited and where changes in water management are relatively small.

Model approaches are generally made in a more or less objective way, but this objectivity risks causing problems. Especially when a large number of indirect site factors are considered, an approach that is to a great extent correlative may lead to accidental results as well as to apparent correlations and, in this way, to unusable probability functions. Broodbakker (1990) describes an example of a wrong prediction founded on apparent correlations. On the basis of ICHORS results he predicted a decrease in the species richness and conservation value of aquatic vegetation in the shallow lake 'Naardermeer' after the inlet of lake 'IJsselmeer' water that had been cleaned of phosphate but still contained high concentrations of chloride and sulfate. In reality, however, the lower phosphate level resulted in higher transparency of the water and the restoration of species-rich vegetation with Characaeae and *Najas marina* (Bouman, 1992). Because of the removal of phosphate, the negative correlations between plant species and high chloride and sulfate contents were no longer applicable.

Apparent correlations can be avoided by exclusively considering those site factors that are most important for the species composition of the vegetation. Logically, these factors influence the vegetation in a direct (operational) way. There is an overwhelming number of publications (for reviews see Runhaar and Udo de Haes, 1994; Runhaar, 1999) showing that the most important operational factors for vegetation are 'light', 'salinity', 'moisture regime', 'nutrient availability' and 'acidity'. Unsurprisingly, nearly all ecohydrological models (DEMNAT, MOVE, NICHE, NTM, NUCOM, WAFLO) make sole use of these factors.

A second problem has to do with the method of field sampling. Models like ICHORS (but also the partly correlative MOVE) predict the occurrence probability P as a function of one or more site factors. However, P is deduced from a database with plant species and site factors and thus merely reflects the occurrence probability in the database instead of the actual occurrence in the field. This may seem a trivial and superfluous drawback, but experience (e.g. Runhaar *et al.*, 1994) shows that it is necessary to stress it. The Dutch landscape is dominated by fewer than 20% of the some 1500 indigenous vascular plant species (i.e. by common species with a rarity class of 'UFK8' or 'UFK9' according to Van der Meijden *et al.*, 1991). So most species are rare in a statistical sense, and therefore these species are easily missed with the random sampling procedure that is required for statistical analyses. What makes the sampling problem all the worse is the fact that conservationists have a special interest in very rare species, since they contribute particularly to the value of an area for nature protection (Witte, 1996, 1998). The only practical way to gather information about such rare species is to search for them and sample non-randomly. In this way, a good statistical tool, like logistic regression, is abused.

The sampling method is likely to also affect ecological amplitude – the optimum and the shape of the probability function. For instance, when a certain species is over-sampled at the low values of a variable, the probability function will tend towards a log-normal shape.

Finally, correlative methods have the drawback that no use is made of available knowledge, and that they contribute little to understanding of ecosystem functioning. This is exemplified by a brief review of research into the relationship between vegetation and hydrology as performed by phyto-sociologists in the second part of the 20th century. Ellenberg (1952) and Tüxen (1954) were the first to systematically study the relationship between groundwater level and the occurrence of vegetation types. Later, Niemann (1963) devised a method to characterize the hydrology of sites on which certain vegetation types occur in the form of 'duration lines': frequency diagrams in which the duration that a certain groundwater level is exceeded is represented. Inspired by the work of these researchers, there have probably been several hundred studies in which vegetation types are related to groundwater levels and duration lines. However, since little or no attention has been given to the underlying processes, these studies have added little to the work of Ellenberg and Tüxen, and brought little or no new insights relevant for ecohydrological modelling.

In summary, a typical correlative approach particularly has a descriptive meaning, not a predictive one. From a scientific viewpoint, a mechanistic approach, in which the whole chain of processes is explicitly modelled, is to be preferred above a correlative approach. Mechanistic modelling reveals more, as it deals with causal processes that take place in nature. Moreover, because of these causal relationships, the avoidance of accidental results and the apparent correlations, more faith may be placed in the predictive capacity of mechanistic models. Finally, mechanistic modelling may be the only solution to an ecosystem whose state depends strongly on its history and feedback mechanisms. This is especially the case with long-term predictions, to take account of processes like soil leaching, the accumulation of organic matter and vegetation succession.

There are, however, also some practical objections against mechanistic modelling. First, mechanistic modelling tends to reductionism: the 'cutting into small pieces' of a real system. The more the knowledge about a real system grows, the more it appears that certain model pieces contain lumped relationships, with correlative features. In order to raise the mechanistic character of the model, such lumped pieces may be split into smaller ones. Each of those pieces may then be modelled satisfactorily in a physical way, with a number of mathematical equations. But each piece also contributes to the uncertainty of the final result and sometimes it turns out that the sum of all pieces leads to inadequate outcomes: wrong results founded on good principles (see e.g. the review of the NTM model by Klijn *et al.*, 1998). Eeles *et al.* (1990) compared 'distributed' and 'lumped' models used to simulate the runoff regime of catchments, and Baird (1999) discussed the question of how complex or how simple a model should be.

Second, mechanistic models often require a large amount of input data, for instance about the growth characteristics of species and the kind of litter they produce. For models that are of general application (e.g. DEMNAT, MOVE, WAFLO), such information usually is not available. Related with this problem is the fact that the number of modelled relationships increases dramatically when more species are involved. For instance, one of the most important processes that determine the composition of the vegetation type is the interaction between species. If this was to be incorporated into a model applicable to all of the 1500 Dutch vascular plant species, there would be more than a million interactions to be modelled (i.e. $n(n-1)/2$ interactions, where n is the number of species)! Models in which the competition between species is explicitly modelled are therefore often restricted to simple systems with only a few species, as in the NUCOM model.

There is the possibility that the number of interactions in the model be limited by working with groups of species which share the same ecological niche; for example, by lumping low herbs, tall herbs, shrubs and trees. In this way it may become possible to model the vegetation succession and its effects on nutrient cycling within the ecosystem. However, in impact assessment the lumping of species in functional groups offers no solution, as general ecosystem parameters such as the structure or the productivity of the vegetation are less relevant than the species composition and its associated conservation value.

For a practical and generally applicable ecohydrological model, neither a purely correlative approach nor a fully mechanistic one offers a solution. A combination of the advantages of both approaches is preferable. In this semi-mechanistic approach, available and reliable mechanistic modules for the computation of site factors are used as much as possible, and they are complemented with correlative relationships between site factors and species (Fig. 6.4).

There are very good mechanistic site modules available for simulation of the groundwater table and the moisture regime, such as SWAP (Van Dam *et al.*, 1997) and MOZART (Vermulst and De Lange, 1999). Presently, modules that are able to simulate chemical site factors – like salt concentration, acidity and nutrient availability – are in full development. A good example of such a module is SMART, which was especially developed for natural ecosystems (Kros *et al.*, 1995). These chemical modules perform well in sandy infiltration soils, but are probably not reliable enough to be applied to peat lands, for example, and to situations with upward seepage of alkaline groundwater. Nevertheless, in the long term these chemical modules may take the place of the simple relationships – often based on empirical data or expert judgement – that are presently incorporated in ecohydrological models like DEMNAT, NICHE and WAFLO.

The only pragmatic solution, also in the long term, is the use of correlative relationships in the response model. The number of species in natural ecosystems is simply too large for mechanistic modelling of biotic interactions, and mechanistic modelling simply requires far too much input data. This correlative module may be a weak point of the semi-mechanistic approach, because of the black-box character. But the processes in the box are not completely unknown. Experimental studies show how site factors influence the functioning of plants species through physiological processes and a reasonable understanding exists of how this, through the competition between species, results in changes in the species composition. An example can be found in the relationship between groundwater levels and plant species, where the relationship between hygrophytes and wet conditions can reasonably be understood from their morphology and physiology, even though not all factors that influence the competition between hygrophytes and other plant species are known (Runhaar *et al.*, 1997a).

Furthermore, to avoid statistical change and apparent correlations, relationships between variables with an expected causal relationship are safer. The relationship between groundwater level and fraction of hygrophytes in the vegetation (Fig. 6.9) is an example of the ideal. On the other hand, relationships between groundwater level duration lines and phyto-sociological vegetation types is more problematic.

Species or ecosystem types?

Not only do ecohydrological models differ in the causality of the modelled relationships, they also differ in the kind of biotic component that is modelled: species, vegetation types or ecosystem types. An argument in favour of a prediction in terms of plant species is that protection of species is often regarded as the main target of nature conservation (Latour *et al.*, 1994). Another argument is that species have the advantage of being clearly distinguishable units that are – unlike vegetation types or ecosystem types – more or less independent of a chosen classification. These are indeed strong arguments to use plants as the objects of ecohydrological modelling. So why do not all ecologists use species?

The first argument against a species approach refers back to the last paragraph of the previous section, where the use of vegetation types in correlative studies was criticized because they have no direct causal relationship with operational site factors. The same argument also holds for plant species: to avoid statistical nonsense, one should not seek relationships between site factors and species, but for relationships between site factors and functional groupings of species.

Second, the previous section also argued that in many cases it is impossible to predict the species composition with sufficient certainty. In fact, models as ICHORS and MOVE do not really predict the species composition, but merely indicate the probability that species will occur. Or rather, they predict the suitability of the habitat for a plant species, because only habitat factors are taken into consideration. Other important factors that determine the future occurrence of plant species – such as the actual presence of a species and the availability of a seed bank – are not included in the model. Therefore, a prediction in terms of plant species suggests more precision than can actually be realized.

There are three practical reasons in favour of an ecosystem approach. One important reason is that it is easier to describe spatial variation by means of spatial units, such as ecosystem types, than by practically 'one-dimensional' units like plant species. Using ecosystem types, model results can be presented in terms of changes in surface area or visually presented on maps.

Another practical reason is that the use of ecosystem types reduces the amount of data needed for the construction of the response module. Generally applicable response functions for all Dutch plant species can only be derived from an extensive database with a representative set of records of site measurements and plant species. At present, such a database is not available. Finally, Witte (1998) showed that an evaluation on the basis of ecosystem types produces results that are acceptable to botanical experts, but the use of plant species in an evaluation module gives an underestimation of species-poor ecosystems, such as bogs and salt marshes. Another drawback of a species approach is that all species add to the total value in a positive sense, including the species that are part of disturbances (a heathland, for instance, scores higher when it contains a weed-covered community of low value).

Major Conclusions

Ecohydrological models serve as decision support tools in water management. This chapter has questioned which modelling approach should be followed to arrive at such practical models.

It was argued that a correlative approach is especially useful in an initial stage of research, to reveal unknown relationships. Correlative models, however, have several disadvantages, like the risk of accidental and apparent results. Moreover, nowadays a correlative approach is superfluous since the site factors which determine the plant species composition of Dutch ecosystems are known sufficiently well from previous research.

From a scientific viewpoint mechanistic modelling is to be preferred, since it shows more about the processes that take place in nature. Moreover, mechanistic modelling may be the only solution to long-term predictions, since in such predictions one may have to take into account the ecosystem's history and feedback mechanisms. However, since mechanistic modelling tends to reductionism and an increasing demand for input data, the practical value of mechanistic models is often limited, especially for ecohydrological applications.

A semi-mechanistic approach was preferred, in which site factors are modelled in a mechanistic way as far as possible. This is then followed by the prediction of the vegetation in a more or less correlative way. To avoid statistical nonsense, the correlative relationships between site and vegetation should be based on ecological knowledge. In practice this means two things. First, those site factors that are known to have a great influence on the species composition of the vegetation via water management are used – in The Netherlands, these factors are 'salinity', 'moisture regime', 'nutrient availability' and 'acidity'. Second, these should be correlated with parameters of the vegetation that are supposed to have a causal relationship with site factors. Such a parameter is, for instance, the fraction of hygrophytes, to be correlated with the groundwater level (as a measure of 'moisture regime'), or the fraction of alkaline species, to be correlated with the soil pH (as a measure of 'acidity').

7 The Benefits and Risks of Ecohydrological Models to Water Resource Management Decisions

J.A. GORE[1] AND J. MEAD[2]

[1]Program in Environmental Science, Policy and Geography, University of South Florida, St Petersburg, Florida, USA; [2]Division of Water Resources, North Carolina Department of Environment and Natural Resources, Raleigh, North Carolina, USA

Introduction

It is a reality that, with the expanding human population, the development and alteration of riverine catchments will continue. In turn, alteration of hydrologic regimes has led to severe degradation of fluvial ecosystems. Over the past decades, competing philosophies have been presented as the means and methods of restoring these aquatic ecosystems. These competing ideas range from a purely legal stance that riparian rights or prior appropriation grants primary standing for water consumption to the most recent concepts in ecohydrology, which mandate the restoration of a 'natural state' (Zalewski *et al.*, 1997).

Water has played an essential role in the development of most economies on this planet. While pioneer economies rely upon water for navigation and transportation of goods, as well as local consumption, agricultural technology and industrial development rapidly expand the demand on water resources. For much of the 20th century, water development has been of pervasive importance among the leading economies of the planet. Billions of dollars, mostly provided by federal governments, have been invested in multiple-use projects that provide in-stream generation of power, flood control, low flow augmentation to assimilate waste water discharge, ease of navigation and recreational opportunities, as well as off-stream supply to agriculture, municipalities and industry.

Predictable hydrographic patterns are disrupted by human activity, which either impounds mainstem flow or diverts that flow for human consumption, leaving a dramatically reduced flow or an altered flow regime in the natural channel of the river. Even impoundments that are not subject to withdrawals lose water by evaporation and groundwater recharge. The downstream flow regime is often regulated, producing a flattened hydrograph with reduced peaks and valleys.

For the most part, significant flow-based alterations in physical and chemical conditions within the lotic environment are caused by impoundment for flood control and irrigation storage, impoundment for hydropower operation (especially hydropeaking), irrigation diversion or abstraction, channelization and afforestation.

Concurrent with the decline in the availability of public capital for such water projects, the 1970s to the 1990s brought an increased consciousness, among many nations, of the value of maintaining base flows for pollution control and the protection of fish and wildlife. In some cases, this has extended to various mandates to protect or preserve some waterways in an undeveloped condition as part of a national heritage or trust of an undisturbed natural ecosystem. In order to meet the conflicting demands of economic growth and the protection of ecosystem integrity, there has been a substantive change in the legal position of humans and the natural resources upon which they rely.

The legal rights to access water resources on this planet have been based on some form of riparian doctrine that viewed flowing water as a basic amenity to the adjacent landowner, whether public or private, until the past few decades. For example, English common law held a 'natural flow' doctrine that allowed the landowner to remove only that portion of flow to be used for domestic purposes while the remainder was to be left undisturbed and unpolluted. This concept works well in widely dispersed populations where water flow is abundant. However, the demands of industrial societies have required that this interpretation be changed to 'reasonable use' by the landowner rather than limited personal consumption (Beuscher, 1967). Most societies, however, have operated on the notion that non-riparian landowners and others wishing to preserve free-flowing waters do not have legal rights to the water nor the standing to pursue such an action.

In the USA, the Public Trust Doctrine, first defined in Illinois Central Railroad versus Illinois (1892) 146 U.S. 387, defined the waters of the State as those which could be used for public benefit. However, this definition has traditionally been extended only to those waters defined as 'navigable'. Designation of navigable waters (those under jurisdiction of the federal and state governments) has been quite vague. Increased public use of waterways for recreational purposes has changed the definition of 'public waters' to include those intended for non-commercial recreational use (Kaiser Aetna versus United States (1919) 444 U.S.). Indeed, riparian landowners have discovered that primacy of ownership no longer extends to non-navigable waters if they are commonly used by the public for such activities as fishing, swimming and non-motorized boating (Arkansas versus McIlroy – Arkansas Sc 1980n – 268 Ark., 595 S.W. 2d 659 cert. denied, 449 U.S. 843, 101 S.Ct. 124, 66 L.Ed.2d 51). This expanded definition of 'public waters' has since been extended to include the habitat necessary to maintain these activities, effectively protecting those species as part of the Public Trust.

The 'environmental movement' of the 1970s and 1980s afforded a chance to change the structure of water rights and, hence, water management. In the USA, the Wild and Scenic River Act of 1968 (16 U.S.C.A. §§ 1271–1287) is the only federal statutory acknowledgement that reserved rights exist for non-human in-stream uses. These are restricted to designated free-flowing systems (or portions) destined for protection as part of the national heritage. It became incumbent on the states to determine the status of in-stream uses in their jurisdiction.

The first steps were based on a reinterpretation of water rights from 'reasonable use' to 'beneficial use'. In theory, beneficial use could include in-stream uses by resident floral and faunal elements. Although essentially without power of enforcement, the Federal Power Act of 1967 and the concurrent Fish and Wildlife Coordination Act required that federal authorities give 'equal consideration' to fish and wildlife when permitting the development of water resources. The precedent-setting case to test this power was a US Supreme Court decision in 1978 (New Mexico versus United States – 438 U.S. 696, 98 S.Ct. 3012, 57 L.Ed.2d 1052), when the Court concluded that the United States is legally entitled to reserve flows necessary to sustain plants and wildlife. Indeed, the Court admonished petitioners that the State had the right to reserve flows against future claims to flows of public waters. This trend in the acknowledgement of in-stream flow reservations as a statutory water right and a basic component of appropriate water resources management continued into the final decade of the 20th century. Many states in the USA and other nations have begun to consider duration, intensity and volume of flow to be critical components of 'water quality' and substantive components of new water quality regulations (Gore, 1997). The State of Montana has authorized both state and federal agencies to reserve flows for in-stream biological use, including an arbitrary limit of 50% of the annual flow of gauged rivers (Mont. Code Ann. § 85-2-316). The State of Colorado has taken this reservation a step further by declaring in-stream flow rights as a prioritized 'appropriation' for the state (Tarlock *et al.*, 1993).

Regardless of the legal status of the resource, there is an increasing global commitment to preserving riverine environments. The greatest problem has been in the conflict between continued economic development and the legal and scientific definition of 'beneficial' use. That is, what are the mechanisms that can be used not only to predict the impact of altered flows on lotic biota but also, at the same time, equate these gains or losses to some sort of economic index which allows a sound management decision on the fate of the resource, the fate of the in-stream communities, and the riparian users? Indeed, the relatively weak legal requirements for protection are the result of the inability of science, the legal system and resource managers to come to agreement on the 'value' of in-stream reservations and human rights to access that same resource.

The ecohydrological approach demands an evaluation of surficial needs, to provide for human and other ecosystem needs of sufficient quantity and quality at the correct temporal sequence. A number of intermediate philosophies have been introduced since the late 1970s in an attempt to address ecohydrological concepts. These concepts range from an engineering approach to river regulation, which provides water for human consumption and a modicum of water for ecosystem demand, to recent proposals for techniques, which advocate a management programme based on 'naturalized' flows (see e.g. Richter *et al.*, 1997).

An array of models, usually referred to as 'in-stream flow' models, is among these management approaches. In general, such models attempt to predict or preserve flows for ecological targets on an equal footing (that are of comparable ecosystem 'value') with human consumptive uses. Most commonly these models are used to evaluate specific stream reaches impacted by flow regulation such as impoundment, diversion or abstraction. Although far from perfect, these models provide an adequate base for development of the ecohydrological goals of catchment-based management. However, in order to understand where these models must be improved, the strengths and weaknesses, as well as uses and misuses, of these models must also be understood. This understanding is an ever-increasing need, since some of these models, despite their shortcomings, have obtained a certain amount of legal confidence. Thus, there is a tendency to accept the results of these predictive models as 'findings of fact' rather than the estimates that any model produces.

While the application of some models remains controversial, this chapter examines the various scientific opinions about the value of hydrological phenomena to riverine communities, traditional management approaches to preserving flows, and the new computer simulations and models that attempt to combine hydrological, hydraulic and biological phenomena into a prediction of gain or loss of ecosystem integrity. Controversies and management decisions often hinge on the appropriate use of these models. We suggest what is the greatest source of error and misinterpretation of the results of these models and indicate beneficial avenues of modification or more appropriate use of these models to better manage riverine resources in the future.

Within the context of making sound management decisions that balance the needs of human consumption and ecological integrity, the ecohydrological models we have examined are not, properly, predictive ecological models. That is, these models are not used to predict changes in ecological dynamics. Instead, these are decision models incorporating hydrological and hydraulic phenomena on which water availability has been traditionally based (such tools as hydrographic analysis, flood frequency analysis and time-series analysis) coupled with the ecological phenomena most closely associated with changes in flow conditions (i.e. density, diversity, life stage behaviour and recruitment). Thus, it is necessary to understand the responses of in-stream and, to some extent, riparian flora and fauna to changes in hydrological conditions within a catchment.

Ecohydrological Models

Regardless of the type of approach to defining ecological flows in riverine ecosystems, five elements must be considered before an adequate decision can be made. These are: (i) the goal (such as restoring or maintaining a certain level of ecosystem structure); (ii) the resource (target fish species or certain physical conditions); (iii) the unit of achievement (how achievement of the goal is measured, such as a certain discharge in m^3/s); (iv) the benchmark time period (over 20 years of hydrographic record, for example); and (v) the protection statistic (a mean monthly flow or a mean daily or weekly flow) (Beecher, 1990).

Unlinked models: hydrographic techniques

Most standard-setting flows are created by statute within a governmental agency and are designed to create a uniform enforceable flow (sometimes an optimal flow but, more often, a minimum flow) for the State's rivers and streams (Lamb and Doerksen, 1990). These techniques assume that gauged hydrographic records reflect flows that support aquatic life in that system (Wesche and Rechard, 1980). However, this assumption only applies where streams are undeveloped or where development has been stable for a sufficiently long period of time to supply an adequate post-development hydrographic record (Stalnaker et al., 1995). Where development is ongoing or will change significantly in the future, it is possible to reconstruct the natural hydrograph from gauging records when accounting for current diversions and withdrawals, but this requires that the water manager make some significant assumptions and speculations about the condition of the fishery prior to disturbance (Bayha, 1978).

Several solutions have been presented to address statutory flow allotments in regulated rivers and streams. One of the most common techniques has been to base allocations on the 'aquatic base flow' (Kulik, 1990). This method assumes that the annual historic low flows are adequate to maintain the ecosystem integrity. In most cases the median flow for the lowest flow month (in North American rivers, usually August or September) is selected. Again, this requires an adequate hydrographic record (10 to 20 years) under near-natural conditions to make an adequate assessment. This is a particularly common statutory standard in the western USA, where pre-development historical records are available to make such predictions. A modification of this technique, requiring at least 20 years of hydrographic record, is to assemble monthly flow records and produce a 'normal' hydrographic record by eliminating the 'abnormal' dry or wet months. From the remainder, flow duration curves are created for each month. The monthly in-stream flow allocation is that flow which is exceeded 90% of the time (10% flow) (NGRP, 1974). The selection of '10% flow' is quite arbitrary, but is based on the assumption that most biota can accommodate a few days of low flows each month.

In heavily populated regions and areas where historical records are poor or inadequate, a fallback statutory allocation has been the '7Q10'; that is, the mean low flow occurring for 7 consecutive days, once every 10 years. This has been a minimum flow allocated to protect water quality standards during drought (Velz, 1984). This is the level required to provide sufficient dilution for sewage effluent in the system. It is not adequate to address the flow requirements of biota. Nevertheless, many US states that have not established statutory flows to maintain biota still enforce the 7Q10 as the standard for minimum flows.

Building-block models

An alternative to creating a flow allocation based on flow duration curves is that of recreating the pre-development or 'natural' hydrograph. The assumptions behind this technique are based on simple ecological theory; that organisms and communities occupying that river have evolved and adapted their life cycles to flow conditions over a long period of pre-development history (Stanford et al., 1996). Thus, with limited biological knowledge of flow requirements, the best alternative is to recreate the hydrographic conditions under which communities had existed prior to disturbance of the flow regime.

The most simple of these allocations models was proposed by Bovee (1982), who recommended that a surrogate of the natural annual pattern of streamflows could be approached by allocating the median (exceeded 50% of the time) monthly flow. Again, this technique requires an extended period of undisturbed flow records or the ability to reconstruct these records.

There is a wealth of research to indicate that hydrological variability is the critical template for maintaining ecosystem integrity. The use of this natural variability as a guide for ecosystem management has been widely advocated since the late 1990s. Thus, even the simplest of monthly allocations based on some sort of restoration of a natural hydrograph is preferred to a standard allocation. Although variability is a key to ecosystem maintenance, some sort of predictability of variation must be maintained. Survival of aquatic communities occurs within the envelope of that natural variability (Resh *et al.*, 1988). Thus, the simplest of the building-block models may not include sufficient variability. In addition to the seasonal pattern of flow, such conditions as time, duration and intensity of extreme events, as well as the frequency and predictability of droughts and floods, may also be significant environmental cues. The frequency, duration and intensity of higher and lower flows can affect channel morphology and riparian vegetation, and therefore change aquatic habitat. Indeed, the rate of change of these conditions is also important (Poff and Ward, 1989; Davies *et al.*, 1994; Richter *et al.*, 1996).

In order to include conditions that reflect greater variability yet maintain some of the natural predictability, Arthington *et al.* (1991) proposed a method that draws on features of the daily flow record for flow allocations in dryland regions such as Australia and South Africa. Four attributes of the natural flow record are analysed: (i) low flows (based on an arbitrary exceedance interval); (ii) the first major wet-season flood; (iii) 'medium-sized' flood events; and (iv) 'very large' floods over a period of record (usually 10 to 20 years). These are progressively summed (as 'building blocks') to recommend a modified flow regime that provides predictable variability in duration, intensity and frequency of flood and drought events.

Richter *et al.* (1996, 1997) have suggested a more sophisticated building-block model, termed the 'Range of Variability Approach' (RVA). This approach is specifically designed as an initial, interim river management strategy that attempts to reconstruct the natural hydrograph. By a statistical examination of 32 hydrological parameters most likely to change ecological conditions (Table 7.1), the RVA establishes management targets for each of these characteristics and then proceeds to establish a negotiation session in which a set of guidelines is established to attain these flow conditions. This requires that more than 20 years of daily streamflow records are available for the analysis. The RVA requires that, during each subsequent year, the hydrograph created by RVA be compared with the target streamflows and new management strategies be created to more closely match the RVA. This process of revisiting the management strategy allows ongoing ecological research to contribute new information that may result in the change of RVA targets. These iterations continue until the management targets are achieved.

Building-block models are particularly useful in situations in which there is, quite literally, no time for development of the calibration models necessary to

Table 7.1. Hydrological parameters used as indicators of hydrologic alteration (IHA) in the Range of Variability Approach (RVA) system. (Adapted from Richter *et al.*, 1996.)

IHA group	Target characteristic	Hydrological parameter
Group 1: Monthly water conditions	Magnitude Timing	Mean monthly value
Group 2: Annual extreme water conditions	Magnitude	Annual mean 1-day, 3-day, 7-day, 30-day and 90-day minima
	Duration	Annual mean 1-day, 3-day, 7-day, 30-day and 90-day maxima
Group 3: Annual extreme water conditions	Timing	Date of each annual 1-day minimum Date of each annual 1-day maximum
Group 4: High/low flood pulses	Frequency Duration	Number of high pulses each year Number of low pulses each year Mean duration of low pulses each year (days) Mean duration of high pulses each year (days)
Group 5: Changes in water flow conditions	Rates of change	Means of all negative differences between consecutive daily values
	Frequency	Means of all positive differences between consecutive daily values Number of rises Number of falls

link biological and hydrological information together. This need is especially high in developing nations where demand on the water supply is increasing at a pace that will soon create great losses in available surface and groundwater. For example, in South Africa, that has a population doubling every 15 years and an increased demand for industrial and agricultural development, river managers and ecologists argue that there is insufficient time to carry out the ecological research necessary to create specific models for this semi-arid region. Yet, there is an increasing demand to rapidly create management strategies that will sustain some amount of development and protect ecological integrity (J.M. King, University of Cape Town, personal communication). The building-block models are the 'first-best approximation' of adequate conditions to meet ecological needs. More often than not, resource agencies have kept hydrographic records for long periods of time when little or no biological data have been maintained. Even when poor hydrographic records have been collected (or for less than 10 consecutive years), Larson (1981) suggested that a surrogate indicator for minimum flows could be assigned as 0.0055 m^3/s for each square km of drainage area during dry months with adjustments for spawning flows.

It must be understood that these models make two important assumptions that must be carefully evaluated. First, they reasonably assume that those biotic communities best suited to the river ecosystem developed in response to the natural hydrograph. As a result, reconstruction of the natural hydrograph will create those conditions. An unwritten understanding of this assumption is that there remains a source of colonists, without exotic contributions, to re-colonize or re-establish populations as the flow variability is returned. Second and more importantly, there is a tacit understanding that land use within the catchment has not changed substantially, so that chemical and sediment inputs remain predictable and at acceptable levels. If these two criteria cannot be met, there is a danger that re-establishment of the natural hydrograph will not necessarily result in maintenance or recovery of the original biotic communities.

Hydrographic and building-block models have the advantage of being easy to explain to the public and decision makers and, because they are rapid and less time-consuming, are frequently chosen to make water resource management decisions. The greatest potential misuse of the building-block models, as in any ecohydrological model, is the institutional assumption that the first answer from a model is the only answer necessary to make adequate management decisions. That is, there is a tendency in regulatory agencies to make long-term management decisions from the first set of output data provided by a model. It is almost always the case that the first iterations of any model are based on the smallest amount of calibration information. With the building-block models lacking any ecological information, it can be quite dangerous to make long-term decisions on the first output from these models. There are no assurances that the goal for the reservation will be met. Indeed, the resource goal may not have been correctly identified. Yet, it often occurs that 'permits' to utilize the resource are issued for a period of 5 years or more; thus, re-evaluation of the strategy can only occur at those intervals. These management strategies must, however, be revisited on an annual basis and modified as ecological research determines more accurate information on flow requirements to sustain ecological processes (Richter *et al.*, 1977). This process is in significant conflict with the resource user who prefers a known release schedule for as many years in advance as possible in order to make sound business decisions about supply to customers. This is a conflict that still must be addressed by the users of all models. However, some agencies have attempted to modify this process by incorporating a building-block process but using other ecologically based hydrologic models to set minimum flows and levels for some low-flow 'blocks'. A particularly successful example of this strategy is adoption of Minimum Flow Level standards (MFLs) to protect seasonal in-stream flows in the State of Florida (Munson *et al.*, 2005; Munson and Delfino, 2007).

Intermediate models

Intermediate models rely upon hydrographic information for decision processes but attempt to incorporate some amount of biological information, usually through consultation with professional fisheries biologists and aquatic ecologists. In some respects, these intermediate models are similar to expert systems.

The Tennant approach

The most renowned of the intermediate models is the Tennant method (Tennant, 1976), sometimes called the Montana method. The original model was based on 10 years of observation of salmonid habitat in Montana and provided a table of flows, expressed as a percentage of the mean annual flow, and the relative 'health' of the habitat under those flow conditions. Tennant chose to divide the water-year into several components and recommend flows for each increment. A modification of this technique substitutes the expertise of the local fisheries biologist or aquatic ecologist. The analyst would observe habitats known to be important to the various life stages of the target biota during various flows approximating certain percentages of the mean annual flow. For each observation, the adequacy of the habitat conditions (ranging from severely degraded to optimal) would be recorded. These data would be inserted into a chart and negotiations or management strategies could begin (Table 7.2). Thus, decisions are based on the hydrographic record and the empirical observations of an accepted biological expert on the species or stream of interest. The Arkansas method (Filipeck *et al.*, 1987) is a modification of the Tennant method for warmwater fisheries and suggests a division of flow allocations across the water year as 60% of the mean monthly flow in the high flow season, 70% during the spawning season and 50% during the low flow periods.

Wetted perimeter technique

Among the most popular techniques to attempt a combination of habitat data and hydrographic information is the wetted perimeter technique (Nelson, 1980). This technique selects the discharge that provides the narrowest wetted substrate estimated to protect minimum habitat needs. Typically, the stream manager or analyst chooses a riffle as the area critical to the stream's function. It is generally assumed that providing and maintaining a wetted riffle promotes secondary production, fish passage and adequate spawning conditions. One or several typical riffle areas are surveyed and a relationship between discharge and wetted perimeter is calculated. The 'break-point' (the inflection point where wetted area declines rapidly) is designated as the discharge supporting minimally acceptable habitat (Fig. 7.1). A modification of this approach is to select a set of cross-sections that represent the range of habitats available. A coefficient of the sensitivity to de-watering may also be applied to each cross-section.

Table 7.2. Flow analysis based on empirical observations and the Tennant method. (Adapted from Stalnaker *et al.*, 1995.)

Habitat condition	% of mean annual flow	
	October–March	April–September
Flushing flow	200	200
Optimum	60–100	60–100
Outstanding	40	60
Excellent	30	50
Good	20	40
Fair	10	30
Poor	10	10
Severely degraded	<10	<10

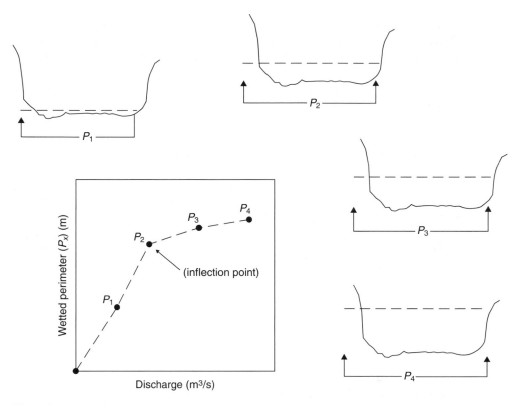

Fig. 7.1. Use of the wetted perimeter technique to estimate in-stream flow values. Dashed lines represent the water surface elevation. (Adapted from Stalnaker *et al.*, 1995.)

As might be expected, the shape of the cross-section of the channel has considerable influence on the ability of this method to be useful in making management decisions. Thus, the wetted perimeter technique is most useful at cross-sections that are wide, shallow and relatively rectangular.

The Texas method

Mathews and Bao (1991) developed a unique intermediate technique for application to warmwater ecosystems after examination of many of the intermediate techniques and understanding that most of these have been developed for salmonid streams dominated by snowmelt hydrology. The ecological focus of this intermediate model is the assumption that, in semi-arid environments with 'flashy' hydrographs, the lowest monthly median flow (designated as the 'Min MMF') results in the greatest metabolic stress to aquatic organisms. These occur as a result of high water temperature, diminished habitat area, and decreased dissolved oxygen and food supply (Matthews and Maness, 1979; Orth, 1987).

The Texas method relies upon examination of the historical flow record and calculation of the ratio (R_j) of the monthly median flow (MMF_j) to the minimum monthly median flow (*Min MMF*), for each month (*i*). Values for R_j are calculated for a suite of streams within a region or catchment. These values are log-transformed and regressed as a percentage (P_j) of the monthly median flow. The recommended flow for each month of the year (RF_j) is equal to the product of P_j and MMF_j. Mathews and Bao (1991) emphasize that this is a preliminary, 'desk-top' estimation of minimum flows that must be revisited and modified as more ecological information is obtained.

Linked models

Intermediate techniques offer a 'snapshot' of the stream resource, as exemplified by records of recent years. Most often the result of the application of these models is the designation of a 'minimum flow' or a regulated set of flows as standards, in perpetuity. Such results make it difficult to negotiate other management strategies. Although the building-block models attempt to give consideration to natural variability in streamflow as a sustainer of ecological integrity, such models poorly address the economic demands of resource evaluation. When the imperatives of negotiation (or legal proceedings) require a 'valuation' of the ecological benefit or a more dynamic approach to evaluation of the ecosystem, other techniques must be employed. Models have been developed that attempt to show the relationship between hydraulic conditions and biotic behaviour, usually reflected as the relationship between flow and habitat condition or availability.

There is a considerable body of knowledge which indicates that virtually every habitat parameter and life stage requirement of lotic biota are linked, either directly or indirectly, to changes in hydrological or hydraulic conditions. The challenge to ecohydrological modelling is the choice of the most appropriate sets of conditions that can be easily simulated, yet strike a balance between ecological reality and sound management choices. The model must be able to equate maintenance of biological integrity with comparable offsets in resource management. However, the model must also serve as a communication tool between water resource managers (often civil and environmental engineers) and wildlife and fisheries biologists. It should be understandable in its underpinnings so that all parties in the process of negotiating the 'beneficial' use of the water resource establish a level of comfort.

There are two groups of models that equate hydrologic conditions and biotic distributions. The first group utilizes statistical analyses to correlate hydrologic features with fish or macroinvertebrate population size. Such models use statistical predictions of change in density or diversity as the governing output for management decisions. The second group links hydrological or hydraulic models, sometimes with the addition of physico-chemical modifiers, with biological preferences for those physical conditions to obtain a prediction of some index of ecosystem change.

Statistical models

Binns (1982) developed the Habitat Quality Index (HQI) for use in the western USA. The HQI is the result of a multiple regression of several habitat variables (including flow conditions) versus the standing crop of fish. Each of 11 habitat characteristics (Table 7.3) is subjectively rated from 0 (worst) to 4 (best). After a series of streams in a catchment are rated and HQI values calculated, a predictive function for that catchment is created. The standing crop of the fish species is predicted as a function of change in HQI. The HQI procedure is fairly labour-intensive, requiring a variety of field measurements in order to create the initial model. There appears to be a direct predictive relationship between flow and standing crop of fish, but this correlation is, in fact, quite stream-specific (having been developed and tested on low- to medium-order, high-gradient streams) and is related primarily to critical low flows. The HQI has been developed for salmonid streams, but Bowlby and Imhof (1989) found that development of a separate regression set was more appropriate for Ontario trout streams than attempting to alter the HQI for those streams. Again, the Ontario Trout Habitat Classification (OTHC) is very stream-specific and must be recomputed for other streams in other regions.

Statistical models to predict changes in hydrological conditions have also been attempted for stream macroinvertebrates. Armitage and co-workers in the UK developed the River Invertebrate Prediction and Classification System (RIVPACS), a system that predicts changes in macroinvertebrate species composition with changes in environmental conditions (Armitage, 1989). This system is based on the multivariate analysis TWINSPAN (Hill, 1979) and uses a combination of chemical water quality, various hydrological conditions and geomorphological conditions to develop the predictive relationships (Table 7.4). The RIVPACS system has proved to be quite effective in the UK, but does rely upon an extensive database of macroinvertebrate collections and water quality measurements from both impacted and unimpacted streams; these data having been collected over many years and collated for the development of the model. The availability of such extensive data sets is quite rare and the acquisition of such data is necessary in order to compare disturbed sites with reference streams. The system is capable of predicting macroinvertebrate community composition downstream of impoundments, but RIVPACS has not been used successfully to predict the effects of changes in discharge at a site proposed for impoundment. Armitage (1989) believes that more research on the relationships between fluctuations in discharge and changes in substrate character will help to make RIVPACS a more successful predictive model.

Systems models

The second group of linked models includes those models that tie open channel hydraulics with measured elements of fish or macroinvertebrate behaviour. The most widely used example of this model is the Physical Habitat Simulation (PHABSIM) (Bovee, 1982; Nestler *et al.*, 1989). PHABSIM is the model most frequently used within the procedure called the Instream Flow Incremental Methodology (IFIM).

IFIM and PHABSIM are often thought to be synonymous. In fact, IFIM is a generic decision-making model that employs systems analysis techniques. IFIM guides stream managers in the process of choosing appropriate targets, endpoints and data requirements to achieve the management goal. At one level or another, IFIM requires a substantive knowledge of how aquatic habitat value

Table 7.3. Stream habitat conditions used to create the Habitat Quality Index (HQI). (Adapted from Binns and Eiserman, 1979.)

Characteristic	Quality rating				
	0	1	2	3	4
Late summer streamflow	CPF < 10% ADF	CPF = 10–15% ADF	CPF = 16–25% ADF	CPF = 26–55% ADF	CPF > 55% ADF
Annual streamflow variation	Intermittent	Extreme fluctuation, but never dry	Moderate; base flow occupies two-thirds of channel	Small; base flow stable	Little or no fluctuation
Max. summer stream temperature (°C)	<6 or >26.4	6.0–8.0 or 24.2–26.3	8.1–10.3 or 21.5–24.1	10.4–12.5 or 18.7–21.4	12.6–18.6
Nitrate-N (mg/l)	<0.01 or >2.0	0.01–0.04 or 0.91–2.00	0.05–0.09 or 0.1–0.90	0.10–0.14 or 0.26–0.50	0.15–0.25
Macroinvertebrate abundance (no./m²)	<25	26–99	100–249	250–500	>500
Macroinvertebrate diversity (Shannon's index)	<0.80	0.80–1.19	1.20–1.89	1.90–3.99	>4.00
Cover (%)	<10	10–25	26–40	41–55	>55
Eroding banks (%)	75–100	50–74	25–49	10–24	0–9
Submerged aquatic vegetation	Lacking	Little	Occasional patches	Frequent patches	Well developed
Discharge (m³/s)	<8.0 or >122.0	8.0–15.4 or 106.6–122.0	15.5–30.3 or 91.4–106.5	30.4–45.5 or 76.1–91.3	45.6–76.0
Stream width (m)	<0.6 or >46	0.6–2.0 or 23.0–46.0	2.1–3.5 or 15.1–22.9	3.6–5.3 or 6.7–15.0	5.4–6.6

CPF, average daily flow during low flow period; ADF, average daily flow.

Table 7.4. Four sets of variables (derived from TWINSPAN analysis) used in predictive equations in the River Invertebrate Prediction and Classification System (RIVPACS). (Adapted from Armitage, 1989.)

Physical variables	Physical and chemical variables
Distance from source (km)	Mean substrate particle size (phi)
Altitude (m)	Total oxidized nitrogen (mg N/l)
Air temperature range (°C)	Total alkalinity (mg CaCO$_3$/l)
Mean of January, April, July and October	Chloride (mg Cl$^-$/l)
temperatures (°C)	
Variables above plus:	
Slope (m/km)	Altitude (m)
Latitude	Air temperature range (°C)
Longitude	Mean of January, April, July and October
	temperatures (°C)
Discharge (average daily flow) (m^3/s)	Mean water depth (cm)
Mean water depth (m)	

changes as a function of incremental changes in discharge. This knowledge must be employed a priori, during the negotiation phases of the decision-making process. Indeed, a thorough knowledge of political science, negotiation and resource law is as important in the negotiation process as knowledge of science is in subsequent studies (Lamb, 1998). The use of replicate habitat sampling, biological sampling for the development of habitat suitability curves, sediment and water routing studies, as well as physical habitat, temperature and water quality simulations, may be necessary to properly depict the condition of the catchment under new operating scenarios (Sale, 1985). In IFIM, habitat suitability is treated as both macrohabitat and microhabitat. Macrohabitat suitability is predicted by measurement and/or simulation of changes in water quality, channel morphology, temperature and discharge along the length of the managed reach. These conditions may have an overriding impact on decisions made at the microhabitat level. Microhabitat suitability consists of individual species' preferences for these same criteria, reflected as depth, velocity, substrate or channel condition and cover. Those individual preferences are incorporated into PHABSIM to obtain predictions of changes in available habitat at a selection of stream segments 'typical' of the reach being managed. In combination, microhabitat and mesohabitat provide the information necessary to adequately negotiate management alternatives (Fig. 7.2).

The microhabitat evaluation within the IFIM methodology is completed through PHABSIM. Through a series of subroutine programs contained within PHABSIM, a prediction of the amount of available habitat (as weighted usable area, WUA) for a target organism over a range of discharges (Fig. 7.3) is created. HABTAT and its associate programs require hydrologic information in the form of transect (cell-by-cell) information on depth, velocity, cover value and/or substrate composition, and biological information in the form of preferences or suitabilities for these conditions by the target organism. Where possible, the

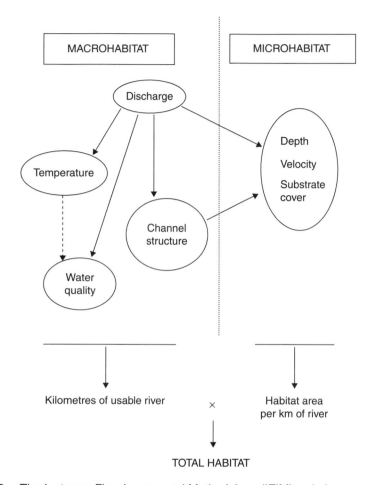

Fig. 7.2.   The Instream Flow Incremental Methodology (IFIM) technique combines both macrohabitat and microhabitat elements to make estimates over the entire length of a river or stream. (Adapted from Stalnaker *et al.*, 1995.)

hydraulic information for each transect should be measured. However, there are several 'desk-top' simulations that can also simulate these data when field measurements are not available or are impossible to measure (in the case of very large rivers or those with rapidly varying, unsteady flow). In addition to simulations within PHABSIM (routines such as WSP, MANSQ and IFG4), other hydraulic simulations are frequently used. These include steady-state models such as HEC-2 (USACE, 1982) and dynamic flow models such as RIV1H (Bedford *et al.*, 1983) and BIRM (Johnson, 1982, 1983). Regardless of how the hydraulic information is provided to PHABSIM, stage–discharge relationships are provided to the hydrologic simulation (usually IFG4), which predicts changes in velocity, cell-by-cell, with changes in water surface elevation. This prediction is accomplished through a series of back-step calculations through Manning's equation or, at the option of the user, Chezy's equation. This calculation assumes that substrate or

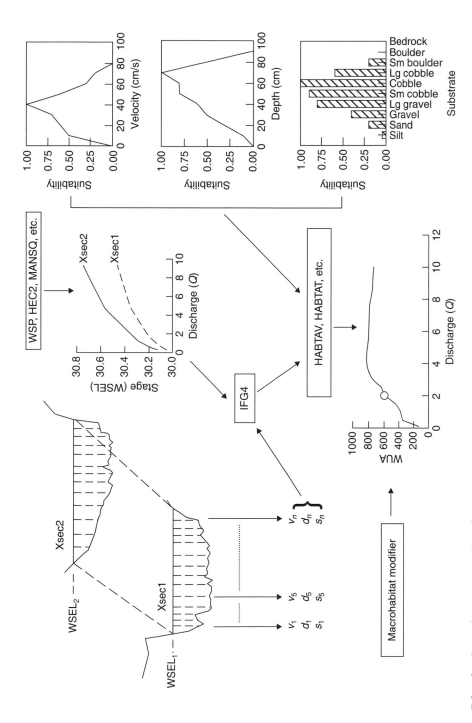

Fig. 7.3. A schematic representation of the basic elements necessary to create predictions of habitat availability (as weighted usable area, WUA) using the Physical Habitat Simulation (PHABSIM). The macrohabitat modifier is derived from external evaluation or simulations.

channel geometry will remain stable over the range of discharges to be simulated. As an alternative, Jowett (1998) has suggested that onsite measurement of changes in hydraulic geometry provide estimates comparable to the back-step predictions contained within IFG4.

Thus the PHABSIM model, in its current form, represents at best a quasi two-dimensional model, since it distributes velocities and discharges laterally along each transect. Cell-by-cell evaluations of WUA (the product of the preference criteria for each of the hydraulic conditions simulated and the total surface area of each cell) are computed through HABTAT and related subroutines (Table 7.5). Although the WUA–discharge relationship can provide information on the potential gains and losses of habitat with changes in discharge and can provide information on the apparent optimum and minimum flows, the output of PHABSIM is often not the product from which flow decisions are made. It will still be necessary to determine the relationship between optimum and minimum flows and their duration during wet and dry conditions. That is, the decision maker must decide what percentage of the time a selected flow is met or exceeded during an average hydrographic year and during unusually wet or dry years (Fig. 7.4).

Table 7.5. The most commonly used subroutines for habitat prediction within the Physical Habitat Simulation (PHABSIM) program. The different subroutines vary in how they treat cell boundaries along the transect.

Subroutine	Output	Common application
HABTAE	WUA WUBA WUV	Recreational boating (canoeing and kayaking) Water contact: wading, cattle crossings, fishing
HABTAM	WUA	Fish movement Allows lateral movement between cells, assumes longitudinal connections
HABTAT	WUA	Most commonly used subroutine Assumes a predictable response by biota to changes in combinations of velocity, depth and substrate
HABTAV	WUA	Assumes that fish habitat is dependent on velocities within the resident cell and adjacent cells (user-specified distance) Also used in lieu of HABTAT in low-order streams with small cell widths
HABVD	WUA	Rapid but less reliable prediction using calibration data from gauging stations Predicts 'threshold' values as minimum acceptable velocity and depth Supplement to HABTAT and others
HABEF	WUA	Competition between two species or life stages Minimum or maximum habitat available at two flows for a life stage Effective spawning area Stranding index analysis

WUA, weighted usable surface area; WUBA, weighted usable bed area; WUV, weighted usable volume.

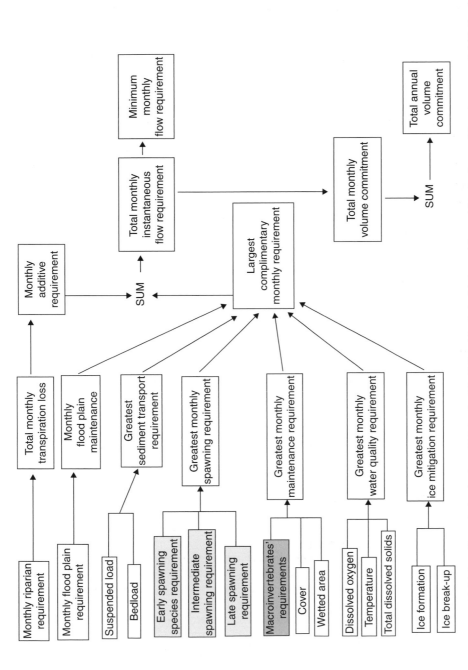

Fig. 7.4. A suggested initial outline of physical and biological 'building blocks' which could combine Instream Flow Incremental Methodology (IFIM)-style evaluations with hydrographic considerations of other physical and chemical requirements of ecosystem integrity. (Adapted from Bovee *et al.*, 1978.)

This is accomplished through the Habitat Time Series (HTS) component of IFIM (Milhous *et al.*, 1990). Such conditions as median habitat value over 10 or 20 years of record, the percentage of available habitat if certain magnitudes of flood are attenuated or enhanced, and the duration of low habitat conditions are typical predictions of an HTS evaluation. As previously mentioned, the benchmark period (10 or 20 years of record) must be carefully chosen to represent the most representative hydrographs under the least impaired conditions. Recently, for example, Kelly and Gore (2008) have reported the effects of changes in the Atlantic Multidecadal Oscillation (AMO) on regional weather patterns in the south-eastern USA and have suggested that two benchmarks must be analysed: a wet weather period and a dry weather period, each alternating period being approximately 30 years in duration.

Traditionally, the IFIM technique has focused on habitat availability of target fish species. Gore and Nestler (1988) believe that habitat suitability curves can be thought of surrogates for the basic niche (Hutchinson, 1959). That is, the derived suitability curves reflect maximized density when preferences approach unity. This should not, however, be interpreted as the equivalent of the carrying capacity of the system. The conversion to WUA is an attempt not to predict density changes, but changes in relative habitat quality and availability.

Habitat suitability information can come from a variety of sources. Most frequently, resource managers use published suitability curve information (the so-called 'Blue Book' series published by the US Fish and Wildlife Service; see Aho and Terrell, 1986). However, onsite development of habitat suitability criteria often produces the most accurate predictions (Bovee, 1986; Gore, 1989a). Habitat suitability among fish species is most often generated for spawning, incubation, fry, juveniles and adult stages. Frequently, when several life stages are involved, several different release scenarios must then be considered to assure the success of all life stages. In salmonid streams, this type of evaluation is relatively simple. However, as the number of species of concern increases, the decision process to provide adequate releases to support all species and life stages becomes quite complex. Competitive interactions between species assemblages can result in significantly different species' preferences among several streams in the same catchment (Freeman *et al.*, 1997), thus making transferability of standard curves impossible. In warmwater streams, where fish communities can be dominated by a variety of species using distinctly disparate habitats, Leonard and Orth (1988) have suggested that 'habitat guilds' are more appropriate than individual life stages or species-specific habitat suitability criteria. These kinds of compromises support Gore and Nestler's (1988) conclusion that the appropriate use of IFIM, in its current composition, is as a predictor of habitat quality rather than as some surrogate of density or productivity.

The in-stream flow requirements for benthic macroinvertebrates received equal attention during the development of IFIM (Gore, 1978; Gore and Judy, 1981). However, most stream managers have largely discounted these considerations because of perceived difficulties in collection (large sample size), taxonomic identification and habitat suitability curve generation, as well as inability to assign 'benefit' to the maintenance of benthic communities. Instead, many regulatory agencies and managers have concluded that enough flow for target fish species (and their individual life stages) is also sufficient for benthic species. Only recently have benthic macroinvertebrate habitat conditions become a frequent component of IFIM analysis. These applications have been quite generic, based on curves created from literature surveys (the Delphi approach; Bovee, 1986) or broadly defined curves (at the ordinal level; Peters *et al.*, 1989). However, Statzner *et al.* (1988) and Gore and Bryant (1990) have demonstrated that different macroinvertebrate life stages require different hydraulic conditions to achieve completion of life cycles, just as fish species have very different spawning, incubation and maintenance requirements.

PHABSIM has been frequently criticized for its simplicity or lack of ecological accuracy (e.g. Patten, 1979; Mathur *et al.*, 1985; Orth, 1987; Scott and Shirvell, 1987). The greatest criticisms have come from those scientists who have attempted to 'force' the model to become a predictor of ecological conditions rather than as a tool for guidance in making a management decision. For example, Irvine *et al.* (1987) rejected the value of PHABSIM because it did not accurately predict densities of rainbow trout in New Zealand streams. Gore and Nestler (1988) responded to the most common criticisms and indicated that the addition of various coefficients of biological interaction and ecological change is probably not justified by the expense required to obtain the data. That is, the management decisions (frequently alternative discharge release scenarios that must balance some amount of economic gain or provision of human consumption with some level of ecological integrity) are often so coarse-grained that the minute adjustments potentially predicted by a more responsive ecological model cannot be created.

Although initially rejected for application in New Zealand streams, Jowett (1989) developed a new set of field techniques for accurately measuring transect conditions in New Zealand streams and created a modified version of PHABSIM, the River Hydraulic and Habitat Simulation (RHYHAB), which places greater emphasis on the hydraulic changes that occur during changes in discharge (Jowett and Duncan, 1990). Jowett and Richardson (1990, 1995) have successfully developed a catalogue of habitat preference information for New Zealand fish and macroinvertebrates for direct application to IFIM and PHABSIM analysis. Further, Jowett and Biggs (1997) have begun to incorporate unique habitat conditions (periphyton and silt accumulations) to the ecohydraulic portions of RHYHAB and PHABSIM in order to implement better management decisions in New Zealand.

There are a number of inherent dangers in the use of PHABSIM and related models. Most of these are related to the misinterpretation of the results or misapplication of the model or its component subroutines. These models provide an index to habitat availability, not a measure of habitat actually used by aquatic organisms; that is, these are not response predictors. There has been a tendency for negotiators to assume that a 25% decline in the WUA index can be directly correlated to a comparable loss in fish density or biomass; thus leading to many early criticisms of the model (see Irvine *et al.*, 1987).

The linked models can only be used if the species or life stage being considered can be demonstrated to have definitive preferences for depth, velocity, substrate/cover and other variables that are incorporated into the model.

Some lotic biota do not respond to changes in channel hydraulics that are reflected in changes in density or location. For example, Layzer and Madison (1995) determined that the traditional application of PHABSIM to mussel species is inappropriate and showed that a more complex analysis, requiring measurement of shear stress and appropriate subroutines, creates a better WUA index. However, a concurrent analysis of the preferences of glochidial host species and the fall velocities for metamorphosed larvae is an even better index for maintenance of mussel communities under regulated flow conditions. Gore *et al.* (2001) have suggested that the extent and location of optimal conditions across a stationary mussel bed, as flow changes, is the most appropriate level of analysis in this circumstance.

Although dynamic flow models can predict the different changes in depth and velocity during the rising and falling limbs of 'flood' events (hydro-peaking generation, for example), the typical application of PHABSIM models assumes a relatively steady-state condition. In their current form, these models are not able to predict changes in channel geometry or condition. Thus, the impacts of extremely high flow events (channel-shaping floods, for example) cannot be predicted and, after such events, transects must be re-measured to reflect the accompanying changes in channel shape and form.

Finally, accuracy and precision are critical components of all modelling systems. That is, when applying the linked models described here, there must be consideration to the effort to accurately assess the underpinnings of the predictive model. All too frequently there is a concerted effort to accurately calibrate the hydraulic component of the model with multiple sets of calibration data at various flows and long and time-consuming iterations of the hydraulic subroutines to get the best estimates of cell-by-cell changes in velocity. Yet, little or no attention is paid to the accuracy of the biological models or to the precision of the data collection techniques that were used to create the models. Despite the fact that Bovee (1986) and Gore (1989a) have recommended the onsite development of habitat preference curves, perceived time constraints on making management decisions 'force' users of PHABSIM-type models to use 'catalogue' (literature-based) biological information. Although they are an acceptable first approximation of potential changes accompanying regulated flows, the generic models are likely to have been developed under different physical, chemical and biological conditions than the system being assessed. In addition, published preference curves reflect unique conditions of competition and predation (Gore and Nestler, 1988). While Thomas and Bovee (1993) suggested a statistical technique to enhance the transferability of habitat suitability information, local biological and physical conditions which significantly alter habitat use may still compromise the ability to utilize generic curves in other catchments (Glozier *et al.*, 1997; Leftwich *et al.*, 1997). Gore and Pennington (1988) expressed similar opinions for the transfer of habitat criteria for benthic macroinvertebrates. It will still be necessary to complete an onsite analysis of each site and rely upon the professional judgement of local biologists to support the use of generic, previously published, habitat information.

Despite their apparent shortcomings, PHABSIM and similar models have been successfully employed in many countries to create adequate management strategies under conditions of new regulated discharges. Indeed, the US Supreme Court has determined that the IFIM procedure is valid and an applicable management tool for negotiating water reservations (Gore, 1989b; Stalnaker *et al.*, 1995). In its current form, the model is sufficiently robust to suggest that it can provide guidance on the restoration of lotic ecosystems (Shuler and Nehring, 1993) and on the management of introduced and endangered species (Gore *et al.*, 1991, 1992). The predictions of linked models must be continually revisited and revised however, as Richter *et al.* (1996) suggested for building-block models, as new hydrological and biological information becomes available. It must be remembered that the production of these management models does not preclude or negate professional opinion and continued research into the interactions between hydrological conditions and biota in the ecosystem being managed.

There remains a lack of ecological prediction and a focus on single species or, at best, a suite of species for management, so there is still a need to more accurately evaluate the complexities of hydraulic conditions within the channel during hydrological changes. In addition, the subtleties of biological interactions within a community, and their potential change after flow alteration, are largely ignored in these models. In some situations, especially where species assemblages are quite complex such as in warmwater streams of the south-eastern USA, or where hydrographic extremes are common and less predictable (arid and semi-arid environments; see Davies *et al.*, 1994), or where a high degree of endemism demands a focus on protection of the entire ecosystem, PHABSIM in its current configuration may not always be adequate (O'Keeffe and Davies, 1991). It will be necessary to continually revise and expand both building-block and linked models as conservation and preservation efforts begin to emphasize ecosystem-level processes.

The Future of Ecohydrological Models

Three issues face the development of better management decision models based on ecohydrological phenomena. These are: (i) the creation of models that more accurately describe both hydrological and biological conditions and responses; (ii) obtaining negotiable results that provide a better estimate of the 'value' of ecosystem maintenance when compared with the economic gain from flow alteration; and (iii) the development of a standardized protocol for model application and management goals prior to negotiation of new flow schedules or flow reservations.

PHABSIM currently uses a one-dimensional hydraulic model to describe depth and velocity changes in a stream. This system is based on some simple energy and mass balance equations both between and across measured transects. A major criticism of this system has been the lack of accuracy when measuring a stream's frequently complex flow patterns such as eddies and divergences (Waddle, 1998). Two-dimensional (2D) models use finite element or finite difference techniques (rather than back-stepping through Manning's equation) to describe fluid motion. Thus, it becomes easier to describe such phenomena as partially wetted side channels, point bars, islands and intermittent inundation; conditions ignored or avoided in current IFIM analysis. The 2D models rely upon explicit areal representations of a stream rather than data derived from individual transects. This makes 2D models quite compatible with the technologies in Geographical Information Systems (GIS). When fully developed, 2D models should be able to allow the evaluation of conditions such as velocity shear zones and side channels, which are critical habitats for various life stages of both fish and invertebrate species.

Since the late 1990s, a number of investigators have been developing 2D models and attempting to couple them with calculations of habitat availability (Leclerc *et al.*, 1995; Bovee, 1996). The initial results are encouraging and seem to provide a more accurate representation of habitat conditions and the location of unusual habitat phenomena, such as eddies. It remains for ecohydrologists to begin the development of new habitat criteria that reflect preference for these conditions and response to changes in eddies or wetting of side channels, for example.

The River Systems Management Section of the Midcontinent Ecological Science Center (US Geological Survey), formerly known as the 'Instream Flow Group', is currently exploring the application of 2D hydraulic models to the IFIM system. Their primary emphasis is on the creation of a 2D WUA subroutine that will link to the existing PHABSIM model and the testing of new habitat metrics to improve the ability to represent more complex biological communities. It is anticipated that the output of the new models will generate a visualization of the stream reaches of concern, using GIS platforms such as ARC/INFO or ARCVIEW (Waddle, 1998).

Of course, one of the greatest criticisms of the IFIM-type procedures has been the inability of these models to predict ecosystem dynamics. Indeed, many regulatory agencies have bemoaned the fact that, although the habitat suitability curves are based on density relationships, the output of PHABSIM – WUA or habitat quality – is not a predictor of density or biomass. Many river managers believe that negotiations would be far easier if a more accurate 'value' could be placed on the species of concern than habitat quality or availability. Density and/ or biomass are easier to translate into economic loss or gain achieved by an altered flow regime. Although it would not create a prediction of ecosystem change, the attempt to create a response model based on changes in density or diversity is a considerable step towards achieving the goal of an ecosystem dynamics model.

Gore and Nestler (1988) have pointed out that fish are very mobile and respond quite rapidly to changes in flow, seeking refugia or moving out of a disturbed area. Niemela (1989) was able to demonstrate that rainbow trout (*Oncorhynchus mykiss*) maintained a refuge as part of their territory when occupying areas impacted by peaking hydropower. Indeed, the trout detected the changes in flow from the rising limb of the peaking wave and moved to cover or left the area prior to detection by human observers. By the same argument, that mobility allows fish to return to these areas rapidly after adequate flow conditions have been restored. Thus, the quality of the habitat (as reflected by an index such as WUA) is probably the more appropriate management target or tool. Some of this migratory ability is compensated for in the subroutine HABTAM. However, Cheslak and Jacobson (1990) have suggested a mechanistic simulation model that can be tied to PHABSIM to evaluate the potential effects of temporal changes in carrying capacity (where WUA is used as a surrogate for carrying capacity) on rainbow trout populations impacted by peaking hydropower generation. The quantitative performance of the model has yet to be compared with the actual performance of fish populations in the field, but the model is quite robust and the principal algorithms should be appropriate to most fish species.

Lamouroux *et al.* (1996) have taken a new approach to predicting the relationship between fish habitat and density and hydrological conditions. These models are based on spatial probability distributions of depth (Lamouroux, 1997) and velocity (Lamouroux *et al.*, 1995) and combining multivariate habitat-use models to predict changes in habitat suitability over the length of a river. In tests on the River Rhone, Lamouroux *et al.* (1996) were able to predict changes in frequency of the occurrence of various size classes of species, such as barbel (*Barbus barbus*) and chub (*Leuciscus cephalus*). Currently, this model has not been applied to other systems and the authors believe that the greatest problem, as in many other ecohydrological models, will be in the transferability of models to other systems.

Predicting changes in benthic macroinvertebrate numbers and density seems to be closer to reality than are the same predictions for fish. Gore (1989a) argues that macroinvertebrates, being less mobile, can only respond by leaving an impacted area (usually through drift). Re-invasion by the same process suggests that an area might support fewer or greater numbers of individuals of a species, depending on the quality of the habitat under the new flow conditions. Gore (1987) was able to demonstrate that the models of Gore and Judy (1981) were adequate to predict the densities of various species of *Baetis* collected from the same ecosystem from which the original preference curves were developed. In addition, Statzner *et al.* (1988) were able to produce statistically acceptable density predictions of the hemipteran, *Aphelocheirus aestivalis*, using both complex hydraulics and the simpler, bivariate models of Gore and Judy (1981). They later showed (Statzner *et al.*, 1998) that the joint preference curves (utilizing velocity, depth and substrate preference curves) provided adequate density predictions at lower sample sizes than required for complex hydraulic models. Similarly, Gore *et al.* (1998) obtained reasonable predictions of macroinvertebrate diversity along artificial riffles in stream restoration projects.

If we continue to employ IFIM or other linked models to manage regulated rivers, it will be necessary to continue to evaluate the appropriate level of biotic target for management. As mentioned before, the traditional level of evaluation has been the individual fish species, being further subdivided into critical life stages. However, as the number of target fish species increases, the resulting evaluation of habitat availability and change in hydrologic condition becomes more complex and muddled. Thus, Leonard and Orth (1988) suggested that habitat-use guilds were more appropriate in complex warmwater fish assemblages. O'Keeffe and Davies (1991), who suggested a similar approach to southern African rivers where endemism is high and all species are targets, echoed this concern for protection and management. Gore *et al.* (1992) suggested that maintenance of a high diversity of flow conditions in a river was a valuable alternative to constructing a great number of habitat preference curves, especially in river systems of southern Africa, where high endemism restricts transferability. Gore (1999) has also suggested that an analytical level of community diversity, developed on a site-specific basis, is a more appropriate target for macroinvertebrates and obviates difficult taxonomic problems and the need to develop large numbers of individual habitat suitability criteria. While the adequacy of other curves is still open to verification, Gore (1999) also suggested that suitability criteria for functional feeding groups (Cummins and Klug, 1979) might additionally serve as appropriate targets for maintaining the integrity of macroinvertebrate communities.

It is necessary to re-examine the management decisions required to maintain the integrity of lotic ecosystems. Currently, most river managers assess maintenance of 'integrity' by flow reservations that preserve one or two target species of concern. This simplistic scheme has been widely criticized, both for its lack of ecological reality and for its apparent inability to preserve whole-system integrity. Indeed, the proponents of the building-block models argue that the widely popular IFIM techniques make no attempt to reconstruct a natural hydrograph. We suggest that this is not the case with the proper application of IFIM. For example, in some of the earliest work on the development of IFIM, Bovee *et al.* (1978) suggested combining IFIM results with other models to create just such a set of building blocks to provide minimum monthly flow requirements which could be combined to produce a manageable annual hydrograph and a total annual volume commitment. Current stream managers do utilize PHABSIM results to allocate different monthly discharges during the year, but the focus remains on the hydrological needs to maintain the biotic component of the system. It is quite apparent, however, that such phenomena as flood plain maintenance, ice mitigation and water quality are also ecological integrity issues linked to maintenance of a certain hydrograph. In that respect, building-block models probably provide better management of the physical integrity of catchment ecosystems. It will be the task of ecohydrological models of the future to link improved ecological and hydrological models with negotiating techniques that focus on both the biological and physical targets. These models then, by combining a more complete model of hydrological change within the fluvial corridor with a sophisticated model of ecosystem response to these flow changes, could be used to assess not only restoration potential, but also the vulnerability of these systems to continued disturbance from catchment alteration.

Research to move in-stream flow models away from management of a single channel segment to entire catchments is underway. Stalnaker *et al.* (1989) had already suggested that certain specialized habitats (cover, edge, backwaters and flood plains) may have a stronger ecological role in large rivers and must, therefore, receive greater focus in the development of future ecological models. The first critical step to incorporate riparian needs has been taken by Stromberg (1993) and Stromberg and Patten (1990, 1991, 1996), who in a series of examinations of the flow requirements of riparian cottonwoods in a variety of basins have determined that the linkage between in-stream flows and riparian success is quite varied. In wide, unconfined valleys, where groundwater usually fluctuates with water surface elevations, there was a strong link between streamflow and tree growth. However, in narrower mountain canyons and other confined catchments the relationships were not as strong, suggesting that the same species

Table 7.6. In-stream flow models appropriate for integrating spatial and temporal scales of habitat analysis with a river or stream catchment. (Adapted from Stalnaker *et al.*, 1995.)

Target for evaluation	Type of model
Longitudinal succession	One-dimensional macrohabitat models: temperature, dissolved oxygen, other dissolved chemicals. Evaluate: degree-day accumulations of temperature, thresholds of tolerance, extent of available acceptable conditions
Habitat segregation or patchiness	Two-dimensional microhabitat models: depth/velocity or complex hydraulics (especially shear for mussels) in association with substrate materials and cover in small streams
Variable meteorological processes	Time-series analysis: total amount of usable habitat in the aggregate over the stream network. Evaluate: seasonal occurrence and duration of ecological bottlenecks associated with flood, drought or human-created water demands

in these ecosystems rely upon other hydraulic connections. Thus new in-stream flow models will have to account for basin morphology when attempting to preserve riparian integrity.

IFIM has developed over the past quarter of a century into a river network analysis that incorporates biotic habitat (especially target fish), recreational opportunities and woody vegetation response to alternative management schemes (Bartholow and Waddle, 1986). Information is usually presented as a time series of flow and habitat at selected points within a river system (Milhous *et al.*, 1990). These are especially useful in predicting the timing and location of habitat bottlenecks that might degrade the integrity of the entire catchment. The HTS is the first step in establishing critical flow periods (monthly, weekly or even daily) to avoid habitat bottlenecks and to allow a more careful examination of restoration of a 'naturalized flow' that supports the dynamic equilibrium observed in most fluvial ecosystems (Table 7.6). Critical habitat questions can then be related to comparable questions regarding available water supply for human consumption during those same time intervals. Cardwell *et al.* (1996) have suggested an optimization model as a planning tool that combines both the size and frequency of water shortages with habitat requirements to suggest appropriate water management schemes. Indeed, Cardwell *et al.* (1996) suggest that if we can express political, economic or other social concerns as a linear combination of storage, release and or diversion in a given time period, these can be used as additional constraints in the model. Such integrated approaches – that link theoretical models, ecological phenomena and institutional concerns – will be the next great step in better allocating water to the demands of an ever-increasing human population and maintaining the integrity of riverine ecosystems.

8 Nutrient Budget Modelling for Lake and River Basin Restoration

G. Jolánkai and I. Bíró

Environmental Protection and Water Management Research Institute (VITUKI), Budapest, Hungary

Introduction

Quantification of the fate of plant nutrients over the whole catchment of water systems is a key issue of ecohydrology, as eutrophication has become the most widespread problem of recipient waterbodies and especially standing waters since the 1940s (Chapter 3, this volume). The design and selection of effective management strategies require the quantification of water and contaminant (e.g. nutrient) flows along all the pathways of the hydrological cycle (Chapter 2, this volume). This control has in some cases been achieved through the use of models to predict the future state under different scenarios. Nevertheless, in the light of recent trends in water-related sciences and of relevant policies (e.g. the European Union (EU) Water Framework Directive), this activity is or rather should be the major scientific tool of River Basin Management Planning (RBMP), which is the major objective of Integrated Water Resources Management (IWRM).

In this chapter we review the options available for these planning activities, first for standing waters and then for catchment processes.

Quantification of Lake Processes

The quantification of lake processes has been carried out in a large variety of ways since the late 1950s beginning with experimental and empirical relationships (Vollenweider, 1969; Chapra, 1975; OECD, 1982). Other simple nutrient (mostly phosphorus) budget approaches were developed as it became clearer that eutrophication was a serious problem for water management (Lorenzen, 1974; Lewis and Nir, 1978; Chapra and Canale, 1991; Rossi and Premazzi, 1991; Salas and Martino, 1991).

Research into lake processes evolved through various phosphorus–phytoplankton models (Thomann *et al.*, 1974; Imboden and Gachter, 1975; Orlob, 1977; Jolánkai and Szöllősi-Nagy, 1978; Jolánkai, 1992) into multiparameter, dynamic, lake ecosystem models (Kelly, 1973; Jörgensen, 1976; Di Toro *et al.*, 1977; Kelly and Spofford, 1977; Park *et al.*, 1979; Bierman *et al.*, 1980; Di Toro and Conolly, 1980; Kutas and Herodek, 1986), including multilayered – epilymnion/hypolymnion – models of deep lakes (Lung *et al.*, 1976; Jörgensen and Harleman, 1977; Knoblauch, 1977; Niemi, 1979; Vincon-Leite and Tassin, 1990). Stochastic approaches were also applied or coupled with conceptual and deterministic models (Canale and Effler, 1989). Some approaches used the techniques of artificial intelligence for interpretation and qualification of the complex hydroecological processes involved (Guerrin, 1991). A separate group of models coupled hydrodynamic transport processes with the contaminant transformation processes of aquatic ecosystems (Paul, 1976; Shanahan and Harleman, 1986).

Drawing a flow diagram of the state variables and processes of a lake ecosystem like the one in Fig. 8.1 is the first stage in modelling the processes of eutrophication. This seeks to establish the cause-and-effect relationship between the measures (indices) of eutrophication and the inflowing phosphorus (nutrient) loads, if one intends to use such models for designing ecohydrological management actions. This diagram describes the lake ecosystem processes in terms of the 'forcing variables', such as inputs of energy, water and substances (nutrients), thus identifying the potential control points.

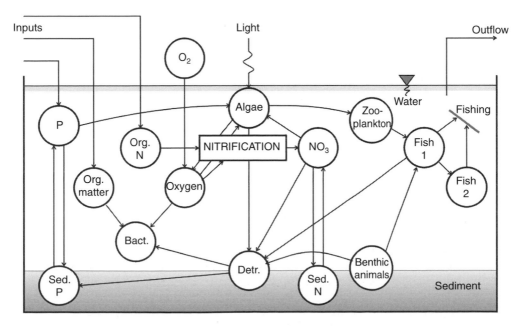

Fig. 8.1. Schematic illustration of some of the major processes in aquatic ecosystems (Bact., bacteria; Detr., Detritus; Org., organic; Sed., sediment).

Empirical and experimental relationships

The easiest approach to model eutrophication is the derivation of experimental relationships, which have the advantages that they are easy to use and (when reliable) can provide insight into the expected state of an aquatic ecosystem as a function of a few parameters and indices of the waterbody and its inputs from the catchment. In this approach, indices of the trophic state of the waterbody, such as the mean or maximum chlorophyll-a value and/or the rate of primary production, are expressed as functions of the flushing-corrected average inflow concentration of phosphorus.

The best example of this was produced for the Organization for Economic Cooperation and Development (OECD) (Vollenwider and Kerekes, 1981; OECD, 1982), which also provides a trophic state classification of waterbodies (Table 8.1). Refinements and adaptations of it were subsequently made (Yeasted and Morel, 1978; Rechkow, 1979; Golterman, 1980; Hoare, 1980; Kerekes, 1983; Mahamah and Bhagat, 1983; Salas and Martino, 1991).

Despite being a simple, robust method, its disadvantage is that the information obtained is very approximate and cannot be used for designing management actions. Nevertheless, predictions can be made about land use or the point-source treatment needed to change from one trophic state to a better one – or, in reverse, the risk of a waterbody deteriorating from one trophic state to another. These empirical relationships have remained indispensable tools in assessing the state of lake ecosystems, especially when quick answers to lake recovery problems are required on the basis of limited data (or now in Europe where, under the Water Framework Directive, classification of a large number of waterbodies is required and a new guidance document on eutrophication assessment is being prepared at the time of writing this chapter). They must be used with due concern for their limitations, perhaps together with parameter sensitivity and error analysis.

Classifications such as the OECD trophic state may be used in association with dynamic nutrient budget models, which express the change of in-lake nutrient concentrations as the result of inputs minus outputs of nutrients. The advantage of such models is that long-term variation of in-lake (and sediment) nutrient

Table 8.1. Fixed trophic state categories of the Organization for Economic Cooperation and Development study. (From OECD, 1982.)

Trophic category	Index of trophic state (mg/m^3)		
	P_L	Chl_{mean}	Chl_{max}
Ultra-oligotrophic	<4.0	<1.0	<2.5
Oligotrophic	<10.0	<2.5	<8.0
Mesotrophic	10–30	2.5–8.0	8.0–25
Eutrophic	35–100	8–25	25–75
Hypertrophic	>100	>25	>75

P_L, total phosphorus concentration in the lake water; Chl_{mean}, mean chlorophyll-a concentration in the lake water; Chl_{max}, maximum chlorophyll-a concentration in the lake water.

concentrations can be described and the predicted future conditions assessed as a function of planned variation of input loadings. They are relatively easy to calibrate, when appropriately reliable records of inflows, outflows and nutrient concentrations are available, but they are limited in their flexibility.

Ecosystem models

Dynamic lake ecosystem models not only include dynamic nutrient budget models but also describe the variation (e.g. growth, predation, mortality, etc.) of various ecological elements of the system (e.g. algae, zooplankton, fish, macrophytes). There is great variety of approaches (Bierman *et al.*, 1974; Larsen *et al.*, 1974; Thomann *et al.*, 1974; Imboden and Gachter, 1975, 1978; Jolánkai and Szöllösi-Nagy, 1978; Alcamo, 1983; Messer *et al.*, 1983; Wang and Harleman, 1983; Watson, 1983) and comprehensive literature reviews (Jacobsen, 1983; Jörgensen, 1988). Some models described very fine details of ecosystem processes even in the early years of the development of such models (Park *et al.*, 1979).

This detail is their advantage, expressing the trophic state with concentration and biomass values of the ecological components, along with the variation of flows and water volumes. Nevertheless, a major disadvantage of these models also stems from their complexity. Dozens, sometimes hundreds, of model parameters and coefficients might be required, with little or no chance of them being calibrated against locally measured data. Thus the reliability of forecasts is questionable. They are better suited for scientific purposes; for example, some important internal pathways of material cycling have been identified with the help of such models.

The processes of ecological components, such as the growth of algae, in many eutrophic lakes may not be influenced by the present rates of external nutrient loads because nutrients are, in many cases, no longer limiting. In such systems growth may be limited by light and temperature, so a good fit of the simulated time series of an ecological variable to measured data might only be due to a good temperature limitation function and thus will not reveal any real cause-and-effect dependencies of the system.

Models aimed at the provision of 'ecohydrological' planning tools and that succeed in quantifying in-lake ecosystem processes need to meet the following criteria:

- The model has to be tailored to the intended use and objectives (e.g. simulation, forecasts, long-term trends, research or management).
- It also has to be tailored to the availability of data, considering possibilities for field studies. It cannot include state variables for which no measurement data exist (at the least, they have to be minimal).
- It has to include hydrology (water levels, inflows and outflows).
- The model parameters have to be calibrated against a set of observed records and verified with another set that was not used for calibration, particularly when intended to serve management/planning purposes.

Some examples of simple and robust ecosystem models are now described briefly.

A model for Lake Balaton, Hungary

The example presented here was built during the course of the activities made for restoring Lake Balaton (Hungary) from severe eutrophication, which it had experienced since the early 1960s. Lake Balaton is the largest shallow lake in Middle Europe (surface area 596 km^2, average depth 3.14 m, catchment area 5776 km^2). External anthropogenic sources of phosphorus were blamed for this process. Eutrophication of the lake was subject to many studies, including the modelling of the eutrophication process (Jolánkai and Szöllössi-Nagy, 1978; Somlyódy and van Straten, 1986). Some remarks on the modelling experiences follow:

1. The OECD equations (OECD, 1982) for 'shallow lakes' were used to estimate the degree of eutrophication. For the highly eutrophic bays of the lake the OECD model somewhat underestimated the actual trophic state, while for the remaining less eutrophic eastern basins the calculated values either matched the measured ones or overestimated the actual state. In short, the empirical literature model gave a fairly good insight into the lake.

2. Phytoplankton growth was simulated (Jolánkai et al., 1994) by a general dynamic model concept (fully mixed reactors of a chain of four subsequent basins, nutrient budget, Michaelis–Menten growth limitation, temperature and light limitation functions) (Fig. 8.2). A special feature of this model was that the variations in lake volume (water depth) were also explicitly considered by an additional model equation written for the water balance for each of the four lake basins. The relatively good fit to measured data was mostly due to the effect of the temperature limitation function, as dissolved reactive phosphorus, i.e. bioavailable phosphorus (BAP), was in abundance in the lake water. Thus this model was able to simulate algae growth (but still cannot be used for management purposes).

3. A simple dynamic lake model of the type given below was utilized in the attempt to relate long-term variations of nutrient loads to trophic state indices of the lake water (the actual model equations are more complex ones and the equations below show only the type of water and mass budgets utilized):

$$\frac{dP}{dt} = \frac{1}{Ah}\left[P_{in}Q_{in} - PQ_{out}\right] - K_1 P + K_2 P_s \frac{d}{h}$$

$$\frac{dP_s}{dt} = \frac{h}{d}K_1 - K_2 P_s$$

$$\frac{dh}{dt} = \frac{1}{A}\left[Q_{in} - Q_{out}\right] + P - E$$

$$\frac{dAB}{dt} = \frac{1}{Ah}\left[Q_{in}AB_{in} - ABQ_{out}\right] + \mu AB - K_a AB$$

$$\mu = \mu_{max}\frac{P}{K_P + P}F_{temp\,lim}$$

$$Chl\text{-}a = aAB$$

(8.1)

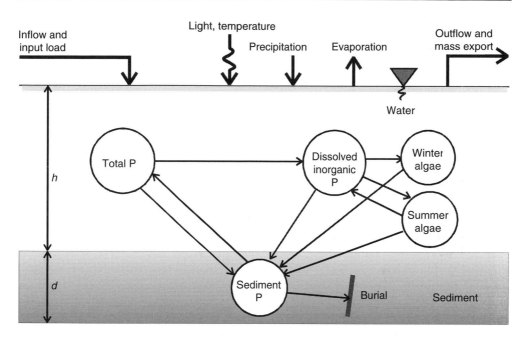

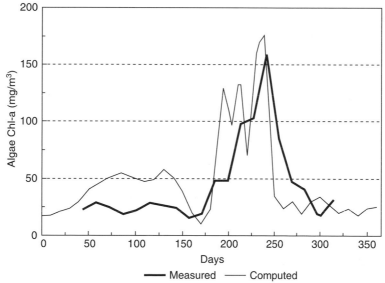

Fig. 8.2. Model scheme (top) and simulation results (bottom) of measured chlorophyll-a (Chl-a) time series of the Bay of Keszthely, Lake Balaton.

where

P = in-lake phosphorus concentration [M/L^3],
t = time [T],
A = average lake surface area [L^2],

h = lake depth [L],
P_{in} = phosphorus concentration in the inflow [M/L^3],
Q_{in}, Q_{out} = inflow and outflow rates of the lake respectively [L^3/T],
K_1 = sedimentation (retention) rate constant of phosphorus [1/T],
K_2 = phosphorus re-suspension (scouring) rate constant [1/T],
P_s = phosphorus concentration in the sediment [M/L^3],
d = depth of (active or interactive) sediment [L],
P, E = precipitation and evaporation on to/from the lake surface respectively [L],
AB = concentration of algae biomass [M/L^3],
AB_{in} = algal biomass concentration in the inflow [M/L^3],
μ = growth rate of algae [1/T],
K_a = lumped algae loss rate constant (mortality and zooplankton grazing) [1/T],
μ_{max} = maximum growth rate of algae [1/T],
K_P = half-saturation constant for algae growth (phosphorus concentration at which the growth rate is half of the maximum) [1/T],
$F_{templim}$ = factor of temperature limitation of algae growth [factor from a function not shown here],
$Chl\text{-}a$ = chlorophyll-a concentration [M/L^3],
a = proportionality factor between algae biomass and chlorophyll-a (set for the model as that of the recent average of Keszthely Bay of Lake Balaton).

This model had a 17-year simulation run, giving the following results:

- Simulated and measured in-lake phosphorus (BAP) concentrations were fairly well matched in the eastern bays but were considerably different in the western ones. One likely explanation for this difference might be the method of estimating BAP of the inflowing streams.

- The exchangeable sediment phosphorus pool (the plateau of the curve) matched the single set of sediment exchangeable phosphorus data (Heródek *et al.*, 1995) relatively well, showing also the same spatial variation (the value of 50 g P/m^3 for Keszthely Bay corresponds to the reported value of 7.8 g P/m^2 if the active sediment layer was 6.4 cm, while the modelled value was 7.5 cm).

- The model predicted an increase in chlorophyll-a concentration until about the end of the 1980s, followed by a decline afterwards. The actual decline of chloropyhll-a concentration started in the early 1990s and is still continuing, in spite of an increase in orthophosphate concentrations of the lake water subsequently. This latter phenomenon has not yet been satisfactorily explained.

A model of the Ráckeve–Soroksár–Danube arm

A dynamic nutrient budget model, coupled with an empirical relationship between phosphorus and chlorophyll-a, was used for simulating the effects of planned water and environmental management strategies for the Ráckeve–Soroksár–Danube arm (RSD), just downstream of Budapest. This 56 km long arm of the river (enveloping the Isle of Csepel with the main Danube) is one of the most important recreational areas of Budapest. It is fed by the Kvassay Zsilip (a river barrage) from the main Danube and is highly eutrophic, being also the recipient of the South-Pest Sewage Treatment Plant (SP-STP). Non-point sources, suburban settlements without sewerage and the large number of summer cottages along the river arm are also known as substantial contributors to the phosphorus load. A project (PHARE, 1998) was launched to plan control strategies. The following model was developed/adopted to the RSD:

$$\frac{dP_w}{dt} = \frac{1}{Fh_w}\left[P_{load} + Q_{feed}P_{feed} - (Q_{out} + Q_{abstr})P_w\right] - k_1 P_w + k_2 \frac{h_s}{h_w}P_s \quad (8.2)$$

and

$$\frac{dP_s}{dt} = k_1 \frac{h_w}{h_s}P_w (k_2 - k_b)P_s \quad (8.3)$$

where

P_w = phosphorus concentration in the water [M/L^3],
F = average surface area of the lake segment [L^2],
h_w = water depth in the lake segment [L],
P_{load} = external phosphorus load into the reactor (lake segment) [M/T],
Q_{feed} = flow of the feed water (or outflow water of the former segment) [L^3/T],
P_{feed} = phosphorus concentration of the feed water [M/L^3],
Q_{out} = outflow rate from the lake segment to the next one [L^3/T],
Q_{abst} = water abstraction (intake) rate from the lake segment [L^3/T],
k_1 = sedimentation rate coefficient of phosphorus [1/T],
k_2 = scouring/re-suspension rate coefficient of sediment phosphorus [1/T],
P_s = exchangeable sediment phosphorus concentration [M/L^3],
k_b = inactivation/burial rate coefficient of sediment exchangeable phosphorus [1/T].

Equations 8.2 and 8.3 describe the temporal variation of phosphorus interaction between water and sediment as a function of loads from various external sources, assuming a 'fully mixed reactor'. The RSD river arm was approximated by a series of such 'fully mixed reactors', interpreted as stream segments corresponding to monitoring stations.

The trophic state of the lake segment corresponding to the phosphorus concentration was simulated with the OECD (1982) relationship relating the mean and maximum in-lake chlorophyll-a concentrations to the in-lake phosphorus concentration (actually by equations of identical form which gave the best fit to the observed values):

$$\begin{aligned} Chl_{max} &= 0.74 P_w^{0.97} \\ Chl_{mean} &= 0.52 P_w^{0.81} \end{aligned} \quad (8.4)$$

The model simulation was made for three segments of RSD:

1. Kvassay Gate–Szigethalom gauge (using the Szigethalom gauge records as the basis of the calibration).
2. Szigethalom station–Ráckeve station (using Ráckeve station's records as the basis of the calibration).
3. Ráckeve station–Tassi Zsilip station (using Tassi Zsilip station's records as the basis of the calibration).

In this simulation the feed water concentration parameters of the first segment were those of the records of the Kvassay Zsilip station. In the subsequent reaches the 'feed water' concentration parameters were those of the calculation results of the previous segment.

Calibration and simulation were both made for phosphate-P because: (i) this is the parameter that can be considered (as an approximation) to be the biologically available form BAP, which drives the eutrophication process; and (ii) relatively consistent continuous records were available (compared with data for total phosphorus).

There were no records available for sediment phosphorus so the calibration run was made against new experimental data for 1998 (Jolánkai, unpublished). These data, 'Danube water leached sediment phosphorus', were used as the approximation of exchangeable phosphorus concentration of the sediment.

It should be noted that any more sophisticated algal model would be meaningless until the biologically available phosphorus (BAP or phosphate-P) concentration in the water is much higher than that which would limit the growth rate to half (approximately 2–5 μg/l, whereas the present 5-year average range is 68–105 μg/l).

Calibration was made for a period of 10 years (1986–1996) against the measured time series. Relatively good fit was achieved for both orthophosphate-P and chlorophyll-a. The multi-annual average values of measured and simulated data differed from each other by between only 1 and 3% for phosphorus and 0.2 and 1.0% for chlorophyll-a. No good fits to peak values were achieved, owing to the fact that an annual single date represented the input data series. The simulated sediment exchangeable phosphorus time series stayed well within the measured data range (the single profile measurement data).

Several alternative control options were analysed with the calibrated model. The results are summarized in Table 8.2.

The main conclusions were as follows:

• No appreciable improvement in the trophic state of the RSD could be achieved by any control strategy because the feed water to RSD from the River Danube already supplies enough nutrients to maintain high trophic levels.
• All of the control strategies contemplated together (Scenario 5) will result in substantial reduction of the values of trophic state variables (by approximately 40%). This means that all control strategies must be implemented, because local loads alone could maintain eutrophic conditions.

To develop the model further, one would need to express the functional dependency of the ecological parameters and state variables (such as algae biomass,

Table 8.2. Summary of the results of eutrophication control strategies of the Ráckeve–Soroksár–Danube arm, as revealed by model experiments.

Control scenario	Average improvement (%)		Trophic state		Improvement in trophic state category	
	PO_4-P	Chl-a	PO_4-P	Chl-a	PO_4-P	Chl-a
Scenario 1: P removal at SP-STP	28.0	22.0	Hypertrophic	Hypertrophic	None	None
Scenario 2: diversion of SP-STP effluent	34.0	28.0	Hypertrophic eutrophic (Tass)	Hypertrophic	In the 3rd segment	None
Scenario 3: sewerage + Scenario 1	33.0	27.6	Hypertrophic eutrophic (Tass)	Hypertrophic	In the 3rd segment	None
Scenario 4: 40% NPS removal + Scenario 3	40.0	33.3	Hypertrophic in 1st and eutrophic in 2nd and 3rd segments	Hypertrophic	In the 2nd and 3rd segments	None
Scenario 5: 60% NPS removal + Scenario 3	43.3	36.6	Hypertrophic in 1st and eutrophic in 2nd and 3rd segments	Hypertrophic	In the 2nd and 3rd segments	None (close to limit value)
Scenario 6: partial dredging	Simulation results in no appreciable changes					
Scenario 7: full dredging	Simulation results in no appreciable changes					

SP-STP, South-Pest Sewage Treatment Plant.

primary production, chlorophyll-a, etc.) on the controllable forcing variables (flows, water stages, input rates of nutrients and other pollutants). A good fit, which is achievable by the use of temperature limitation functions, will not be sufficient.

This would need reliable records on all important state and control variables (flows, water stages, nutrients in both the water and the sediment, input discharges of nutrients for both point and non-point sources, etc.). Such data on non-point source inputs are usually missing or are only estimated by rough methods, which will impair the usefulness of most of the more sophisticated tools.

The same applies to the lack of knowledge on the quantitative use of ecohydrological control options. These are of two major types: (i) how nutrient input loads of diffuse source origin can be curtailed by ecological methods (e.g. by natural and man-made wetlands, ecotones); and (ii) how the appropriate control of flows and water levels can alter the processes (e.g. growth, uptake, decay, mortality) of the ecological components.

Oxbow lake modelling experiences in the River Tisza flood plain

One of the major objectives of the international, EU-supported Tisza River Project (Jolánkai *et al.*, 2005; www.tiszariver.com) was to review knowledge on the state and problems (critical issues) of, and the revitalization options for, the high number of unique wetlands along the lower reaches of the River Tisza and its tributaries (a map of the Tisza river basin is given in Fig. 8.13 below). The ecohydrological revitalization strategy also means the integrated catchment management approach in which the objectives are multiple ones, including pollution control, recreation, fishery enhancement and flood control, etc.

The ecohydrological partners in the project were UWIEN.IECB (University of Vienna, Institute of Ecology and Conservation) and VITUKI (Environmental Protection and Water Management Research Institute, Budapest). Several of the hydrological–ecological lake models developed by VITUKI (the project coordinator) were considered for use in the project. Finally, owing mostly to a lack of data, it was decided to use a relatively simple, dynamic water and mass (nutrient) budget model, which related the resultant in-lake nutrient concentration to ecological parameters (such as chlorophyll-a concentration) by means of experimental expressions. In this latter context this model also made use of the OECD (1982) modelling concept and experimental equations; an approach that is frequently used worldwide for the assessment of trophic state in the lack of detailed measurement data. As this model was a kind of new development (mostly because of the hydrological–hydraulic part of it), it was given a new name: EcohydSim. Model equations were very similar to those given in the above examples with the exception of a very detailed hydrological model that accounted for time-varying inflows to the oxbow lakes during flood and of the rising and falling water levels of the lake.

The model was applied for one of the pilot oxbows of the project, the Mártély oxbow (Fig. 8.3), where the most data (longest records) were available.

Fig. 8.3. Location and map of Mártély oxbow lake.

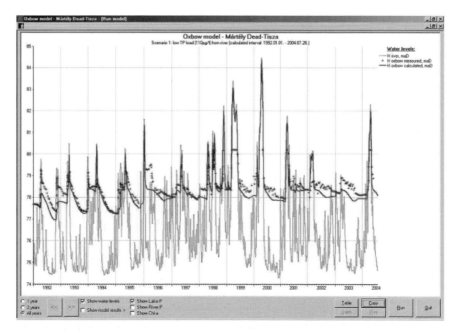

Fig. 8.4. Simulation of the water level of Mártély oxbow lake with the hydrological sub-model of EcohydSim, also using records of the nearest monitoring station of the River Tisza (for inundation inflow).

Calibration results of the model against Mártély data are shown in Figs 8.4 and 8.5, while some scenario simulation results are discussed below. Considering the complexity of the situation (especially of the flood-induced inflows) and the relative simplicity of the models, one may state that the calibration fits are acceptable for both hydrology and chlorophyll-a.

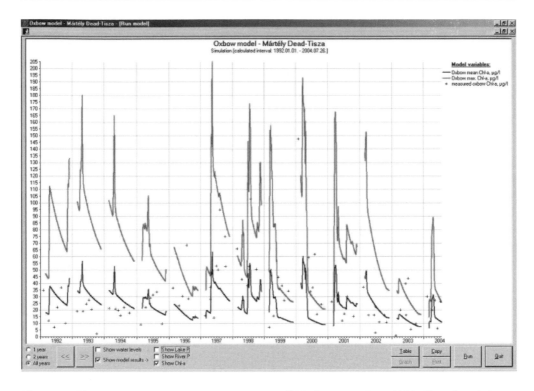

Fig. 8.5. Measured and simulated chlorophyll-a (Chl-a) time series of Mártély oxbow lake using the EcohydSim model.

The major findings of the project, with respect of the ecohydrological management strategies of the unique wetlands (oxbow lakes) of the River Tisza (and tributary) flood plain corridors, are summarized below:

1. Results of detailed field investigations of some 30 wetlands (oxbow lakes) of the several hundred along the River Tisza and its tributaries indicated that they are in various degrees of silting up, drying and high trophic state; in a rather late stage of succession, close to semi-terrestrial. Their long-term future is uncertain (resulting also from heavy anthropogenic impacts on the main river and the local sources). Many of them are likely to end up as a single macrophyte (*Trapa natans* or *Lemna minor*)-dominated, uniform 'green desert', or – in the case of macrophytes losing the competitive struggle with phytoplankton – just an 'algal desert', in the sense that both terminal states result in extremely low biological diversity, low ecological and conservation value, and bad ecosystem health.

2. The overall floristic results of five oxbows indicated that all are in a serious in-filled situation characterized by extreme water quality conditions (e.g. low concentrations of dissolved oxygen, selective fish and mollusc deaths, overwhelming production of single species of macrophytes). The biodiversity of these dead arms is relatively low with dominating eutrophic/polytrophic species groups in the different locations.

3. Out of a spectrum of potential revitalization options that were intensively discussed with – in part – the application of respective models, reconnection is out of the question as this will result finally in complete loss of the biodiversity of the wetlands, turning them into a riverine ecosystem equalling that of the main Tisza channel.

4. The strategy of providing 'refreshing' flows from the main river might have some advantage (halting the drying process), but will result in excessive inputs of nutrients and suspended solids (stemming from the hypertrophic concentrations and turbidity loads of the main river). This then may even accelerate eutrophication (as indicated by model results). This strategy may be effective only in the case of a full cleaning-up of the nutrient sources of the whole basin (see section on 'Integrated system models' and Figs 8.12–8.14 below), likely a very long-term procedure.

5. Ecologically sensitive dredging may have a favourable impact, but the recovery or build-up of bottom sediment and its nutrient content will likely be a rapid process (as indicated by model results, Fig. 8.6) in the oxbows of the inner flood plain, subjected to frequent flooding.

6. The most promising strategy (illustrated by Fig. 8.7) seems to be the slow raising of the water level ('filling up') of the oxbows, using hydraulic structures (gates, sluices, siphons, inverted siphons, etc.), without throughflow, in periods

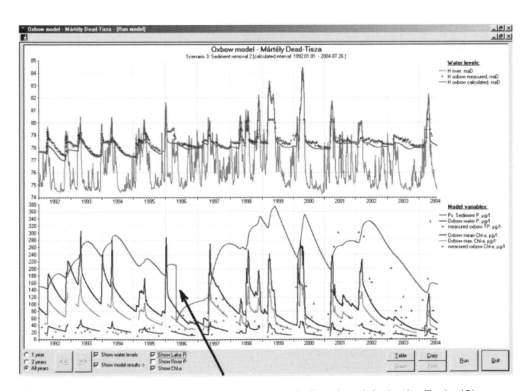

Fig. 8.6. Simulation of the dredging scenario for Mártély oxbow lake by the EcohydSim model. Note the relatively rapid build-up of sediment phosphorus (arrow).

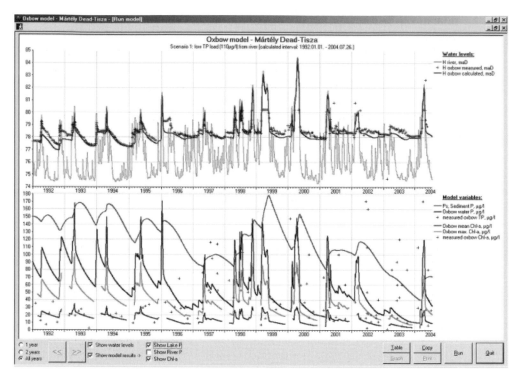

Fig. 8.7. Screen printout of the EcohydSim model, showing all simulated components for low total phosphorus (TP) input from the River Tisza. Note the relatively high improvement of the trophic state.

when the river has relatively low sediment and nutrient concentrations. However this solution is likely to be very expensive and cumbersome, needing also careful and target-oriented monitoring of the main river, and may be possible (with respect to hydraulic conditions) only in special cases and in favourable discharge periods.

7. For the already dried oxbows and flood basin depressions outside the flood levees there is an option of returning to the traditional land-and-water management of earlier times, by letting out flood discharges into the formerly inundated wetlands and by reconnecting the depressions. This is a programme also included in the flood control programme of Hungary (using these depressions, flood basins, also for flood water storage) and could bring many benefits (especially in enhancing fisheries, traditional animal grazing, re-creation of habitats for rare and endangered species, and enhancing ecotourism and recreation). More details on this technique are included in the final project reports (www.tiszariver. com), which describe a running programme of this kind.

8. Probably the most important conclusion is that the monitoring of these oxbows should be much improved (for better learning about their state and to provide data for supporting decisions; also with the help of the models described above). To this end the VITUKI team has prepared for the Hungarian authorities a detailed programme that is also in compliance with the EU Water Framework Directive.

9. Another important aspect is that all Danube countries have signed the UN Convention on Biological Diversity (CBD). This document obliges nations not only to protect, but also to enhance biological diversity wherever possible, and to save from further deterioration those landscape units still having high quality regarding biological elements. Certainly, wetlands of the quality found at the Tisza and Latorica rivers are subject to the intentions borne in the CBD. Therefore, the aspect of biodiversity within the scope of the Tisza River Project is more than just a matter of conservation or protection in legislature and practice: it is a prime issue of any modern concept of landscape development and land-use change. And certainly it must be considered in strategies of flood plain management which are foreseen in the Tisza and Latorica river corridors.

10. An additional matter must be pointed out: the ecological part of ecohydrological management relies much on vegetation structures, which influence – in a reciprocal way – the hydrology and habitat hydraulics of waterbodies. Aquatic ecotones formed by macrophytes are essential for numerous other aquatic organisms. Preserving and enhancing the abundance of valuable, usually submersed, aquatic plant species is one of the best ways leading to good ecosystem health in aquatic ecosystems. An option to do this is discussed in some detail in the final report of the project (www.tiszariver.com), in the light of the 'Nagykörű' case study.

Catchment Models: Quantification of Catchment Processes

The major ecohydrological question – the quantification of the sources and fate of nutrients – can only be addressed by models at catchment or whole basin scale. Quantification options of catchment processes are illustrated schematically in Fig. 8.8.

As shown in Fig. 8.8, the main tools of quantification and modelling include: (i) the determination of overall indices, such as runoff coefficients and unit area loading rates, as derived from observed flow and concentration time series; through (ii) the development of bi- and multivariate regression models (between flow, concentration, load and their affecting factors); to (iii) sophisticated catchment models that describe some details of the hydrological events of runoff generation processes, along with pollutant (particle) detachment, transport and transformation processes, in a lumped or spatially distributed way.

Figure 8.8 and its brief description suggest that an intermediate group of modelling options is missing between the second and third ones: that is, at a larger catchment scale, knowledge and data on the hydrological and ecological processes which govern the fate of water, its constituents and the ecosystem are generally poor. One usually measures (monitors) 'some' of the inputs and the responses (e.g. precipitation, flow hydrographs, pollutant concentrations, indices of the state of the ecosystem, etc.), but not the processes. And certainly

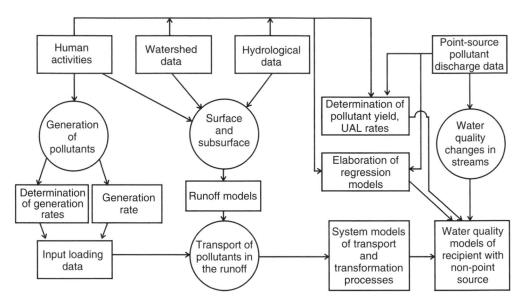

Fig. 8.8. Options of quantifying (modelling) catchment processes.

not the processes that need to be described for the purpose of a reliable planning tool.

Consequently, owing to the lack of data and experimental evidence, many or rather most of the quantification techniques are severely limited to some simple methods. Even worse is the case when the method is sophisticated, but there are no data to support the derivation of model parameters and coefficients (making the model a so-called 'scientific computer game or toy').

Ecohydrological processes can be implicitly considered in the simple methods (such as runoff coefficients and/or unit area loading rates corresponding to various land-cover types) or explicitly accounted for in the more sophisticated ones (calculating species-dependent interception and evapotranspiration rates, nutrient uptake rates, sediment and pollutant detention/delivery factors, etc.). This means that the representation of eco(hydro)logical processes is rather poor in all available modelling/planning options.

Before trying to elucidate a management and practice-oriented catchment basin modelling tool with case study examples, let us briefly review the major properties of the various modelling options available.

Major properties of available modelling tools of various degree of sophistication

Runoff and washoff indices as the simplest catchment modelling tools
The advantage of the simple 'overall indexing' method is that available data often allow the derivation of such indices. The disadvantage is that these indices

are usually too general and cannot provide reliable planning for pollution reduction or runoff modification. Nevertheless they can or could be improved, especially when the data for a number of different catchments can be obtained and analysed. One of the 'ecohydrological improvements' could be, for example, when unit area loading rates are defined for various different land-cover types including various crops and various natural plant community types.

To this end it is important to know how the unit area loading rates (in mass per area $[M/L^2]$ or in mass per area per time $[M/L^2/T]$, e.g. kg/ha or kg/ha/year) are derived. The common method of deriving such unit area yield values is the continuous (or very frequent) measurement of flow Q and concentration C in the outflow section of a selected drainage area. The product of Q and C is calculated to obtain load, then integrated over a selected period of time and divided by the drainage area A and the period of concern T. Thus:

$$Y = \frac{1}{AT} \int_{T_1}^{T_2} Q(t)C(t)\,\mathrm{d}(t) \tag{8.5}$$

where T_1 and T_2 are the time points of the beginning and end of period T (i.e. $T = T_2 - T_1$).

Considering discrete values and also the presence of point-source loads in the catchment basin, the yield rate is obtained as:

$$Y = \frac{1}{A}\left[\sum_{i=1}^{n} 86.4 C_i Q_i \Delta T - T \sum_{j=1}^{m} L_j\right] \tag{8.6}$$

where

Y = unit area yield rate (over time period T) (kg/ha),
C_i = measured (or estimated) concentration value in time step i ($i = 1, \ldots, n$) (g/m^3),
Q_i = measured rate of flow in the ith time step (m^3/s),
A = area of the catchment basin (ha),
ΔT = time step of the discretization (i.e. $n\Delta Tt = T$) (days),
L_j = jth point-source loading rate ($j = 1, \ldots, m$), considered here as a constant (kg/day), of the same quality constituent.

The yield value in kg/ha dimension corresponding to time period T can be converted into appropriate units such as kg/ha/year or t/ha/year as desired.

If one derives this value for different ecosystems, i.e. for catchments characterized by different vegetation cover, then certain ecohydrological indices of contaminant (nutrient) washoff yield rates can be obtained.

In possession of the data for many catchments and/or experimental plots, one could introduce further variables into the tables or expressions of contaminant yield rate. Such further variables could be, for example, the runoff (in mm), which induced the export process. Further parameters can be the average slope, the soil moisture (or antecedent precipitation index), the rainfall intensity, the size of the catchment, etc.

The final result is a set of multivariate expressions (regression models), which enable the estimation of pollutant (nutrient) loads from catchments of different land use and ecological character.

This, however, belongs already to the second group: the regression models.

Regression models

The advantage of regression models is that they are relatively easily derived when monitoring records of catchments are available. The disadvantage, however, is that regression models are usually derived from the data of a single catchment (such as flow–load correlations) and then the formula is valid only for the catchment in question.

The case when data of a larger (statistically interpretable) number of catchments are available is different. In this case, generalized runoff–yield functions, tables and expressions can be derived. Owing to the very high labour and monetary costs of this type of work, very few general models of this type are available. The well-known and widely-used Universal Soil Loss Equation (USLE) and its modifications (Wischmeier and Smith, 1978) is an example of such experimental models.

Paying more attention to ecological parameters and/or patterns is highly desirable and can be done by developing expressions as, for example, a function of the forested fraction of the catchment, repeating it for various forest types (e.g. based upon soils, geology). A characteristic early example can be found in the study by Betson and McMaster (1975), who analysed the data of 66 watersheds. Many more such models were reviewed in Jolánkai (1992) and Jolánkai et al. (1999).

Integrated system models

More sophisticated complex catchment models have been developed for all the dominant hydrological, ecological, material transport and transformation processes. There is a large variety of such models, most of them now available as computer software (Novotny and Chesters, 1981; Jolánkai, 1983, 1992; Giorgini and Zingales, 1986; Donigan, 1988; Thornton et al., 1999). The common feature of these models is that they integrate several sub-models from various disciplines, such as hydrology, agricultural and soil sciences, water quality management. Many of them can also handle ecological processes, such as evapotranspiration and the uptake of nutrients by various plants.

Nevertheless, they are usually not practicable as a planning tool (because they were not designed for such purpose; thus they are also called conceptual models). Depending on their degree of sophistication, they involve from a dozen to several hundred parameters and coefficients whose values cannot be obtained from measurement data (or with much expense and for very small catchments only). Most importantly these conceptual models cannot be calibrated. Thus they remain 'scientific tools'. Nevertheless they offer promise for the future, especially those based on Geographical Information Systems (GIS), which make use of digital maps of topography, land use and soil, along with the records from monitoring systems.

It is necessary to take advantage of all existing useful databases and information (including GIS digital maps, all types of monitoring records and the data that characterize anthropogenic impacts), while minimizing the details of processes where no data or information is available.

An example of a model which seeks to do this is presented through two case study applications.

Catchment modelling case studies using the SENSMOD model

The major steps of this model are shown as a flow chart in Chapter 2 (Fig. 2.6) for the River Zala, a tributary of Lake Balaton in Hungary, to which it was applied in an EU-supported project (VITUKI, 1998; Jolánkai and Bíró, 2000). The model, applied in many case studies (Jolánkai, 1986, 1992; Jolánkai *et al.*, 1993), works on a steady-state basis and calculates flow and concentration profiles along any selected, branching, sub-river system of the entire catchment for which the model network was created. It is aimed at allowing calibration against measured (monitoring) data of the river system. The basic principle is the water and material budget at overall and river section/sub-catchment level.

The model calculates diffuse-source pollutant (nutrient) input (washout export from an area unit, e.g. cell or pixel), with a unit area loading rate formula, as a function of the land-use form, runoff and fertilizer input. Parameters of the model are estimated on the basis of runoff–load experiments (literature data or own field experiments).

In the second step the model calculates the load reaching the next, model-network-defined recipient stream with a delivery formula. The parameters of this delivery formula are calibrated by fitting the simulated stream profiles of flow and concentration to measured ones.

The next sub-model model calculates the in-stream processes with a simple first-order reaction kinetic model (a water quality stream model), which considers point-source inputs and distributed (non-point source) inputs.

Essentially the model relates point- and non-point source inputs to the measured in-stream conditions and thus allows the analysis of load reduction scenarios. This means that ecohydrological management strategies aimed at flow and/or nutrient concentration reduction/alteration over any part of the catchment can be predicted. Ecohydrological processes are highlighted by weighting total observed runoff according to vegetation-dependent runoff coefficients and when estimating delivery factors across different land-use patches on the basis of literature values.

When fertilizer input time series are available at sub-catchment scale, the model also utilizes a soil nutrient (usually phosphorus) build-up and emptying sub-module, which can be tailored to the degree of data availability. If no such data are available, a literature estimate is used for the estimation of the unit area load–runoff function.

The capabilities of SENSMOD are illustrated by two examples: the catchment of the River Zala and the whole basin of the River Tisza, a major tributary of the Danube.

Application of SENSMOD to the Zala river basin
The Zala river, the largest tributary of Lake Balaton, Hungary, 1200 km^2 area, has comparatively good data availability, including fertilizer application and soil

phosphorus data. The model was applied as part of the Project INCAMOD (Development of Integrated Catchment Models for supporting water management decisions). The key goal was to reduce the eutrophication of Lake Balaton through nutrient load reduction (and, within this, phosphorus load reduction). The bulk of phosphorus loads originates from non-point sources. Therefore the major objective of the study was to construct a GIS-based hydrological and pollutant transport model of the catchment of the river, upstream of Zalaapáti, in order to simulate the effects of natural (climate) changes and man-made control scenarios on the quantitative and qualitative properties of the basin. The River Zala (Fig. 8.9) drains roughly half of Lake Balaton's catchment, contributing half the total phosphorus load.

Orthophosphate-P concentrations had decreased by about an order of magnitude in the period 1985–1997, for two reasons. The first was the introduction of phosphorus removal facilities at the sewage treatment plant of Zalaegerszeg, the largest discharge from the catchment. The second was the drastic decline of fertilizer usage as a consequence the political–economic change when Hungary ceased to be communist-ruled, when fertilizer prices rose as subsidies were stopped.

There were relatively detailed data available for the loads of various pollutants, including point sources and estimates of some of the non-point sources (urban runoff and from rural sewage from small communities or individual dwellings). An overall summary of these data for the period 1995–1997 is shown in Table 8.3.

The average total phosphorus load at the Zalaapáti river monitoring station was 46.2 t/year in the above period, indicating that about 62% of the total phosphorus load into streams must have stemmed from other sources, mainly agricultural ones. It follows from this that the major environmental strategy should be the control of agricultural sources of phosphorus.

Nevertheless, point-source management strategies were also considered on the basis of very detailed village-to-village sewerage and sewage treatment development plans of the local environmental authority for the 21st century (Kővári et al., 1997). The control strategies examined were as follows:

- Strategy P1 – sewerage and sewage treatment development strategies, assuming adherence to effluent requirements of 0.7 mg total P/l for plants of special importance and 1.8 mg total P/l for the rest (complying with national standards).
- Strategy P2 – as Strategy P1 but assuming adherence to effluent limit of 0.7 mg total P/l at almost all plants (the optimum solution).
- Strategy P3 – an unfavourable forecast based on present experience of non-compliance with effluent standards (effluent concentration generally 3.0 mg total P/l).

The formulation of agricultural control strategies had to rely on international literature and experience, as no detailed development plans were available. The fertilization strategies were:

- Strategy F1 – pessimistic scenario; return of application rates to those of the mid-1980s.

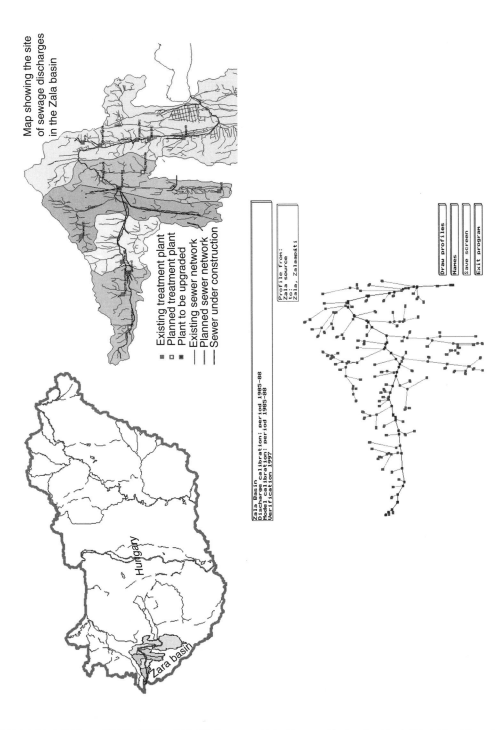

Map showing the site of sewage discharges in the Zala basin

- Existing treatment plant
- Planned treatment plant
- Plant to be upgraded
— Existing sewer network
— Planned sewer network
— Sewer under construction

Fig. 8.9. The Zala river system, Hungary (left, above) and its model replica (below), the network of the modelled river system with computation nodes.

Table 8.3. Summary of phosphorus loads to streams in the Zala catchment.

Sewage discharge (m³/day)	Urban runoff (t/year)	Sewage from rural settlements without treatment (t/year)	Total (t/year)
6,784	7,510	3,007	17,301

- Strategy F2 – realistic or intermediate scenario; average rates of the years before and after the dramatic drop in fertilizer use.
- Strategy F3 – do-nothing scenario; use of present (1997, very low) rates.

Agricultural management strategies included:

- Strategy A1 – contour tillage in orchards and plough land.
- Strategy A2 – buffer strip cropping.
- Strategy A3 – terracing.

In the model these scenarios were considered with multiplier factors of the calculated total runoff export load from the unit area (land-use patch A_{ij}) of concern. The basis of the estimation was the USLE tables (Wischmeier and Smith, 1978). The multiplier factor was equivalent to the 'conservation practices' P value of the USLE method and varied with slope category.

Two other control strategies were considered in the model runs. These were two different strategies for reducing urban runoff load (Strategy U1 – 30% reduction of the estimated load; Strategy U2 – 70% reduction of the estimated load) and three scenarios for phosphorus load reduction by reservoirs (Scenario RV1 – 10% reduction; Scenario RV2 – 20% reduction; Scenario RV3 – RV2 plus new reservoirs). A reservoir retention sub-model, similar to that outlined above, was developed for the purpose.

A relatively detailed impact analysis of climate change on precipitation runoff change was also carried out in this project. The local (regional) climate scenarios were developed by making use of Global Climate Model (GCM) results and the results of international and national literature. The flow generation technique, based on Monte Carlo analysis, yielded runoff changes of –13%, –2% and +7% in an average year, respectively, for the three precipitation scenarios (P1–P3). The result was 12 time series, each of 200 years length. These generated time series were used for running the reservoir operation model, which was also developed for analysing the effects of climate scenarios on water resources management.

Calibration of the water quality sub-model against the measured total phosphorus conditions was made for the period 1985–1988, when detailed fertilization and soil investigation data were available, using 37 stream sections and the corresponding sub-catchments of the model network. Phosphorus input data were used in the form of digital maps. This was before the introduction of phosphorus removal at the sewage treatment plant of Zalaegerszeg, and thus the subsequent effects of a major load control strategy could be analysed.

Results of the calibration run are shown in Fig. 8.10 (the line showing the points of measured concentrations). Relatively good fit to the measurement data

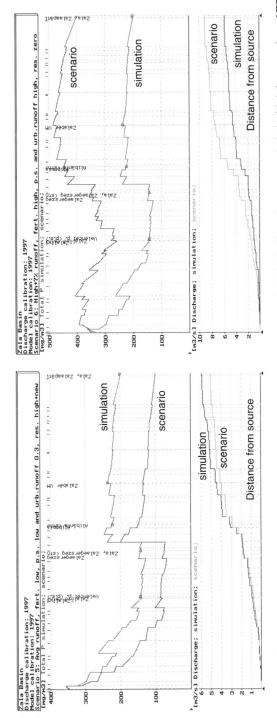

Fig. 8.10. Total phosphorus (TP) calibration results and simulation of the most optimistic (left) and most pessimistic (right) scenarios of TP load reduction measures.

was achieved, the largest deviation between measured and observed data was only 3.5%. It is worthwhile to note that the overall average of the 'virtual diffuse load' was 0.4 kg/ha/year (or 67.6 t/year) and this refers to the larger part (58.2%) of the multi-annual total load.

Partial verification of the model was also obtained by simulating the conditions of 1994 (a much drier year), while keeping the same model parameters as those of the calibration run. No full verification was possible since the detailed changes of P fertilizer reduction could not be followed. Heródek *et al.* (1995) suggested that the use of 230–280 kg fertilizer/ha in the 1980s had been reduced to 19 kg/ha by 1992. The model was verified to some extent by the small (1.8%) difference between measured and calculated loads of Zalaapáti outflow station (33.5 t/year and 34.1 t/year).

An intermediate result of the calibration runs was a map of the unit area loading rates of phosphorus. An example of the latter is shown in Fig. 8.11, with highest values corresponding to orchards and vineyards on highly sloping cultivated land (realistic quantities in the light of the relevant literature).

Natural conditions, the variation of runoff, were considered in three scenarios:

- Scenario R0 – the multi-annual mean runoff.
- Scenario R1 – the highest recorded annual runoff.
- Scenario R2 – the annual runoff corresponding to the highest positive climate scenario (+7%).

These were considered by driving the model with the respective modelled runoff map.

Land-unit P export map
for scenario R0 (1985–1988)
kg P/ha/year

6.127
5.518
4.909
4.299
3.690
3.081
2.472
1.863
1.253
0.644
0.035

Fig. 8.11. Land-unit phosphorus export map of the Zala basin for climate scenario R0.

Anthropogenic impacts and control scenarios were considered as above. Representative combinations are shown in Table 8.4, together with simulation results for the outflow of the catchment at Zalaapáti.

The runs of the model analysed the following management options:

1. Scenario 1 was the low fertilizer but assuming high runoff. The results indicate that runoff increase alone will not result in appreciable concentration increase when input to agricultural land has been nearly zero for several years. Load increases will be directly proportional to the increase in runoff.

2. Scenario 2 was high runoff associated with high fertilizer application over a long time. The results suggest that the phosphorus load from the catchment will nearly double if fertilization returns to the high rates of the 1980s (if no other control action is taken).

3. Scenario 3 was high runoff and long-lasting, medium fertilizer application (as 1989–1991). The results show that return to a medium level and lasting fertilization will increase the present phosphorus concentration by about 0.1 mg/l (or by 50%) if no other changes are made in the catchment.

4. Scenario 4 was the average condition with medium point-source and non-point source control efficiencies. The interesting result is that the effects of higher point-source loads and higher-than-present fertilization rates are alternately counterbalanced by the effect of contourline tillage and urban runoff load reduction, but the final output load of the catchment will change but slightly compared with the present situation.

5. Scenario 5 was the 'theoretically achievable' best combination of point- and non-point source management strategies, with the lastingly long present very low fertilization levels. This optimistic scenario results in total phosphorus concentration of about 0.1 mg/l in the outflow at Zalaapáti station, which is highly desirable (falling into water quality class II (good) in terms of the Hungarian standard).

Table 8.4. Summary of the results of water quality management scenarios.

Scenario	Description of scenario combination	Total P concentration (mg/m^3)	Total P load (t/year)
Calibration (1997)	Measured (matching calculated) values in 1997	202.1	37.62
Verification (1985–1988)	Measured (calculated) values	556.0 (486.9)	116.26 (101.83)
Scenario 1	Present condition with high runoff	198.7	53.54
Scenario 2	Present condition with high runoff, high fertilizer	384.1	103.48
Scenario 3	Present condition with high runoff, medium fertilizer	291.4	78.51
Scenario 4	Average conditions with medium control efficiency	186.9	29.63
Scenario 5	Theoretically achievable best condition	99.3	15.74
Scenario 6	Most pessimistic condition	403.3	116.26

6. Scenario 6 was the 'most pessimistic' one, where high runoff (including runoff increase by 7% due to the positive climate scenario) is combined with lastingly high fertilization rates, low efficiency point-source control, no control of urban runoff, zero reduction in reservoirs and no agricultural management strategies. The result is also dramatically bad: 0.403 mg P/l or 116.26 t P/year at Zalaapáti, exceeding the loads of the base period (1985–1988).

Eventually, many other combinations of control and management strategies can be investigated with the model, which can be considered as calibrated and verified. We have strongly suggested that local and high-level managers and decision makers make use of this model in their actual work. For this purpose we prepared a Hungarian version of the original English study and submitted it to all parties interested.

What are then the lessons of this project from the ecohydrological point of view? The major lesson is that the role of various land-use forms (and thus their vegetation cover) in the transport and retention of phosphorus was estimated on a relatively weak, subjective basis, making use of literature values or their proportions only. Results of experimental plots on the water, sediment and nutrient budgets of various land-use forms and 'ecosystem units' could have improved the estimates substantially, and thus the ability of the planners to design ecological management options. Even the effects of well-known options, such as the use of vegetation buffer strips on reducing (filtering) sediments and nutrients, can be considered as being on a rather subjective basis in such models (making use of the conservation practice parameters of the USLE method, for example).

Use of SENSMOD to model nutrient loads in the Tisza river basin

The catchment area of the River Tisza is a very large international river basin of 157,000 km^2 (Fig. 8.12) where the modelling faced serious data shortages (www.tiszariver.com). The project's general objective was to conserve the basin's water resources and ecological values with the help of integrated catchment management tools and to secure their sustainable use. The scientific objective was to develop a 'real-life scale' integrated catchment model system. This means the development (and selection) of a practical set of tools (set of computer models for water flow, water quality and ecosystem functioning) that are exactly tailored to the issues to be solved and the availability of data. The establishment of an international database of the Tisza river basin was also an objective. There was a special emphasis on an ecohydrological approach to revitalization of the many large unique wetlands – oxbow lakes formed by 19th century canalization – along the Tisza river and its tributaries.

There were two ecohydrological objectives supported by modelling tools. One was the basin-scale application of SENSMOD for phosphorus budgeting and the other analysis of wetland (oxbow lake) conditions by two different ecological models (see section on 'Oxbow lake modelling experiences in the River Tisza flood plain' above).

There was a serious lack of data in applying SENSMOD to phosphorus transport in the basin. Large parts were without water quality monitoring data and practically no data on diffuse-source phosphorus inputs (e.g. fertilizer application

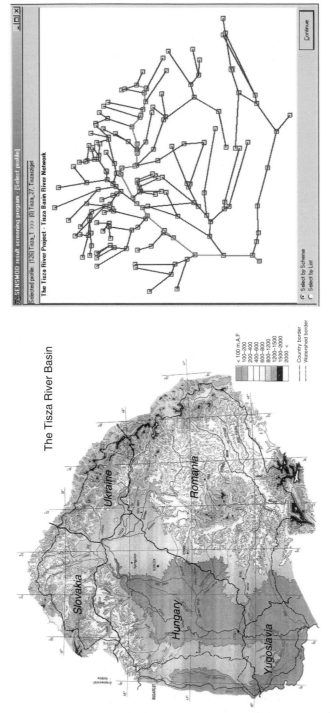

Fig. 8.12. The Tisza river basin (left) and the model network of the river system with its computation nodes (right).

rates) were available. The vertical scale of the Digital Terrain Map (DEM) was very rough and hydrological monitoring data as well as channel morphological data were missing along many river sections of the network.

The main problem, however, was that measured (monitored) in-stream loads were not explained by the point-source loads of the database but made a smaller fraction of the measured loads only (Table 8.5).

Consequently only very rough pollutant budgeting calculations were possible. The whole model system was restructured for the purpose of identifying missing load components by calibrating the model against in-stream monitoring data in order to obtain the gross, diffuse, unit area loads of sub-basins of the entire catchment.

The use of the model for calibration against the data of the selected base period (year 2000) was made in such a way that the diffuse loads from different land-use patches of different runoff coefficient values were increased (by the calibration algorithm) until the load (concentration) value of the next down-stream monitoring point was matched. That is, the model was calibrated for the non-point (unidentified) source inputs of total phosphorus (and also of the rest of the modelled parameters, not discussed here). For this procedure the land-use maps, the digital elevation models (DEM) and the generated maps of the runoff coefficients and times of travel were utilized. This means that the original model structure of SENSMOD had to be abandoned and a new rough budgeting algorithm developed. This type of calibration needed a fixed input of the decay (detention, decomposition or 'loss') rate coefficient (K) for each model stream reach (between

Table 8.5. Total phosphorus loads of the main Tisza and its tributary rivers, showing the contribution of point-source loads.

River name	Computed load at mouth station (t/year) [A]	Point-source load upstream of the station (t/year) [B]	Diffuse load (t/year) [A–B]	Contribution of point-source load (%) [1–(A–B)/A]
Túr	56.3	0.0	56.3	0
Szamos/Somes	547.6	230.7	316.9	42
Kraszna/Krasna	550.7	33.6	517.1	6
Lónyai Canal	95.0	25.1	69.9	26
Bodrog	545.3	0.0	545.3	0
Sajó-Hernád/ Slana-Hornad	560.7	29.5	531.2	5
Eger	115.8	0.5	115.3	0
Zagyva-Tarna	215.2	36.4	178.8	17
Körös system/Crisuil	1151.2	724.6*	426.6	63
Maros/Mures	2065.8	244.7	1821.1	12
Tisza/Tiszabecs station, entering Hungary	283.9	0	283.9	0
Tisza at Tiszasziget, leaving Hungary	5331.1	1515.5	3815.6	28

*Of this, the load of the city of Debrecen, Hungary, is 503 t/year (i.e. the point-source contribution of the larger part of the sub-basin is reduced to 20%).

junctions). This was done in such a way that a relatively low literature value was chosen (so as not to put a too high, non-justifiable, share on non-point pollution sources). This fixed K value was increased only in stream reaches where the 'elimination' of a higher point-source load needed it for matching the next monitoring-point load measured downstream. The results were obtained for any combination of the model stream network as illustrated in Fig. 8.13. The output map shows the diffuse loads of total phosphorus for various sub-basins in units of kg/ha/year. These are unrealistically high unit loading rates, characterizing large parts (sub-catchments) of the basin. Any loading rates above 10 kg P/ha/year are very high (characteristic only of highly fertilized agricultural units) and any values above 20 kg P/ha/year are absolutely unrealistic. From the point of view of controlling ecological conditions of the Tisza river basin, this yielded a dramatic conclusion: that large, unknown or non-registered sources of pollution dominated the phosphorus load conditions.

A clearer picture is provided by Fig. 8.14, which shows the model-simulated phosphorus concentration profiles for the main Tisza river between the upstream, Black Tisza in Ukraine and the downstream Hungarian station, Tiszasziget. The two profiles on this figure are as follows:

• The upper curve is the calibration result and thus runs through the measured data of monitoring stations (indicated by small squares). Total phosphorus concentrations are in the hypertrophic range along nearly the entire river.

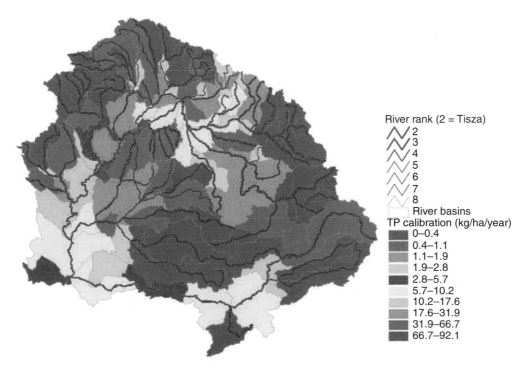

River rank (2 = Tisza)
2
3
4
5
6
7
8
River basins
TP calibration (kg/ha/year)
0–0.4
0.4–1.1
1.1–1.9
1.9–2.8
2.8–5.7
5.7–10.2
10.2–17.6
17.6–31.9
31.9–66.7
66.7–92.1

Fig. 8.13. Calibration results of SENSMOD for total phosphorus (TP) load conditions of the Tisza river basin in the year 2000.

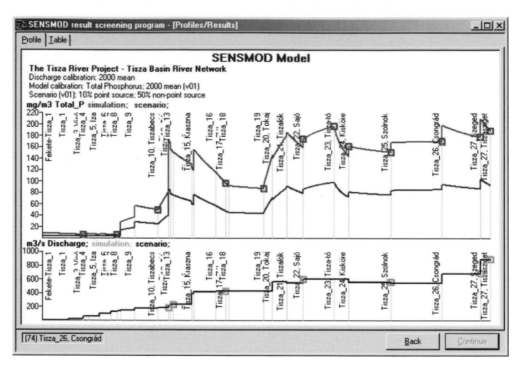

Fig. 8.14. Calibrated total phosphorus (TP) and flow profile of the River Tisza from the junction of the Black and White Tisza (in Ukraine) to Tiszasziget (in Hungary), for the mean flow and load conditions of the year 2000, showing also the most optimistic clean-up scenario.

- The lower curve shows the results of the most optimistic clean-up scenario, i.e. 90% removal of all point-source loads and 50% removal of all non-point source loads. It is evident that roughly half of the entire Tisza remains in the eutrophic–hypertrophic range even after such a technically feasible (but optimistic) reduction of input loads.

The assumption of 50% load reduction of all non-point sources is based on the ecohydrological management strategies outlined in Chapter 2. They span from contour tillage, through vegetation buffer strips, catch basins, grassed channels and macrophyte ponds to other ecological 'filtering' techniques, and all possible means of controlling fertilizer and manure usage. It is highly probable that the diffuse sources which cause these extremely high unit area loads are mostly small point sources, so the most effective management would be to clean them up individually.

This result becomes even more dramatic when one looks at the multi-annual average concentration data of the streams entering Hungary from upstream neighbours (Table 8.6). This table proves that every tributary falls into the hypertrophic domain under the OECD classification and the state of ecological (ecohydrological) conditions matches this finding. Any revitalization of the whole or part of this aquatic ecosystem, for example the globally unique oxbow lake

Table 8.6. Multi-annual mean phosphorus concentrations in the tributaries of the River Tisza entering Hungary.

River name	2002–2005 (μg/l)	
	Total P	Orthophosphate-P
Szamos	123.3	66.4
Kraszna	919.8	616.8
Bodrog	120.2	68.1
Sajó	188.9	106.0
Ronyva	1296.0	996.5
Hernád	393.5	295.3
Fehér-Körös	121.2	37.8
Fekete-Körös	103.7	31.6
Sebes-Körös	172.4	100.4
Maros	218.6	37.1

wetlands, needs the upgrading or cleaning up of the entire river basin, with much international effort.

The overall conclusions are that the Tisza river basin is (and will be) facing serious problems even in the case of implementing national and international (EU Water Framework Directive) point-source clean-up strategies, owing to the high amount of non-accountable (diffuse or small point) sources of pollution. For total phosphorus and chemical oxygen demand (and possibly also for other parameters) even the most optimistic clean up of point and diffuse sources would not result in 'good' status of this aquatic environment. Ultimately the only solution could be to identify all yet unregistered sources of pollution and clean them up to some extent, which is higher than the 50% diffuse load reduction assumed in this study.

Ecohydrological conclusions from the model applications

1. The quantitative tracing or routing of the transport and transformation processes of plant nutrients through the flow pathways of larger river basins is still a very problematic issue, which can be solved only with simple and robust models that require multiple and uncertain assumptions.

2. There is a long way to go until reliable decision-support or planning tools for nutrient reduction by ecohydrological means will be achieved. The unavoidable consequence is that much research and field experimentation will be needed, a fact which is frequently overlooked and neglected by the responsible financing authorities.

3. Plant nutrient loads originate mostly from diffuse or non-point sources, among which agricultural and urban sources are dominant. More importantly, the available databases do not allow accurate identification of point-source loads and thus model-identified non-point source loading rates can reach unrealistic ranges (implying unidentified point-source loads).

4. Catchment models, planning tools for nutrient load reduction strategies by ecohydrological and other means, must be continuously kept updated, maintained and re-calibrated against new measurement data and field experimental data. This is not achieved in the case of most of the larger, international catchments and thus project results will not be efficiently utilized. Many national and international efforts on follow-up activities of such projects are needed to solve this contradiction of international (e.g. all-European) dimension, which jeopardizes the successful implementation of relevant directives such as the EU Water Framework Directive.

9 Ecohydrology Driving a Tropical Savannah Ecosystem

E. Gereta[1] and E. Wolanski[2]

[1]Tanzania Wildlife Research Institute, Arusha, Tanzania; [2]Australian Institute of Marine Science, Townsville, Queensland, Australia

Introduction

Water is increasingly a scarce and vital resource over much of semi-arid East Africa. The problems are becoming acute in view of rapid human population increase, overgrazing and deforestation. The impacts on wildlife are staggering. In Rwanda the need for land and water for an additional 1 million-strong herd of cattle imported in the country in 1995 resulted in de-gazetting most of the 6000 km^2 Akagera National Park with a 2000 km^2 wetland, the largest protected wetland in Africa after the Okavango Delta. Most of the wildlife and natural vegetation have since been removed, the land is overgrazed, erosion scars are now a common sight and siltation is already apparent in the wetland (G. Gerin, personal communication). In the Ruaha National Park in Tanzania in the late 1990s, water diversion for irrigation of rice fields essentially removed much of the river flow available for wildlife in the dry season in the park. This has resulted in massive disruption to wildlife, including over-population of hippopotamus and crocodiles in the remaining small water holes in the dry season, increased mortality of fish and mammals, overgrazing and erosion of surrounding lands, and disruption to dry-season wildlife migration patterns (Mtahiko *et al.*, 2006). In the lower Tana river in Kenya, the freshwater wetlands have been reduced by probably about 50% since the river flow has been regulated by hydroelectric dams, and this has also significantly affected wildlife and fisheries as well as resulting in silting the river (E. Martens, personal communication). The impact on wildlife does not seem to have been considered seriously at the planning stage of such schemes, nor are follow-up studies undertaken or remedial measures implemented.

If the lessons from such schemes are not studied and published, the same mistakes will be repeated in other river catchments. In the absence of such data, no progress will be made to find solutions that can integrate water engineering and ecohydrological principles to ensure sustainable developments in the future. The problems are pressing because hydroelectric dams and water abstraction and diversion schemes are planned for a number of other semi-arid, wild African areas, including the Sudd wetlands on the White Nile in Sudan, the Okavango Delta in Botswana and the River Mara in Kenya (Gereta et al., 2002; Crisman et al., 2003).

All these waterbodies flow through wildlife-rich areas. The impact on wildlife of such water management schemes needs to be carefully evaluated and managed. The Mara river, for instance, is the main dry-season water supply for the Serengeti ecosystem. The plans (only temporarily on hold) for hydroelectric power in Kenya from the Mara river call for flow diversion to another river which is outside the Serengeti ecosystem.

The difficulty with managing human and wildlife needs for water in semi-arid East Africa is exacerbated by the pronounced seasonal and inter-annual variability of the water. Semi-arid conditions prevail in the dry season when water is scarce. In the wet season excess water is available and wildlife disperses away from the rivers. Seasonal wetlands form and these are important for wildlife at the start of the dry season. The use of the ecosystem by wildlife thus fluctuates spatially and temporally with the seasons as the animals migrate between wet- and dry-season pastures. It also fluctuates from year to year because of the pronounced, inter-annual fluctuations in rainfall. As a result, the wet-season pastures in a wet year are different from wet-season pastures in a dry year. Similarly, the dry-season pastures in a wet year are also not the same as dry-season pastures in a dry year.

Managing wildlife and human needs for water in the face of such variability is the key problem to be addressed in semi-arid East Africa. This requires a sound understanding of the ecohydrological factors driving the ecosystem. Such an understanding is now emerging for the Serengeti ecosystem and thus the lessons may well be applicable to other river catchments.

Geographic Setting of the Serengeti

The Serengeti ecosystem has an area of 67,000 km^2 located primarily in Tanzania. Its core is the Serengeti National Park in Tanzania (Fig. 9.1) and this is surrounded by buffer zones including hunting and game reserves and managed conservation areas in Kenya and Tanzania (Fig. 9.2). The Mbalageti, Grumeti and Mara rivers drain the Serengeti National Park, all flowing westward to Lake Victoria. The River Mara drains a large area (10,300 km^2) in Kenya, flows through the park and drains the far north region at 1300-1500 m elevation. The Mara river in the Masai Mara Game Reserve and the northern part of Serengeti National Park is the only permanent source of drinking waters for migrating animals, especially during the dry season in a drought. Previous studies of the ecological functions of the Mau forest (Fig. 9.2) showed that the forest plays an important hydrological role in preserving and maintaining a flow in the Mara river even in a drought (Dwasi, 2002). By regulating infiltration and runoff and thereby controlling water availability downstream, the Mau forest maintains dry-season flows of water in the Mara river; and prevents or limits river bank erosion, downstream siltation and sedimentation; and prevents or limits flooding. The River

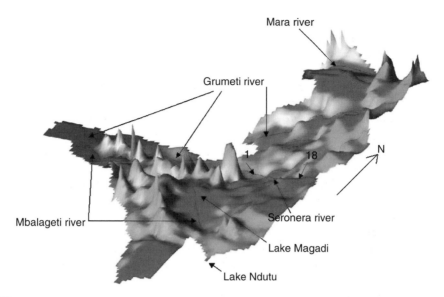

Fig. 9.1. A three-dimensional visualization of the topography of the Serengeti National Park ecosystem.

Grumeti drains much of the central and northern area; most of it is hilly and wooded savannah, with a catchment area of 11,600 km^2, much of it in the park. The River Mbalageti drains 2680 km^2 of the southern open, treeless grassland, nearly all in the park. Most of the grasslands are at an elevation of 1600–1660 m (Gereta and Wolanski, 1998).

The park is famous for its annual, mass migration of 1.2 million wildebeest and about 200,000 zebra. These herds in turn support the largest lion and other mammalian carnivore populations on Earth. The ungulates migrate seasonally. In the wet season they aggregate in the grasslands, which are located in the south. As the dry season sets in, the animals leave the grasslands and migrate northward to the wooded savannah of the lower Grumeti river. Later, in the middle of the dry season, most herds aggregate in the Masai Mara area where water and grazing are available. While migrating, the animals select grazing sites as a function of their minerals (McNaughton, 1979, 1985, 1988, 1990; McNaughton *et al.*, 1988; Sinclair and Arcese, 1995). In the dry season the southern plains are arid and offer no forage or water to the herbivores; the herds return there only in the wet season.

Rainfall

Rainfall over the Serengeti has large but seasonally predictable annual fluctuations (Fig. 9.3) with a marked dry season in June–September, a small wet season in November and a marked wet season centred on April, and the heavy wet season immediately thereafter (SMEC, 1977; Brown *et al.*, 1981; Gereta and Wolanski, 1998).

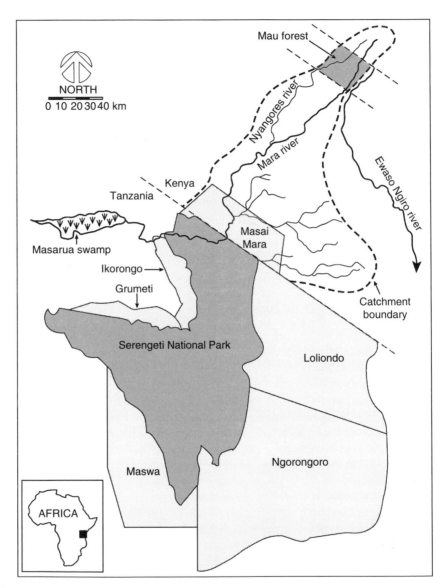

Fig. 9.2. A map showing the Serengeti National Park, Tanzania and its relationship with the various, surrounding managed areas. This map also shows the watershed of the River Mara, much of which is outside the conservation areas. In the park, the elevation varies from a minimum of about 1140 m in the western corridor to a maximum of about 2000 m in the north-east, hilly corner of the Grumeti river catchment. The northern area encompassing the Grumeti river catchment is wooded savannah. The southern area encompassing lakes Ndutu and Magadi and the upper Seronera river is open grassland, and the transition between these two habitats occurs midway along the Seronera river.

The rainfall varies spatially because of orographic effects; it is the highest in the hills to the north (the upper Grumeti river catchment) and in the south-east (Ngorongoro mountains). Rainfall is the lowest in the southern grasslands, which are located in the rain shadow of the Ngorongoro mountains (Fig. 9.2). This general pattern is found in both dry and wet years (Fig. 9.4).

However, there is also a large inter-annual variability (variation by a factor of 3 from year to year) in the annual rainfall (Fig. 9.5). This inter-annual variability also affects the seasonal rainfall distribution as is shown by the large error bars in monthly rainfalls (Fig. 9.3).

Only during a strong El Niño–Southern Oscillation (ENSO) event is the annual rainfall correlated with the Southern Oscillation Index (SOI). The factors affecting these sub-Saharan climate anomalies and the resulting inter-annual rainfall variability are believed to be atmospheric teleconnections. These include the Quasi-Biennial Oscillation (QBO) signals and the interaction between the global ENSO signal and the tropical Atlantic and Indian Ocean coupled processes, including monsoon convection, with resulting time lags of typically 12–18 months. ENSO-based predictability of rainfall is high only for southern Africa. As a result, the rainfall over the Serengeti ecosystem, and East Africa in general, is largely unpredictable (Semazzi *et al.*, 1988; McIntyre, 1993; Jury *et al.*, 1994; Sasi, 1994; Tourre and White, 1995; Jury, 1996; Nicholson and Kim, 1997).

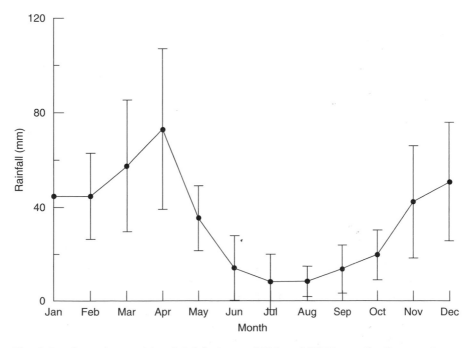

Fig. 9.3. Average monthly rainfall (between 1960 and 1997) over the Serengeti ecosystem. The seasonal variation is apparent. The error bars reflect the huge inter-annual variability in rainfall.

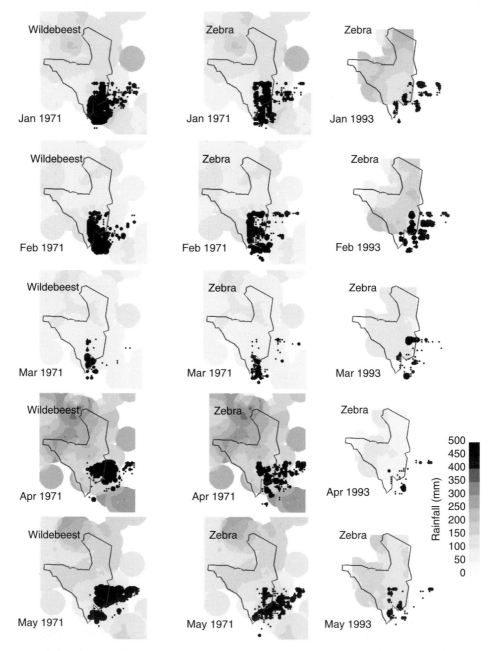

Fig. 9.4. Monthly distribution during the wet season of rainfall and of the herds of wildebeest and zebras in 1971 (wet year) and 1993 (dry year) in the Serengeti grasslands. Shading shows monthly rainfall. The herds of animals are shown by circles, the centre of the circle indicates the location of a herd and the size of the circle is related to the size of a herd using a logarithmic scale. The smallest circle (a dot) indicates 1–25 animals in that one herd; the next size circle indicates 26–250 animals; the next size circle 251–2500 animals, and the largest circle more than 2500 animals.

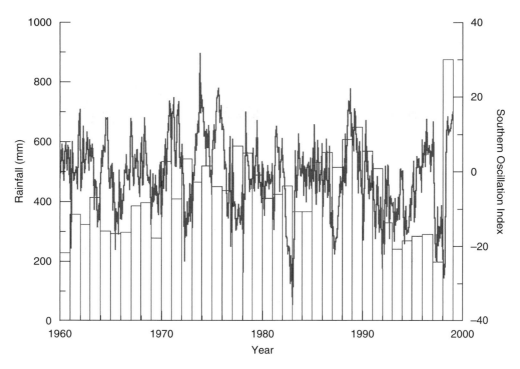

Fig. 9.5. Time series from 1960 to 1998 of the monthly Southern Oscillation Index (line) and the annual rainfall (white bars) over the Serengeti ecosystem.

Salinity and Migration

Salinity of surface waters, the only water wildlife has to drink, varies spatially and temporally (Gereta and Wolanski, 1998). In the south of the Serengeti ecosystem, water is available only in the wet season and for a few months afterwards. Salinity of surface waters in the grasslands at the end of the wet season in a normal year (e.g. 1996 and 1997) is high; commonly more than 5 ppt in the headwaters of the Mbalageti river, 10–15 ppt in the headwaters of the Seronera river and more than 20 ppt in the alkaline lakes. When the salinity is more than 4 ppt no grass grows along the banks of the rivers and lakes. The salinity decreases rapidly with distance downstream, being typically less than 1 ppt in the lower reaches of these two rivers. In the north, salinity less than 1 ppt prevails all year.

At the wet-season salinity of about 2 ppt, the vegetation changes from grassland upstream to wooded savannah downstream (coincidentally, this limit is also that for the maximum salinity of water that can be used to irrigate rice). Thus salinity at decadal time scales may determine the discontinuity between grassland and wooded savannah.

Mammals cannot survive indefinitely drinking water of high salinity. A salinity of 10 ppt appears to be the upper limit for sheep for a few months. In the arid Kalahari-Gemsbok National Park, South Africa, groundwater is pumped to ponds for wildlife (Child *et al.*, 1971). Where this water is hard and saline, the wildebeest migrate; where it is sweet the wildebeest are stabilized. Excessive salinity may be the trigger that starts the annual mass migration of wildebeest and zebras near the end of the wet season northward away from the southern grassland. The timing of these migrations varies annually by as much as 3 months; the onset of the migration is thus not driven by a biological clock. The migrations start when there is still plenty of edible forage and surface water in the southern grasslands, however at that time the water is excessively saline. The wet-season salinity in the grasslands varies inter-annually by as much as a factor of 3, being inversely related to rainfall.

In this hypothesis, that migration is triggered by excessive wet-season salinity, salinity can be used to predict the onset of the northward migration. The salinity of an alkaline lake (Lake Magadi; see Fig. 9.1), located at the vegetation discontinuity between the grassland to the south and the wooded savannah to the north, can be used as the overall indicator of salinity in the grassland. The Seronera and Mbalageti rivers and small ephemeral ponds scattered in the grasslands are the main source of drinking water. Their salinity is less than that of the lakes, but proportional (with a factor of <1) to that of Lake Magadi. The lake has a natural spillway at its mouth so that excess water flooding the site flows out readily, water that remains is trapped by the sill. In the dry season after mid-April the lake becomes an evaporation pond. Lake water salinity increases with time in a way that can be predicted from a mass balance model, since salt is a conservative substance. This requires the measurement of the salinity in mid-April near the end of the wet season. Other parameters in the model are the lake initial mean depth (≈ 0.8 m), the decrease in lake depth due to evaporation (≈ 0.007 m/ day) and the residual rainfall after mid-April (i.e. after the wet season). In this model the migrating herds leave the grasslands and enter the wooded savannah when the salinity exceeds a threshold value, 30 ppt.

This simple model was tested on 1996–2000 data. The prediction of the timing of the migration, which varied from year to year by up to 3 months, was correct for each year with less than a week's error (Wolanski and Gereta, 2001).

Inter-annual Variability of the Size of the Ecosystem

Associated with the inter-annual variability of the rainfall, the size of the ecosystem (i.e. the area used by the migrating herds) varies from year to year. In wet years (e.g. 1970 and 1971), the wildebeest, zebras and Thomson's gazelles dispersed in a similar manner to each other as early as November in the southern grasslands and remained there until May (Fig. 9.4). In dry years (e.g. 1993) the animals entered the grasslands later, in much smaller numbers, little of the grasslands were used and those few animals that did enter the grasslands also left earlier, in March (Fig. 9.4).

Similarly, the dry-season wildlife dispersal areas also vary from year to year. In very wet years (e.g. 1989 and 1998; Fig. 9.5) the migrating herds did not enter the Masai Mara in Kenya. In very dry years (e.g. 1993) they entered the Masai Mara as early as July (S. Mduma, personal communication). Thus the size and the use of the ecosystem by wildlife vary from year to year, in a pattern determined by rainfall.

Water Quality

In dry seasons, the surface waters are used for drinking and are a scarce, heavily used resource. Most are grossly eutrophicated with anoxic conditions on the bottom. When fringing wetlands are present, water quality is improved because of the wetlands' filtering effect (Wolanski and Gereta, 2000). By stirring and oxygenating the water, hippopotamus are vital in preventing anoxic conditions (Fig. 9.6), thus maintaining aquatic life and making the water drinkable by wildlife.

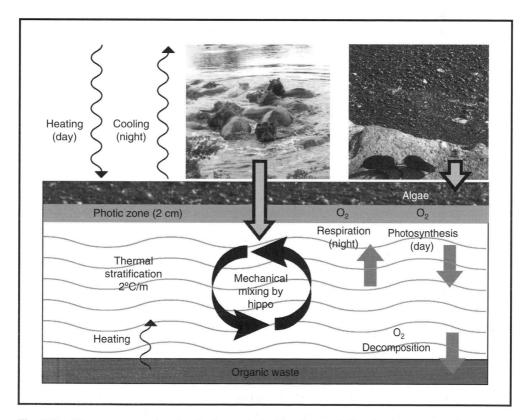

Fig. 9.6. The processes leading to thermal stratification and dissolved oxygen cycling in a typical water hole inhabited by hippopotamus, in the dry season. Left panel: the high turbidity results in solar heating being restricted to the top few centimetres, resulting in a strong vertical gradient in temperature, which prevents aeration of the water column. Right panel: a permanent algae bloom near the surface oxygenates near-surface waters in daytime and results in near anoxic conditions at night. Middle panel: decomposition of organic matter near the bottom removes dissolved oxygen from the water. Hippopotamus stir and aerate the water and prevent the frequent occurrence of anoxic conditions. (Adapted from Wolanski and Gereta, 2000.)

The effects of poor water quality on wildlife have not been studied and may be important if there is an analogy between cattle and wildlife husbandry. Indeed, the performance of cattle drinking from dugouts as opposed to troughs is severely diminished as a result of harmful organisms, the palatability of water, and the release of methane, ammonium or hydrogen sulfide from the muck disturbed by the animals (Willms *et al.*, 1994, 1996; Chu *et al.*, 1998). Similar effects could control wildlife in the Serengeti National Park and deserve detailed investigations. Indeed, poor water quality may explain the observations (S. Mduma, personal communication) that 70% of the deaths of wildebeest are not accounted for by predation or poaching and occur in the dry season, and thus may be due to diseases from drinking polluted water.

Population Dynamics

Ecohydrological concepts were used to develop a biomass model to predict population levels in the Serengeti. This model is forced by rainfall for which data are available at monthly intervals from 1960 to 1999. The model has three trophic layers (Fig. 9.7). The bottom trophic layer is the grass, which grows when watered and withers in the absence of rainfall. The grass is grazed by herbivores. The herbivores calve once a year. The herbivores can die from poaching (for which

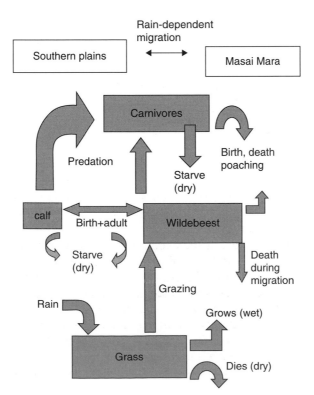

Fig. 9.7. The rainfall-driven, three-trophic-levels biomass model of the Serengeti ecosystem.

data are available from the game warden), starvation (in dry seasons) and disease (mainly in the dry seasons, for which data were made available by S. Mduma). The carnivores prey upon the herbivores. In the model the ecosystem is divided in two areas: (i) area A (the southern grasslands), which are used in the wet season; and (ii) area B (the Masai Mara, Kenya) that is the dry-season refuge. The animals migrate from area A to area B when salinity is excessive in area A (1 month after peak rainfall); they return to area A at the start of the wet season. The migration results in an additional mortality of the herbivores. The model equations are of the Lotka–Volterra type for biomass at each trophic level; they express mass conservation and are given in the Appendix.

The model is successful (Fig. 9.8) in hind-casting many of the changes observed between 1960 and 1999 in the number of wildebeest and lions in the Serengeti ecosystem. The most sensitive parameter turns out to be the herbivore death rate in the dry season; this rate needs to be set to a high value in the model to reproduce the observations of population numbers over those 39 years. The high value of this parameter seems to be related to poor water quality. Predation by lions turns out to be of secondary importance for the herbivore population dynamics. The model suggests that the population of herbivores is limited by the availability of water and forage, and thus fluctuates inter-annually as a result of rainfall.

Conclusions

Ecohydrology thus seems to control much of the dynamics in the semi-arid Serengeti ecosystem:

1. Rainfall determines the salinity of surface waters and in turn salinity may control the discontinuity between open grassland and wooded savannah.

2. Rainfall variations are large at decadal time scales; salinity may shift southward or northward (i.e. upstream or downstream) with the location of the salinity threshold determining this discontinuity. In turn this introduces changes in vegetation at decadal time scales.

3. Excessive salinity, and not available forage and water, appears to be the trigger starting the annual migration of wildebeest, zebras and Thomson's gazelles at the end of the wet season. The onset of migration may be predicted on the basis of a simple threshold value for salinity.

4. Rainfall determines the extent of the occupancy by the migrating herds of respectively the wet-season dispersal area (the southern grasslands) and the dry-season refuge area (the Masai Mara). Because annual rainfall varies by a factor of up to 3 from year to year, the ecosystem size thus varies markedly from year to year. The inter-annual rainfall variations, hence the migration areas, are only weakly correlated with the ENSO phenomenon and remain unpredictable.

5. Water quality is extremely poor in the dry season; stirring by hippopotamuses and filtering by fringing wetlands improve it. The poor water quality in the dry season apparently may be responsible for the increase in mortality of wildebeest.

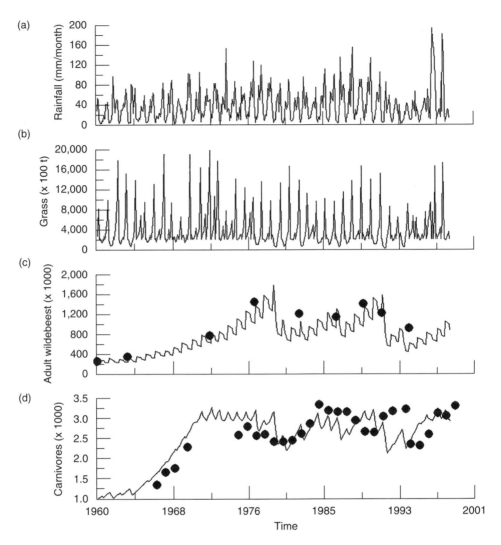

Fig. 9.8. Time series plots from 1960 to 1999 of (a) the observed monthly rainfall, (b) the predicted grass biomass, (c) the observed (circles) and predicted (line) number of wildebeest, and (d) the observed (circles) and predicted (line) number of lions in the Serengeti ecosystem.

6. Hippopotamuses thus play a key role in the ecosystem by preventing the occurrence of anoxic conditions in the dry-season water holes that are the only source of drinking water for wildlife.

There are several implications for managing the Serengeti ecosystem in view of a proposed hydroelectric and water diversion plans for the River Mara:

1. To maintain a viable ecosystem it is particularly important to maintain water quantity and quality in the dry season. A sustainable, dry-season water flow in the dry-season refuge area, in this case the River Mara, and a rich population of hippopotamus appear vital to maintaining a viable ecosystem.

2. The ecohydrological model can be vital for the management of the Serengeti National Park. Indeed, it may enable one to separate the effects on both migrating and resident wildlife of rainfall-driven variability, over which management has no control, from those with other causes (e.g. bush fires, storing surface waters for tourism), over which management does have control. It can also be used to predict the impact that diverting water from the River Mara would have on the entire Serengeti ecosystem.

These lessons are probably also applicable to other wildlife-rich, river catchments in semi-arid East Africa, such as the Okavango Delta in Botswana, the Sudd wetlands in Sudan and the Ruaha National Park in Tanzania, where existing or proposed water diversion schemes may negatively impact on wildlife, which is an important resource of these countries.

Acknowledgements

This study was supported by the TANAPA, AIMS, FZS and the IBM International Foundation. We thank Ernest Sitta, Ephraim Mwangomo, Felicity McAllister, Simon Spagnol, Craig Packer, Markus Borner, Brian Farm and Simon Mduma.

Appendix: The Serengeti Ecohydrology Model

The model equations are of the Lotka–Volterra type for biomass at each trophic level; they express mass conservation. The model considers separately zones A and B and the migration pathways. The equations are improved from Gereta *et al.* (2002).

1. Zone A

Grass (G):

$$dG/dt = \begin{cases} gGR/20, & \text{if } R > 20 \\ 0, & \text{if } 5 < R < 20 \\ -2gG(5-R)/20, & \text{if } R < 5 \end{cases} - hG_0H/H_0$$

Adult herbivores (H):

$$dH/dt = \begin{cases} s_AH(G-G_0)/G_0, & \text{if } G < G_0 \\ 0, & \text{if } G > G_0 \end{cases} - d_1H - pHC/C_0$$

Young herbivores (Y):

Before calving: $Y = 0$

In the calving month: $Y = 0.5rH$

Thereafter: $\quad dY/dt = \begin{cases} s_A Y(G - G_0)/G_0, & \text{if } G < G_0 \\ 0, & \text{if } G > G_0 \end{cases} - d_Y Y - pYC/C_0$

Carnivores (C):

$$dC/dt = \begin{cases} 0, & \text{if } H > H_{01} \\ -d_4 C(H_{01} - H)/H_{01}, & \text{if } H < H_{01} \end{cases} + (b - d_3)C$$

2. Migration

When migrating from A to B:

$\quad H(\text{new}) = q(H(\text{old}) + Y/r)$

$\quad Y(\text{new}) = 0$

When migrating from B to A:

$\quad H(\text{new}) = qH(\text{old})$

$\quad Y(\text{new}) = 0$

3. Zone B

Grass:

$$dG/dt = \begin{cases} gGR/20, & \text{if } R > 20 \\ 0, & \text{if } 5 < R < 2 \\ -2gG(5 - R)/20, & \text{if } R < 5 \end{cases} - hG_0 H/H_0$$

Adult herbivores:

$$dH/dt = \begin{cases} s_B H(G - G_0)/G_0, & \text{if } G < G_0 \\ 0, & \text{if } G > G_0 \end{cases} - d_2 H(1 + H/H_s) - pHC/C_0$$

Young herbivores:

$\quad Y = 0$

Carnivores:

$$dC/dt = \begin{cases} 0, & \text{if } H > H_{01} \\ -d_4 (H_{01} - H)C/H_{01}, & \text{if } H < H_{01} \end{cases} + (b - d_3)C$$

where

$\quad R$ = rainfall (mm/month),
$\quad G$ = biomass of grass (tonnes),
$\quad H$ = biomass of adult herbivores (tonnes),
$\quad Y$ = biomass of young herbivores (tonnes),

> r = calf to adult herbivore weight ratio in April,
> C = biomass of carnivores (tonnes),
> G_0 = equilibrium G preventing starvation of herbivores when $H = H_0$,
> s_A = starvation rate of herbivores in zone A if $G < G_0$,
> $s_B = s_A(1 + H/H_0)$ = starvation rate of herbivores in zone B if $G < G_0$,
> h = rate of removal of grass by herbivores at equilibrium,
> g = growth rate of G,
> p = predation rate of the herbivores at equilibrium,
> q = fraction of herbivores that survive the migration,
> d_1 = death rate of herbivores in region A,
> d_2 = death rate of herbivores in region B,
> b = birth rate of carnivores,
> d_3 = death rate of carnivores (excluding starvation),
> C_0 = equilibrium carnivore biomass,
> H_0 = equilibrium herbivore biomass when $C = C_0$,
> $H_{01} = 0.3\,H_0$,
> H_s = saturation capacity in zone B,
> d_4 = death rate of carnivores at equilibrium,
> t = time (starting from January 1960 when monthly rainfall data are available),
> dt = model time step.

The list of parameters values is shown in Table 9.1. In the herbivore equations, $s_A > 0$ and $s_B > 0$; nevertheless the first term on the right-hand side of the equation is negative (i.e. there is a starvation die-off) because $G < G_0$.

Table 9.1. Value of parameters used in the model. All rates are expressed in month^{-1}; animal population in thousands.

G_0	2000
H_0	200
C_0	2.8
H_s	1200
s	0.15
r	0.7
g	0.6
h	0.05
s_A	1.2 s
p	0.003
d_1	0.01
d_Y	$3d_1$
d_2	0.01
s_B	3 s
b	0.01608
d_3	0.0016
d_4	0.2
q	0.9
dt	1 month

There is no discontinuity in the herbivore and the carnivore model because the starvation rate is proportional to the deficit. Indeed, for the herbivores the rate vanishes when $G = G_0$, and for the carnivores the rate vanishes when $H = H_{01}$.

Observed monthly rainfall data from 1960 to 1999 were used to drive the model. Hence $dt = 1$ month. The model thus utilizes the historical data without averaging them or decimating them. It was not possible to simplify the model into wet and dry seasons because the duration of the wet and dry seasons varies greatly from year to year. The Lotka–Volterra equations are robust because the amount consumed per herbivore depends not only on the abundance of grass but also on a saturating function parameterized by the threshold function $(1 + H/H_s)$.

A model with several rainfall thresholds is not aesthetic compared with a smooth response rainfall–grass function; however, the IF statements in the grass dynamics sub-model reflect our practical experience on the various rainfall thresholds for the grass dynamics.

10 The Mid-European Agricultural Landscape: Catchment-scale Links Between Hydrology and Ecology in Mosaic Lakeland Regions

A. HILLBRICHT-ILKOWSKA

Centre for Ecological Research, Polish Academy of Sciences, Lomianki, Poland

Introduction

Landscape is the spatial composition of different patches and corridors linked and integrated through the transport and transformation of matter and nutrients, and the exchange of biological information (Forman and Gordon, 1986; Forman, 1995). Many indices of landscape pattern or structure and their changes in time and space (Hulshoff, 1995) are used for scientific comparison across types, for evaluating different management and fragmentation effects and for recognizing long-term changes. A diversified landscape has a high density of patches differing in origin, soil conditions and vegetation, each with a relatively small size and irregular shape. The index of patchiness of a diversified landscape is higher than that of a homogeneous landscape. This in turn affects flow and intensity of surface erosion, the retention of water and nutrients in different patches, and the losses of nutrients. Generally, the patchy structure of a landscape influences the variation of the above indicators of catchment functioning. Johnson *et al.* (1997) found that slope and patch density accounted for most of the observed variation of nutrient contents in catchment waters.

The mid-European landscape in northern Poland is diversified (Fig. 10.1). This is a fragment of the young (12,000–10,000 years BP), postglacial landscape which prevails over large lowland areas of North-Eastern Europe. The region which this chapter addresses is the Masurian Lakeland in north-eastern Poland. Two types of landscape occur in the region: hilly and outwash plain, usually bordered with a range of moraine hills (Bajkiewicz-Grabowska, 1985; Kondracki, 1988).

This landscape patchiness is the product of the formation of geological deposits (substrate patchiness), historical and actual geomorphological processes shaping the land forms (relief), and human management in the area. The latter

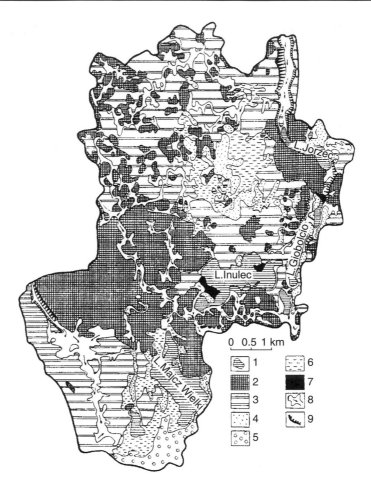

Fig. 10.1.   A relief map of the River Jorka watershed (Masurian Lakeland, north-eastern Poland): 1, lakes; 2, ridges, hills and terminal and dead-ice moraine mounds; 3, ground and ablation moraine plains; 4, esker ridges, hills, ridges, kame mounds and kame terraces; 5, sandy outwash plains; 6, kettles; 7, clear lake terraces made up of lacustrine clays and calcareous gyttya; 8, meltwater stream-way valleys; 9, channel slopes. (After Bajkiewicz-Grabowska, 1985.)

has led to fragmentation of forest and wetland complexes, formation of field plots (with permanent or temporary ploughing in crop rotation) and urbanization (residential and farming buildings, tourist sites). Such changes have resulted in intensification of nutrient input (fertilization, farming, range management, waste water discharge) and of harvest output. This diversified spatial system of patchy substrate and land cover has created a particularly complicated matrix for water movements by infiltration, percolation, surface (overland), subsurface (root layer) and underground discharge plus retention in catchments and lakes.

Landscape Pattern in Selected Lacustrine Catchments

The catchment of the River Jorka (53°45′N–53°53′N and 21°25′E–21°33′E) flowing through five lakes (total length 25 km) is a good illustration of the patchiness that occurs (Hillbricht-Ilkowska, 2002a). The landscape in this region is connected with the Masurian lobe of Baltic glaciation. Terminal moraine hills, kames and eskers reach 160–206 m above sea level and a gradient of 5–20°. Smaller gradients (up to 5°) occur on ablation moraine plains, ground moraine plains as well as Holocene accumulation plains found in land depressions. Typical for this area are very frequent, sometimes very small (0.1 ha) wetland patches (potholes) formed in land depressions without surface runoff, with their subsurface drainage covering about 28% of the whole catchment area.

The distribution of Quartenary (Pleistocene and Holocene) deposits is a very important factor in water movements in the Jorka catchment. They determine the rate and range of infiltration of precipitation, surface and underground retention and, indirectly, the erosion potential and surface runoff. The very fine sandy loam and loam with boulders occupies about 34% of the area, sandy soil about 21% of the area, and peat about 10%. All types of deposits have different infiltration capacity and almost half of the area (43%) has poor conditions for water percolation (<5% of annual precipitation) (Bajkiewicz-Grabowska, 1985), which is thus sensitive to surface erosion. These deposits are located mainly in the middle and lower parts of the catchment (which are the main arable parts). The headwaters are of different character; almost 75% of this area can infiltrate more than 25% of the annual precipitation. The average value of infiltration for the whole River Jorka catchment is estimated at 88 mm from about 580 mm of annual precipitation, but ranges from <50 to 138 mm.

Arable land covers about 48% of the catchment, forests 30% (including swamp forests), grasslands 12% (grazed and cultivated, also wet meadows), lakes 5% and residential areas 5%. However, in different lake sub-catchments (area 1.3–19 km^2) the percentage cover ranges from 15 to 60%, 10 to 68%, 2 to 28%, 1 to 20% and 1 to 7%, respectively, for arable, forests, grassland, waters and residential land (Bajkiewicz-Grabowska, 1985; Rybak, 2002).

Mozgawa (1993) described four lake catchments by their structural indices, divided into three categories according to their potential for nutrient and water retention (Fig. 10.2). Wetlands have the highest retention potential, natural and managed grasslands and forest fragments moderate potential, and arable land has negligible retention potential. The pattern found for all four lake catchments was more or less similar. Wetlands constituted about 10% of the catchments, with patch size varying between 1 and 2 ha, patch density between 4 and 12 per km^2, and the contact zone between them and other patches (length of ecotones) between 2 and 4 km/km^2 of catchment. The dominant patches (more than 50%) were of moderate retention capacities, such as forest fragments and grasslands with larger patch sizes (2–5 ha), higher density (up to 23/km^2) and longer ecotone zone around them (up to 14 km/km^2 of catchment). The arable and residential areas in these particular lake catchments do not cover more than 30% of the area with rather small-sized patches (2–4.5 ha) of relatively low density, up to 10 patches/km^2 (Fig. 10.2).

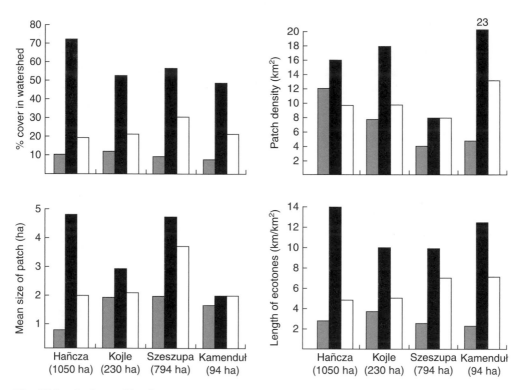

Fig. 10.2. Indices of landscape structure in terms of size and dispersion of three types of patches – 'strong' (wetlands; ▦), 'moderate' (woodlots, natural and cultivated meadows, pastures; ▦) and 'low' (ploughed arable land, urban sites; ☐) – in four lake watersheds in north-eastern Poland (Lake Hańcza, Lake Kojle, Lake Szeszupa and Lake Kamendul) of different surface area (as shown). (Data of Mozgawa, 1993 in Hillbricht-Ilkowska *et al.*, 1995, modified.)

Routes of Nutrients in Relation to Precipitation and Runoff

There is a significant difference between the main routes of transport and transformations of the two major compounds involved in eutrophication: phosphorus and nitrogen. Both nutrients are transported by overland, subsurface and groundwater flows, but the rate and changes of their concentration and forms along these routes are different.

The sources of particulate phosphorus in a landscape are primarily the eroded surface soil layer, stream banks and beds, plant fragments and detritus. The main sources for dissolved phosphorus (DP) are the dilution and desorption processes from soil complexes (including mineral fertilizers), decomposition of organic material and biotic excretion. All these forms are mainly mobilized with water erosion. The amount of phosphorus transported with groundwaters is relatively small due to the sorption capacities of glacial soils (clay loam); therefore the surface and subsurface transport are crucial for the movement of this element

(Hillbricht-Ilkowska, 1995; Hillbricht-Ilkowska *et al.*, 1995; Kronvang *et al.*, 1997). Groundwater flow is much more important for nitrate-N, which is the most mobile compound in agricultural landscapes (Herlihy *et al.*, 1998).

Smolska (1993) and Smolska *et al.* (1995) measured the intensity and range of geomorphological processes in the headwaters of the River Szeszupa, in Suwalskie Lakeland (north-eastern Poland). The surface erosion of topsoil was relatively extensive and occurred on one-third of the area. The amount of eroded material varied greatly between 36 kg/ha/year and 7 t/ha/year depending on the gradient (9–24°) and profile (convex or flat) of the slope, the substrate (sandy or sandy-loamy) and the type of arable activities. Highest values were found for steep, short and ploughed slopes. However, the transport distance of this material appeared to be rather short (Fig. 10.3). Mostly, it was accumulated at the slope base just below the erosion zone and only a small part was moving further. The more intensive erosion, although on a relatively small area (about 15%), occurred through linear washing (along the valley, field paths and baulks in parallel or diagonal to the slope) in the form of rills, sometimes as deep as 30–50 cm.

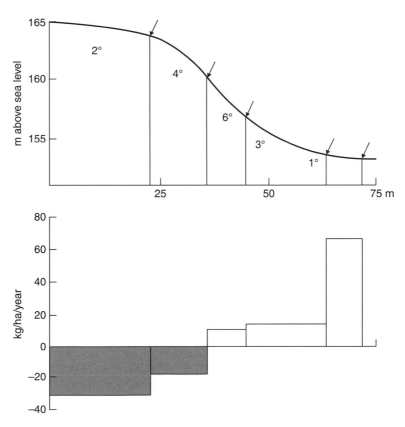

Fig. 10.3. The example of surface washing (■) and deposition (□) of topsoil on cultivated slope (cereals, clover and root crops in crop rotation) in Suwalski Lakeland (north-eastern Poland). (Data from Smolska *et al.*, 1995.)

Between 7 and 24.4 t/ha/year was moved downslope. Thus in hilly, postglacial, arable areas, the mass of topsoil material containing mostly particulate nutrients mobilized with surface erosion is relatively high, but most of it is deposited over a short distance, not usually reaching a lake or river. Rytelewski *et al.* (1985) showed that the amount of phosphorus in the calcium-rich soils of the Jorka catchment oscillated between 120 and 360 kg P_2O_5/ha (average 270) for 0–20 cm of soil layer. The amount of phosphorus that could be washed down annually from 1 ha of arable slope is thus quite considerable. The stream density is as high as 5 km/km^2 in these catchments (Hillbricht-Ilkowska and Wisniewski, 1993) but most of these are usually short, shallow temporary streams, active only in vernal, freshet periods and only rarely in other periods following long and heavy rains. There is a relationship between the stream density gradient and phosphorus losses in the summer period in the sense that the sub-catchments with stream density higher than 2 km/km^2 tend to export more phosphorus per unit area (Fig. 10.4).

The above peculiarities of lake catchments indicate the close relationship between local hydrology and total phosphorus (TP) losses. As an illustration, several Masurian lake catchments compared between years with different amounts of precipitation showed losses of 0.04–0.13, 0.23–0.76 and 0.30–1.24 kg TP/ha/year for dry, moderate and wet years, respectively (Giercuszkiewicz-Bajtlik, in Hillbricht-Ilkowska *et al.*, 1995). There is little, or at best a weak, relationship between the TP concentration and its bioavailability and discharge. The correlation is best for catchments heavily polluted with farm and communal waste waters or in periods of fertilizer spreading (Hillbricht-Ilkowska and Bajkiewicz-Grabowska, 1991; Hillbricht-Ilkowska *et al.*, 1995). The increased discharge from these types of land use temporarily enhances the release and outflow of soluble phosphorus. The best predictor of TP losses from an unpolluted, mosaic landscape

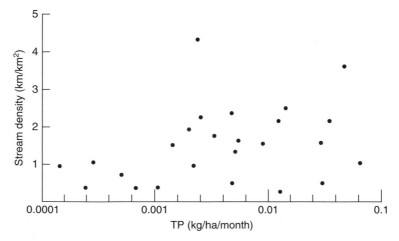

Fig. 10.4.   Stream density from 24 small lakeside sub-catchments in Suwalski Lakeland (north-eastern Poland) plotted against total phosphorus (TP) export rates in the summer period. (Data from Hillbricht-Ilkowska and Wisniewski, 1993.)

is the annual discharge. This is shown by the average TP concentration and load for 16 sub-catchments (Fig. 10.5). Despite the considerable variation of TP concentration (0.01–1.0 mg/l) it is the variation in discharge which determines the final load (Hillbricht-Ilkowska *et al.*, 2000).

In mosaic landscapes with patches of different permeability there is no direct correlation between precipitation and average discharge (Hillbricht-Ilkowska *et al.*, 2000; Rybak, 2000, 2002). The highest discharge is always connected with vernal freshet in the lakeland region but is not quantitatively related with the amount of precipitation accumulated over the winter period. Only during the vernal period is the whole stream network active (Table 10.1) with over 25 m^3/day (average discharge for all streams in April is more than 2000 m^3/day). For this

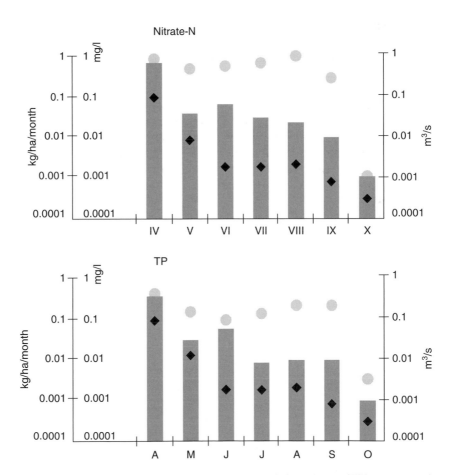

Fig. 10.5. Seasonal changes of nitrate-N and total phosphorus (TP) concentration (mg/l; ◯) in stream water and average monthly export rates (kg/ha/month; ▣) for 16 lakeside sub-catchments in the River Jorka system (Masurian Lakeland, north-eastern Poland), plotted together with data on the average stream discharge (m^{3}s/; ◆). Log-transformed data for 1996. (From Hillbricht-Ilkowska *et al.*, 2000; Rybak, 2000; 2002.).

Table 10.1. Precipitation cumulated for 7 and 14 days before measurement of discharge in 16 monitored streams in a lakeland area (north-eastern Poland), the number of 'active' streams (discharge >26 m^3/day) and their average discharge. (Data for 1998 from Rybak, 2000.)

Month	Precipitation over 7 days (mm)	Precipitation over 14 days (mm)	No. of active streams	Discharge (m^3/day)
April	16 (123)[a]	31	14	2376
May	10	11	7	2145
June	11	66	6	4122
July	11	22	0	–
August	37	48	2	8817
September	6	13	2	1106
October	9	30	1	1261

[a]Value in parenthesis is sum of precipitation for the winter period: January, February, March and April.

reason the TP load is much higher in vernal periods (cf. Fig. 10.5) than in summer, irrespective of how intensive the summer rains are; it usually constitutes 70–80% of annual TP load as well as total nitrogen load which is accumulated during this period (Hillbricht-Ilkowska, 1988; Hillbricht-Ilkowska *et al.*, 1995; Rybak, 2002). This strong seasonality seems to be the feature of these landscapes.

During stream transport the concentration of TP usually decreases downstream owing to sedimentation of particles, although with higher discharge rate sediments are re-suspended and moved downstream. Also marked changes in bioavailability occur during transport due to the absorption process of DP on soil and humus particles in transported sediment. The ratio of dissolved to particulate phosphorus decreases along the stream course (Fig 10.6) when the discharge is below 100 l/s (Hillbricht-Ilkowska, 1988). In this way, fine-sized particles of eroded material are constantly enriched with phosphorus and transported further, into a lake. For example, Kufel (1990) followed this phenomenon over 100 m between stream inflow and lake littoral (Fig. 10.7), showing that the contribution of organic phosphorus in the suspended and sedimented particles increased considerably as velocity decreased from 50 to 3 cm/s. The cumulative effect of this phenomenon is that the amount of phosphorus reaching the lake ecosystem is considerably lower, but relatively richer in the easily-desorbed fraction of phosphorus.

The concentration of DP in groundwaters can occasionally be high. Nowakowski *et al.* (1996) described the distribution of DP concentration in groundwater over a catchment of 600 km^2. In the greater part of this area the concentration was lower than 40 μg/l (i.e. close to the hydrogeochemical background), but in isolated small areas (about 10% of the whole catchment) was higher than 120–180 μg/l. Rzepecki (2000, 2002) found the DP concentration in groundwaters leaving the field edge to be close to 500 μg/l. Generally the concentration of DP in groundwaters in lakeland areas overlaps with the range found for TP concentration in stream waters, but values higher than 500–600 μg/l are rare. However, the final assessment of the phosphorus input in groundwaters to lakes

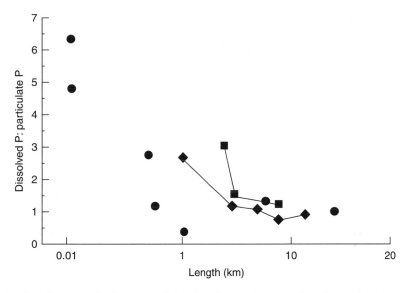

Fig. 10.6. Changes in the ratio of dissolved to particulate phosphorus in the annual total phosphorus load along the length of several small streams (●) and two rivers (■, River Jorka; ◆, River Baranowska Struga) in the River Jorka watershed on Masurian Lakeland (north-eastern Poland) (Adapted from Hillbricht-Ilkowska, 1988.)

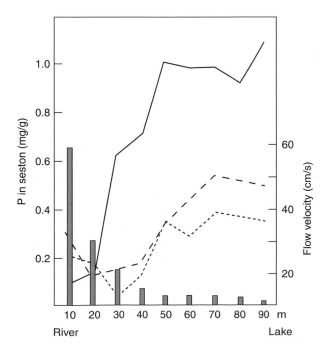

Fig. 10.7. Comparison of phosphorus content in seston transported in river waters (———), the upper (0–2 cm) layer (– – –) and the lower (2–5 cm) layer of sediments (------) along a distance of 100 m between river mouth and lake littoral. The decrease of flow velocity (bars) over this distance is indicated. (Adapted from Kufel, 1990.)

in terms of unit lake surface made by Nowakowski *et al.* (1996) revealed rather low values (0.2–0.01 g/m^2 lake surface/year) compared with input from surface and subsurface waters (1.6–25.0 g/m^2 lake surface/year) (Hillbricht-Ilkowska and Kostrzewska-Szlakowska, 1996).

The stream network is also an important carrier of nitrate-N input to lakes. Figure 10.5 shows that the concentration of this compound varied greatly among a dozen lakeside sub-catchments (between 0.01 and 2.0 mg/l). The maximal losses occurred during the vernal freshet period (as with TP). The concentration of nitrate-N in groundwaters is usually higher than in stream waters. Nowakowski *et al.* (1996) found values between 1 and 10 mg/l and similar ones were reported by Rzepecki (2000, 2002) in the Jorka catchment in groundwaters between field edge and the riparian zone (0.1–7.0 mg/l). The rough estimates made by Nowa-kowski *et al.* (1996) and based on the data on groundwater discharge into the Krutynia lakes (0.01–0.2 m^3/s) revealed that the possible annual load of nitrate-N from groundwaters can vary between 0.4 and 9.0 g/m^2 lake area. Values higher than 1.0 are 'permissible' and those higher than 2.0 g/m^2 are 'dangerous' for shallow lakes according to Vollenweider's (1968) criteria.

Export Rates of Nutrients in Relation to Human Impact and Discharge Rates

The data from different lake catchments in Masurian Lakeland (Table 10.2) and for 2 years of different discharge in freshet periods for a dozen lakeside sub-catchments indicate the effect of human land-use practices. The maximal values of export rates of TP are related to agriculture as well as urban and tourist activities, but generally the range of variation is great (over four orders of magnitude) and overlaps between the three categories of sub-catchment (Fig. 10.8). The values of annual export rates for extremely patchy areas dominated by wetlands (pastures, wet meadows, swamp forest islands) are usually below 0.15 kg/ha/year. These sub-catchments constantly have a low concentration of chloride in their waters, <10–15 mg/l (Fig. 10.8). Higher values occur in sub-catchments with higher contribution of arable land and/or tourist sites, in which chloride concentration usually does not fall below 20 mg/l. A range of TP export values between 0.01 and 1 kg TP/ha/year is common and representative of this type of rural landscape. They are similar to those reported for Lake Paajarvi, Finland (Hakala, 1998), although values between 2 and 5 kg TP/ha/year can occur occasionally (Table 10.2). However, the range of variation of TP concentration is large (0.01–0.5 mg/l) and strongly overlaps between all three groups of sub-catchment. There is no tendency for higher values in sub-catchments with higher human impact (Fig. 10.8). These data indicate that the differences between the TP export rates from sub-catchments of different human impact are mainly the product of differences in discharge and the amount of eroded sediments. Allan *et al.* (1997) found that the sediment yield in sub-catchments is strongly positively correlated with the proportion of arable area and negatively correlated with forested area. Soranno *et al.* (1996) found high differences between the TP losses from arable land in high and low discharge years.

Table 10.2. Range of annual export rates of total phosphorus (TP) in a hilly lakeland region (north-eastern Poland) from small (0.3–22 km²) lakeside sub-catchments (drained by streams) with different arable and tourism impacts.

River catchment	Land cover and use	Tourism impact	TP export rate (kg/ha/year)
River Jorka (n = 16)	wetlands, pastures, swamp forest islands	small	0.009–0.15 (1.3)[a]
	pastures, wet meadows, arable fields	small	0.002–0.9 (1.2)
	arable, ploughed land, pastures, village	small	0.01–1.14 (5.36)
River Szeszupa (n = 28)	wetlands, pastures, wet meadows, arable fields, forest islands	moderate	0.01–0.7 (1.3)
River Krutynia (n = 11)	forest	strong	0.06–2.2 (4.1)
	ploughed land	strong	0.5–2.1 (3.6)

Collected data from different sources in Hillbricht-Ilkowska and Wisniewski (1993). Data for the River Jorka system concern the range for three years 1996–1998 according to Rybak (2000, 2002).
[a]Value in parenthesis is the maximal one reported.

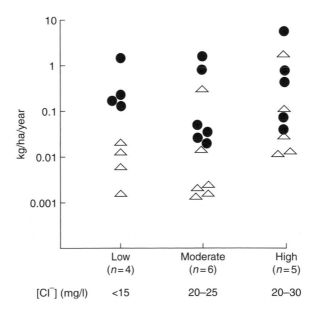

Fig. 10.8. Log distribution of the annual values of total phosphorus export rates (kg/ha/year) from 16 lakeside sub-catchments (River Jorka catchment, Masurian Lakeland, north-eastern Poland) of low, moderate and high arable/urban impact (as indicated by chloride concentration) and for two years of different annual discharge (Q) (●, 1996, Q = 0.014 m³/s; △, 1997, Q = 0.096 m³/s), which determines annual load. (Data from Rybak, 2000; 2002 and unpublished.)

The annual export rates of nitrate-N estimated in surface and subsurface runoff were between 0.002 and 5 kg/ha/year and tended to be higher in sub-catchments with higher agriculture/residential impact as well as in years of higher discharge (Fig. 10.9). However, most values are in the range of 0.01–1 kg/ha/year, similar to those found by Hakala (1998) for Lake Paajarvi. Much higher values for arable land lie within several kg/ha/year (Rekolainen, 1989; Moldan and Cerny, 1994; Burt *et al.*, 1993; Haycock *et al.*, 1997). The differences between nitrate-N export rates from the three groups of sub-catchment here (Fig. 10.9) seem to be the product not only of differences in discharge but also of differences in concentration. The nitrate-N concentration varies greatly, but instantaneous values for two sets of sub-catchments give lower values (71 and 67% of the data for two respective years) for low arable impact with higher values (56 and 63% respectively) for higher arable impact.

The difference between the export rates of TP in patchy landscapes compared with those from more homogeneous areas of agriculture is smaller than the same difference in nitrate-N export rates. This is probably the consequence of a higher retention potential in diversified glacial landscapes for nitrate-N than phosphorus, particularly because of the wetland patches and land depressions.

The discharge and export rates of both nutrients in spring (freshet period) determine the annual loss values (Fig. 10.5). The spring values are up to 20 times higher than summer values for both nutrients, so about 70–80% of the annual load reaches the lake system during a short (about 2 months, April–May) period of time. Using the spring values of export rates of both nutrients (see Fig. 10.11 below), there does not appear to be any trend in annual changes in average loss rates from sub-catchments of low or high agriculture/urban impact (Rybak, 2000, 2002). The mosaic pattern of patches effectively mitigates the possible changes caused by arable activity.

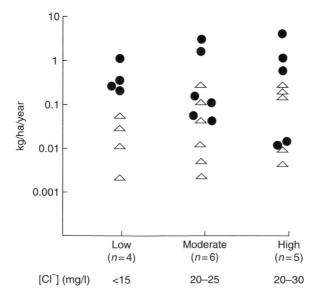

Fig. 10.9. Log distribution of the annual values of nitrate-N export rates (kg/ha/year) from 16 lakeside sub-catchments (River Jorka catchment, Masurian Lakeland, north-eastern Poland) of low, moderate and high arable/urban impact (as indicated by chloride concentration) and for two years of different annual discharge (Q), (●, 1996, $Q = 0.014$ m^3/s; △, 1997, $Q = 0.096$ m^3/s), which determines annual load. (Data from Rybak, 2000; 2002.)

Buffer Functioning of the Wetland Patches: Relationship between Hydrology and Nutrient Loading

Wetland patches are usually small (0.5–7 ha, average <1 ha), dispersed among the fields and meadows with total area around 10% of the lake catchment (Kruk, 1990; Wilpiszewska, 1990; Wilpiszewska and Kloss, 2002). They occupy land depressions between moraines and hills (kettleholes) as well as shallow and larger depressions formed by irregular accumulations of glacial drifts (Kloss, 1993). ^{14}C dating indicates their bottom peat layer was formed 11,430±120 years BP in the older patches and 7370±140 years BP in younger ones (Kloss, 1993). There is no surface runoff from these small wetland patches. The majority of them are fed with subsurface and groundwater discharge (soligenic), others are fed by precipitation (ombrophilous). Two types of vegetation succession (mesotrophic and dystrophic) and 15 plant communities relating to these hydrological conditions were distinguished by Wilpiszewska (1990) and Wilpiszewska and Kloss (2002). The mesotrophic series follow reed and sedge communities, willow scrub and then alder fens. The dystrophic series follows bog moss swamp, bog moss pine forest and then bog pinewood. The wetland patches of the first type are usually smaller in area (below 1 ha) but their basins are four or five times larger. They are more productive, having higher nutrient concentration in waters and plant tissues (Wilpiszewska, 1990). Their soils are organic-rich and hydromorphic. This type of wetland patch dominates in the lakeland area and their role in transformation of nutrients transported with subsurface and groundwaters is particularly important. These patches are continuously disappearing from the region now, due to artificial drainage as well as a decrease in the water table on arable land.

Kruk (1987, 1990, 1991, 1996, 1997) has studied the retention of nutrients and other elements passing with subsurface and groundwaters through the wetland patches of the above two types. He found that the ratio of the patch area to its drainage area is a good predictor of its functioning; patches with an area smaller than 20% of their drainage area are more productive in terms of dissolved organic matter and nutrient concentration but highly variable in terms of intensity and length of water saturation conditions. They are more exposed to the fall in water level between spring (freshet period) and the summer drought (Fig. 10.10). However, during the period of higher water level, wetland patches situated in the hollows on the arable slope have high potential for nitrate-N removal (Fig. 10.11); retention is close to 95% of input in the case of nitrate-N and 45% in the case of ammonium-N (Kruk, 1996, 1997). Wetland patches export soluble phosphorus (negative retention –55%) as well as potassium (negative retention –255%) during high water level conditions. Drained wetland patches lose the retention capacity of nitrogen and export it (negative retention –71%) (Kruk, 1997).

Wetland belts formed along the lake shore and river banks constitute an ecotone isolating the main aquatic ecosystems from arable uplands. Historically, they are the remains of former flood plains of low-order streams and/or the former part of the lake littoral flooded during the higher state of water. Their present state is the product of a more stabilized water level over a wide area by canals connecting the lakes and artificial drainage. Typically, the width of the wetland

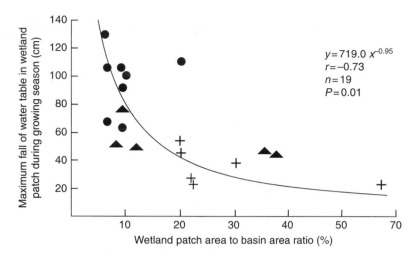

Fig. 10.10. The relationship between the maximum fall of the water table in wetland patches (during the period from April to August) and the ratio between wetland area and its basin area for seasonally flooded (●) and permanently flooded (▲) wetland patches of minerotrophic type, and permanently waterlogged wetland of ombrotrophic type (+). (Adapted from Kruk, 1987.)

belt along the majority of lakes and streams is less than 100 m, sometimes as narrow as 15 m or less (Klosowski, 1993; Klosowski and Tomaszewicz, 1993; Kloss and Wilpiszewska, 2002). The composition of the vegetation includes fragments of tall rush communities, sedge and moss–sedge communities, phytocenoses of nitrophilous and hygrophilous perennials (such as *Urtica dioica*), riverside herb communities, some meadow–pastural communities and forest communities of alder ash and alder carr. These communities often create a corridor along the river bank and the zonal pattern along the lake shores (Fig. 10.12). The community with tall rushes and reeds occurs close to the lake edge while the zone occupied by nitrophilous plants is at the edge of the field. The soil conditions are different in particular zones and layers; the upper layer is prone to temporary drying while that close to the lake could be temporarily flooded. The lower layers of soil are usually organic-rich and highly humid, the waterlogging conditions controlled mainly by the surface and groundwater runoff from uplands and by seeping or flooding from the lake and stream waters. There are very small waterbodies (several m^2), small deposits of lake detritus and sediments (including freshwater mussel shells) and terrestrial litter found here, creating a series of microsites and making the whole habitat extremely patchy. The main difference in the functioning of the riverine and lacustrine wetlands is that the former are periodically disturbed by the floods which scour away the greater part, leaving bare areas for new sediment deposition and colonization of plants.

The ecotones function as barrier zones, which remove and/or transform chemical compounds and organic matter transported here from the arable uplands in surface and subsurface water runoff as well as groundwaters (Correl, 1997). The compounds undergo significant change in mass due to the contact with

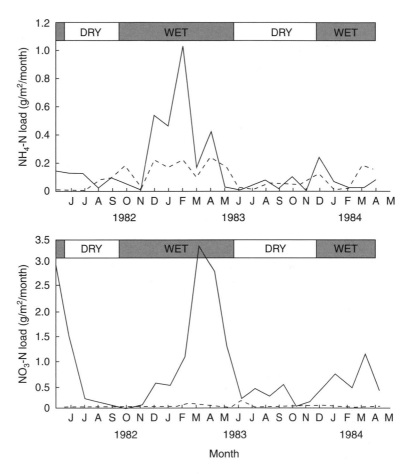

Fig. 10.11. Seasonal dynamics of inflow (——) and outflow (– - – - –) of ammonium-N and nitrate-N for a wetland patch of minerotrophic type in wet and dry seasons. (Adapted from Kruk, 1997.)

wetland plants and soils. There are several papers on the role of buffer zones (Burt *et al.*, 1993; Hillbricht-Ilkowska *et al.*, 1995; Haycock *et al.*, 1997; Naiman and Decamps, 1997). Generally the removal rate for nitrogen supplied with groundwaters is higher and more predictable than for phosphorus (Kruk, 1991; Hillbricht-Ilkowska, 1995; Hillbricht-Ilkowska *et al.*, 1995). The general opinion is that the wetland zones could be permanent sinks for nitrogen and temporary sinks for phosphorus (Naiman and Decamps, 1997; Uusi-Kämppä *et al.*, 1997). Generally, this difference is linked to the different mechanisms for effective removal of both nutrients as well as the seasonality of site conditions enhancing the removal of nitrogen in a different period compared with the removal of phosphorus (Hillbricht-Ilkowska, 1995).

The most powerful mechanism for nitrogen removal from nitrate is the process of denitrification, which occurs in carbon-rich, low-redox, water-saturated

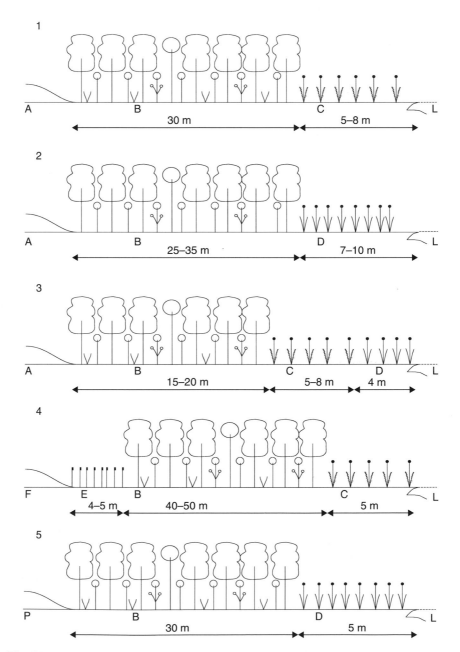

Fig. 10.12. Selected profiles of lake shore plant communities, Masurian Lakeland, north-eastern Poland: A, mineral shore; B, alder carr community (*Carici elongatae–Alnetum*); C, tall rush community (*Thelypteridi–Phragmitetum*); D, sedge community (*Caricetum acutifermis*); E, communities of nitrophilous and hydrophilous perennials with *Urtica dioica*; F, ploughed field; P, pasture; L, lake. (From Klosowski and Tomaszewicz, 1993.)

soil/sediment layers with constant nitrate supply in subsurface and groundwater runoff and is temperature-controlled. The other mechanisms like plant and microbial uptake are mechanisms for temporary removal. The sedimentation of phosphorus-rich suspended and dissolved material, which eventually reaches the wetland zone with surface and subsurface runoff, seems to be the effective mechanism for removal of TP from upland runoff (Mitsch *et al.*, 1995; Uusi-Kämppä *et al.*, 1997). The same is true for suspended organic nitrogen (Gilliam *et al.*, 1997). The removal via deposition is related to the available soil and plant surface area in wetland sites well as the oxic, air-saturated conditions in the soil layer which in turn binds compounds of phosphorus to aluminium, iron, calcium, clay and humic particles. When the soil is water-saturated and low in oxygen, desorption processes start and inorganic phosphorus compounds are released into the soil water (Szilas *et al.*, 1998). The efficiency of these processes in the removal of both nutrients is related to the retention and flow velocity of water in the wetland (Hillbricht-Ilkowska *et al.*, 1995) as well as to the level of incoming load, being higher when the load is smaller (Mitsch *et al.*, 1995). The plant uptake of both nutrients from the wetland soil can be recognized as temporary removal as the nutrients return to the system after the vegetative period. The amounts of both nutrients which possibly could be removed with plant harvest are usually below 10–20% of the total loss (Hillbricht-Ilkowska *et al.*, 1995). The processes for effective removal of nitrogen and phosphorus are thus, to some extent, conflicting and the consequences of this for the functioning of wetland systems are important.

This situation was illustrated well by Stachurski and Zimka (1994), who investigated transformation of nitrogen and phosphorus forms in the load passing through a wetland patch of 20 ha (reed belt and wet meadow close to lake shore). In spring floods the concentration of phosphate-P increased almost three times after passing through the patch, while the concentration of inorganic nitrogen decreased at almost the same rate (Fig. 10.13). The average annual

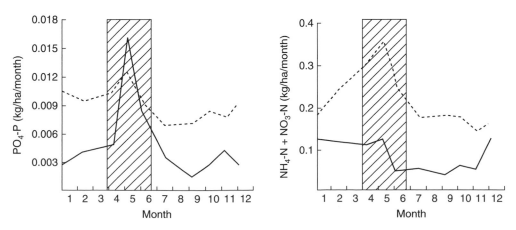

Fig. 10.13. Seasonal dynamics of loads of phosphate-P and sum of nitrate-N and ammonium-N leaving part of a watershed without (------) and with (———) a 24 ha wetland patch. (Adapted from Stachurski and Zimka, 1994.)

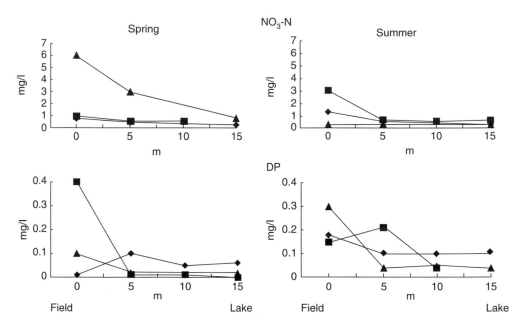

Fig. 10.14. The changes in nitrate-N and dissolved phosphorus (DP) concentrations along the 0–15 m transect through the wetland zone isolating the ploughed field and lake shore. Data for the spring (freshet) and summer (drought) periods, in three successive years (♦, 1997; ▲ 1997; ■, 1998). (Data from Rzepecki, 2000 and unpublished.)

retention of nitrate-N and particulate nitrogen was close to 73%, while that of phosphate-P was only 43%.

Generally, the rate of decrease of nitrate-N in subsurface flow is greater than in surface flow and occurs closer to the field edge when the input of nitrate from upland is higher. It has been shown to be more effective in spring freshet periods for three successive years. In both spring and summer, the input concentrations of nitrate were high (1–7 mg/l) but the decrease in the first 5 m of the wetland zone was considerable on all occasions (Fig. 10.14) (Rzepecki, 2000, 2002). The decrease rate of phosphate-P is more variable seasonally as well as laterally across the ecotone. In the summer period, when the extent of the air-saturated layer in wetland zones is the highest, the retention of DP could be the most effective.

Catchment-scale Processes and Lake Eutrophication

The question arises whether the export rates of nutrients from a patchy landscape are sufficient to control the eutrophication of lake ecosystems. The phosphorus load from the direct lake catchments (g/m^2 lake area) is compared with the permissible load calculated according to Vollenweider's (1976, 1989) criteria in Fig. 10.15 (Hillbricht-Ilkowska, 1999). In a dozen lakes the catchment load was quite small by comparison with the permissible load. In others the catchment

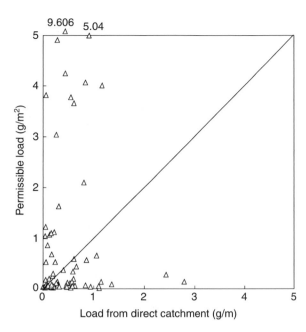

Fig. 10.15. The annual permissible load of total phosphorus calculated according to Vollenweider's criteria (1989) versus the actual load supplied from that lake's direct watershed (lake terrestrial surroundings). (Data from Hillbricht-Ilkowska 1999.)

load was several times greater and is probably the major factor in the eutrophication of these lakes. However, the majority of cases indicate that both actual and permissible loads were similar and below 1 g/m^2 lake area. This means that the phosphorus inputs from direct lake surroundings in this postglacial, arable landscape of Poland are generally lower than from other types of landscape. Long-term study of the phosphorus input and eutrophication rate of the River Jorka system revealed that in the hilly lakeland region the river–lake systems draining small mosaic catchments are rather resistant to further eutrophication (Hillbricht-Ilkowska, 2002b).

Management Implications

The basic guidelines for sustainable management in these lake environments are as follows (Hillbricht-Ilkowska, 2002c):

- Because the landscape patch diversity is the main moderator of water and nutrient access to lake ecosystems, the conservation and/or restoration of the appropriate pattern of patches of different buffering potential is very important; in particular the protection and restoration of small, dispersed wetland patches. These systems function as the point-scale spots of nutrient reduction along their route to lake ecosystems.
- Because the reduction in nutrient load is greater the longer is the transport route in a catchment and ecotone, the drainage and stream network in the lake catchment should be preserved in its natural state – ramification, meandering, small wetland patches and pools. The wetland patches, as well as

wetland strips along lake shores, should be excluded from agriculture and pastoral activities and drainage measures.

- The optimal size and structure of ecotonal zones along the lakes should consider the stabilization of moisture conditions, by enhancing the biomass of low-height vegetation with dense root systems and leaving at least a 20–30 m wide belt of wetland vegetation, connected freely with the littoral zone (macrophyte belt) of the lake itself.

11 The Ecohydrological Approach as a Tool for Managing Water Quality in Large South American Rivers

M.E. McClain

Department of Environmental Studies, Institute for Sustainability Science, Florida International University, Miami, Florida, USA

Introduction

The ecohydrological approach (Zalewski *et al.*, 1997) stems from our understanding that aquatic ecosystems contain intrinsic water purification systems wherein physical, biological and chemical processes interact to maintain water quantity and quality within ranges acceptable to the majority of organisms. These intrinsic services of water purification function to some extent throughout river corridors, but they tend to be accentuated in specific features such as riparian zones, wetlands and flood plains. Ecohydrologists strongly believe that by improving our understanding of natural water purification processes, humans can make use of these features as explicit tools in larger water management programmes, serving to complement and enhance engineering plans. This improved understanding would also enable humans to better preserve critical ecosystem components and to ensure that they continue to fulfil their ecological roles.

In this chapter the utility of the ecohydrological approach to managing water quality in large neo-tropical river basins is examined. Ecohydrology should play an expanded, and in some cases commanding, role in water quality management in the humid tropics. Severe economic problems in most tropical countries preclude the widespread installation of expensive water supply and waste treatment systems. What money is available from international aid agencies is targeted for high-priority urban areas and water-starved areas on the margins of the tropics and in other developing nations. Vast areas of the more humid tropics must rely on low-cost water management systems that make rational and complementary use of the inherent assimilation capacity of riverine ecosystems. Fortunately, large portions of Earth's major tropical river basins retain more or less natural hydrological cycles and geomorphological features. These areas are especially favourable for making an ecohydrological approach the primary approach to water quality management.

This chapter presents an overview of the condition of Earth's large tropical rivers. It then reviews the intrinsic water-purifying capacity of river systems and presents an example of this capacity in the Amazon basin. In the final section it discusses the integration of an ecohydrological approach into tropical water management programmes.

The Condition of Earth's Large Tropical Rivers and Limitations to Their Management

Earth's most intense and persistent band of rainfall encircles the globe at the equator, bathing much of the tropics in abundant fresh water. Consequently, several of the planet's great rivers are born in this region. Earth's largest river, the Amazon, follows the equator eastward across the South American continent, draining the largest remaining tract of tropical rainforest and discharging nearly 20% of all continental freshwater runoff into the western Atlantic ocean (Table 11.1 and Fig. 11.1). In Africa, the Congo river collects a full one-third of that continent's runoff and delivers it into the eastern Atlantic ocean. Across Asia,

Table 11.1. Major rivers of the tropics.

River basin	Basin area (×10^6 km^2)[a]	Average discharge (km^3/year)[b]	Countries traversed[a]
Amazon	6144	6300	Brazil, Peru, Bolivia, Colombia, Ecuador, Venezuela, Guyana
Congo	3807	1250	Congo, Central African Republic, Angola, Zambia, Tanzania, Cameroon, Burundi, Rwanda
Orinoco	954	1100	Venezuela, Colombia
Mekong	806	470	China, Laos, Myanmar, Thailand, Cambodia, Vietnam
Magdalena	264	237	Colombia
Zambezi	1332	223	Zambia, Angola, Zimbabwe, Tanzania, Mozambique, Malawi, Botswana, Namibia
Niger	2262	192	Mali, Nigeria, Niger, Algeria, Guinea, Chad, Cameroon, Burkina Faso, Benin, Côte d'Ivoire, Chad
São Francisco	618	97	Brazil
Nile	3255	30	Sudan, Ethiopia, Egypt, Uganda, Burundi, Tanzania, Kenya, Congo, Rwanda, Eritrea
Fly	61[b]	77	Papua New Guinea
Purari	31[b]	77	Papua New Guinea

Sources: [a]World Resources (1998); [b]Milliman and Meade (1983).

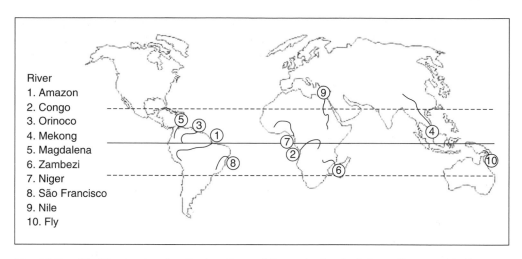

River
1. Amazon
2. Congo
3. Orinoco
4. Mekong
5. Magdalena
6. Zambezi
7. Niger
8. São Francisco
9. Nile
10. Fly

Fig. 11.1. World map showing the locations of the major tropical rivers discussed in the chapter.

where no continental landmass lies on the equator, the countless smaller rivers of the region's island nations exhibit the highest water yields on the planet, and deliver these waters and entrained sediments to the Indian and western Pacific oceans.

In contrast to the humid landscapes through which these rivers flow, other large tropical rivers such as the Zambezi, Niger, Orinoco and São Francisco pass through more arid landscapes characterized by vast savannahs or even deserts. As a special case, the Nile river flows as a singular source of water through one of the most arid parts of the planet, making possible all life within its valley. The Mekong river is interesting in that it presents the reverse case of the Nile, originating on the arid Mongolian Plateau and descending into the humid landscape of Indochina. In all, the rivers listed in Table 11.1 account for 32% of total continental freshwater runoff.

The diversity of life and other natural features in tropical river basins is truly remarkable. While the rainforests of Amazonia, the Congo and South-East Asia occupy little more than 7% of Earth's land surface, they are home to more than half of its species (Wilson, 1988). This phenomenal biodiversity extends into the rivers as well. In the Amazon, approximately 1700 fish species have been described so far, and Goulding *et al.* (1996) estimate that the total fish species diversity will approach 3000. The Amazon river also contains many curiously unique species such as the pink dolphin (*Inia geoffrensis*), piranha (Subfamily Serrasalminae) and giant river turtle (*Podocnemis expansa*).

Over the coming decades, more than 90% of population growth is expected to occur in the tropics. Currently, the human populations occupying tropical river basins vary greatly in density and the extent to which they have altered the river basins (Table 11.2). Population densities are highest in the Magdalena and Zambezi and lowest in the Amazon. Population densities in all of these river basins, however, are low when compared with major European basins such as the Rhine

Table 11.2. Population, land use and environmental conditions in selected large tropical rivers. (From World Resources, 1998.)

River basin	Population density (/km^2)	Land use Cropland (%)	Land use Forest (%)	Land use Developed (%)	Original forest loss (%)	Arid area (%)	Wetlands area (%)	Protected area (%)
Amazon	4.3	14.9	72.9	0.5	13.2	4.0	8.3	7.0
Congo	14.5	7.8	43.4	–	45.9	0.0	9.0	4.7
Orinoco	13.1	18.8	49.6	2.8	22.5	8.5	15.3	23.7
Mekong	77.6	37.9	41.5	2.2	69.2	0.0	8.7	5.4
Magdalena	78.8	38.6	36.6	10.0	87.5	7.2	0.2	4.0
Zambezi	17.7	20.4	4.1	–	43.1	8.7	7.6	7.7
Niger	31.2	5.0	0.3	1.1	95.9	65.4	4.1	4.9
São Francisco	17.6	61.4	0.8	2.4	64.4	32.0	9.7	0.5
Nile	42.7	10.3	2.0	1.1	92.1	67.4	6.1	4.4

(304 people/km^2) and the Danube (103 people/km^2). The Nile and Niger basins have seen the greatest reduction in their original forest cover, but all basins with the possible exception of the Amazon suffer from severe deforestation. Of course the Amazon basin is currently undergoing rapid deforestation, and its percentage forest loss is expected to increase incrementally over the coming years. Percentage cropland is relatively low in each catchment, with the exception of the São Francisco basin. The Magdalena is the most developed from an urban and industrial standpoint.

A common denominator in nearly all tropical river basins is poverty. From the standpoint of water resources management, poverty may be seen as a crippling obstacle in tropical basins. The average per capita gross national product of all countries listed in Table 11.1 is US$837. Arranged according to continent, average per capita gross national products are US$1892 in South America, US$569 in Africa and US$897 in South-East Asia (World Resources, 1998). Significant national investments in water resources have therefore been minimal historically, and they are likely to remain low for the foreseeable future. Nearly all capital for major water works projects has come from international financial aid, which is now at a 50-year low. At least 18 of the countries making up the major tropical river basins are classified as 'severely indebted' by the World Bank (1999) and another five are classified as 'moderately indebted'.

A second common denominator between these tropical basins is a direct and generally unbuffered dependence of the local populations on water and other aquatic resources from the river systems. Although data are not available basin by basin, local people take their drinking water from nearby streams and rivers and consume it with little or no form of treatment. Similarly, human and other wastes are commonly disposed of in rivers without treatment. The effects of poor waste-disposal practices and drinking untreated water are reflected in health statistics from countries of the region. In 16 of the countries for which data are available, 21% of rural children below 5 years of age suffer from chronic diarrhoea (World Resources, 1998). Individual country data range from 13% in Brazil and Egypt to 29% in Kenya. Of the countries considered, Egypt and Brazil are the two in which the greatest percentage of rural households have piped water, 50% and 13%, respectively. On average, only 8% of rural households in the remaining countries have piped water. Cases of children's diarrhoea are not much better in urban parts of these countries (average 19%). In fact, in Bolivia and Côte d'Ivoire there are more documented cases of diarrhoea among urban children than rural children. Inadequate systems for human waste disposal explain much of this poor urban health. Only 32% of urban households in these countries are connected to sewer systems.

The area of large-scale water infrastructure that has received the greatest investment in tropical countries is irrigation and hydropower. Considerable international financing has gone towards the construction of dams to provide for these services. At least 15 major dams (greater than 1000 MW installed capacity) and countless small-scale projects have been constructed in the major tropical basins (Gleick, 1993). In Brazil, a full 95% of the country's electricity comes from hydropower, and the majority of the country's untapped potential lies in the Amazon. Brazil thus has ambitious plans to develop an additional 25 dams in this region. These plans are complicated, however, by the exceedingly low elevation and flat relief of the Amazon. The Balbina Dam, for example, forms a reservoir covering 2300 km^2 but generating only 250 MW of electricity (Smith *et al.*, 1995). This generation efficiency of 2 kW/ha of reservoir is extremely low and means that large areas of rainforest must be flooded to generate modest amounts of electricity. During the construction of the Tucurui Dam in the Amazon, between 20,000 and 30,000 people had to be relocated. With respect to water for irrigation, reservoirs built in tropical river basins generally service only small percentages of land under cultivation. With the exception of Egypt, where 100% of cultivated land is irrigated, other tropical African nations generally apply irrigation waters to less than 5% of their cultivated lands. Although slightly higher, the same may be said for percentages of cultivated land irrigated in South America. In Indochina, between 4% and 52% of cultivated land is irrigated. Thus, water subject to expensive and engineering-intensive management schemes in tropical basins may be said to play a generally small role in overall regional water use.

The quality of water in large tropical river basins is not well known. Dismal health statistics from throughout the region suggest that water quality is poor in most urban and many rural locations. According to the findings of the First Assessment of Global Freshwater Quality published by the World Health Organization and United Nations Environment Programme (Meybeck *et al.*, 1989), organic matter in the form of domestic sewage, municipal wastes and effluents from agricultural industries makes up the most ubiquitous source of water pollution. Organic pollution is made worse by the presence of pathogens. Indeed, in the Amazon basin, alarmingly high bacterial concentrations occur in the La Paz river as it exits the city of La Paz, Bolivia, and extending more than 40 km downstream (Ohno *et al.*, 1997). This bacterial contamination has been linked to a high incidence of diarrhoeal disease in the city of La Paz. The Oropouche virus has been detected near the city of Iquitos in the lowland Amazon of Peru and was found to be transmitted to the population of the city and surrounding areas (Watts *et al.*, 1997). High levels of organic pollutants in tributaries of the Niger river are attributed to runoff from agricultural fields and animal lots (Nwokedi and Obodo, 1993).

Throughout the tropics, inappropriate land management practices and over-exploitation contribute to increased erosion and elevated sediment loads in streams and rivers. The extent of degradation of water quality from erosion is variable and not well known. Large sections of land within the São Francisco, Niger and Nile river basins have been classified as being at very high risk of erosion and desertification. Parts of these basins already show evidence of severe erosion (Seager, 1990). In the more humid basins of the tropics, widespread erosion is less prevalent and generally not a problem so long as an adequate litter and vegetation layer stabilizes the soil (Bruijnzeel, 1990). There are, however, more localized occurrences of severe erosion when the protective layer is absent and when extreme precipitation events occur. The consequences of additional erosion include siltation of reservoirs, possible damage to other engineering infrastructures, smothering of aquatic habitats, and diminished water quality for drinking and other domestic uses. Sediments may also have associated chemical contaminants, which when dissolved or desorbed further degrade water quality.

Contaminants such as pesticides, petroleum products and heavy metals also pollute tropical river basins, but generally in a more localized form. Mining activities in the Papuan highlands have contaminated sediments of the Fly river with copper, lead and zinc (Baker and Harris, 1991). Soltan *et al.* (1996) also reported high levels of lead in sediments above the High Aswan dam on the Nile river. This lead is largely attributed the gasoline spills from the 200 ships and 300 boats operating on that section of the river. Oil spills have been reported in oil-producing areas throughout the tropics and, although there are have undoubtedly been several disastrous spills, there are few quantitative data on which to quantify the impacts of these spills. Pesticide use is generally low throughout the large tropical river basins, as is the use of artificial fertilizers. This is especially true in Africa and South America, and less so in Asia. The explanation for low agrochemical use is largely economical, as poverty within these nations limits the purchase of these products.

Based on the preceding paragraphs, the general condition of large tropical river basins may be summarized as follows. Climatically, geologically and biologically they are extremely diverse, ranging from high-yield rivers draining young terrain covered by dense tropical rainforest to low-yield rivers draining ancient terrain covered by savannahs or deserts. However, they share a common set of problems, which are linked to poverty and a dependence on outside capital to finance large engineering operations. They are also similar in that they contain large rural populations in intimate contact with river water and whose numbers are expected to grow exponentially over the coming decades. In general, water management schemes are simple and controlled by the end users (not large water supply and waste treatment infrastructures). The most prolific contamination problems take the form excess organic waste, pathogens and sediment. These seemingly low-grade problems are accentuated by a lack of proper water treatment prior to domestic consumption. Occasional more severe contamination occurs from pesticides, heavy metals and petroleum wastes in areas of industrial activity and hydrocarbon exploration. Expensive and highly engineered solutions to water resource problems are not practical today, and are unlikely to become practical for several decades.

Ecohydrology and Riverine Ecosystem Resistance and Resilience

Riverine ecosystems possess an intrinsic capacity to regulate flows and quality of water passing through them. This capacity stems from finely tuned ecohydrological interactions between organisms, water and the river's material load. Interactions are biological, physical or chemical in nature, and in concert they lend resistance, resilience and adaptability to riverine ecosystems. Consequently, riverine ecosystems are able to withstand frequent low-magnitude disturbances (i.e. storms, small landslides), recover from less frequent high-magnitude disturbances (i.e. hurricanes, major landslides), and constantly adapt to changing short-term and long-term environmental variables. These disturbance regimes, and the river's response to them, are natural components of healthy riverine ecosystems (Naiman *et al.*, 1992).

In general, riparian zones and wetlands are the most important natural components of riverine ecosystems regulating the quantity and quality of water entering from uplands. Together they function as effective buffers against extreme flooding and excess sediment loads linked to runoff from the landscape. In developed areas, they are also effective buffers against excess erosion, nutrient inputs and contaminant runoff from agricultural fields, pastures and residential areas (Haycock *et al.*, 1997). Riparian zones and wetlands are most effective in the headwater portions of watersheds, where flow is distributed among a large number of smaller streams. In downstream portions of rivers and streams, water quality and fluxes are regulated by riverine wetlands and flood plains. These drainage-system components also dampen extreme runoff events and effectively strip sediments and excess nutrients from river water. A final ecosystem component which is less well understood but potentially quite important is the hyporheic zone underlying and adjoining river and stream channels (Bencala, 1993). River water maintains a direct hydraulic connection to pore waters of the sediments composing the river bed. Depending on the physical properties of the bed sediments and the geometry of the river channel, river water may repeatedly move from the river channel to underlying sediments in a coupled form of downstream flow.

The following sections provide an overview of the ecohydrological processes operating in each of these ecosystem components. Emphasis is placed on the buffering capacities of these processes and ultimately to their potential role as tools of water resource management in tropical landscapes. The great majority of material presented here comes from outside the tropics due to a lack of tropical data. In a separate section it is integrated with a specific example from the tropics.

Riparian zone processes

Riparian zones border streams, rivers, lakes and wetlands and are characterized by plant species associations that are more tolerant of a shallow water table and frequent flooding. Although their geomorphological setting may vary greatly, they commonly occur in valley bottoms and are thus more level than surrounding hill slopes. Their soils are commonly waterlogged and anoxic, creating a chemical environment that is distinct from the drier, more aerated upland soils. Frequent disturbances within the riparian zone also tend to select for fast-growing plant species. Across the landscape, riparian zones act as natural barriers between aquatic environments and terrestrial uplands. In this setting, their qualities of low relief topography, anoxic soil columns and fast-growing species make riparian zones effective traps for particulate and dissolved material moving from terrestrial to aquatic environments.

Riparian zones are capable of removing sediment, nutrients, acidity and toxic chemicals from runoff waters. These removals result from processes of sedimentation/trapping, binding/sorption, plant uptake and denitrification/organic breakdown. Each of these processes is commonly accentuated in riparian zones. Forested and grass riparian zones in the south-eastern coastal zone of the USA have been shown to remove 80–90% of sediment in runoff from adjoining croplands (Cooper et al., 1987; Gilliam, 1994). Here, riparian vegetation slowed the flow of runoff waters, thereby decreasing energy levels and promoting sedimentation. Riparian vegetation also promotes increased infiltration, which captures runoff waters and retains sediments. Riparian zones exhibit a marked capability to remove excess nitrate from runoff waters. Even in agricultural settings where groundwater nitrate levels may exceed $700 \mu M$, riparian zones have been shown to reduce nitrate concentrations by as much as 90% (Jacobs and Gilliam, 1985; Haycock and Pinay, 1993). Nitrate is removed through a combination of plant uptake and microbe-mediated denitrification. Both mechanisms are effective means of cleansing runoff waters, but denitrification is more useful in that it completely expels the excess nitrate from the ecosystem as N_2 gas. Nitrate taken up by plants is only temporarily stored in plant tissue and may be released again to soils and groundwater. Plants may facilitate the denitrification process, however, by taking up nitrate from deeper groundwater and delivering it to surface soils via litterfall (Hanson et al., 1994). Denitrification potentials tend to be greater in surface soils because of abundant organic carbon, which acts as the electron donor in the reduction reaction.

The retention of phosphate in riparian zones is much less predictable than retention of nitrate (Uusi-Kämppä et al., 1997). Phosphate concentrations are regulated by plant uptake and partitioning reactions with soil mineral surfaces and are only indirectly impacted by oxidation–reduction reactions. Under most natural conditions in upland areas, subsurface waters are oxic and phosphate is relatively immobile. In riparian zones anoxic conditions frequently prevail. Anoxic conditions favour the reduction and dissolution of iron oxides and consequently the liberation of sorbed phosphate. Hence, under natural conditions riparian zones may act as a source for soluble phosphorus. Several investigators have shown, however, that riparian zones may strip phosphate from contaminated runoff waters. Madison *et al.* (1992) documented a 90–99% reduction in phosphate concentrations in runoff waters passing through a grass riparian buffer strip during simulated storms. Dillaha *et al.* (1989) similarly documented a 69–83% reduction in phosphate concentrations in agricultural runoff from a Virginia catchment. With regard to particulate phosphorus associated with sediments, retention is proportional to the quantity of sediments retained and the concentration of phosphorus in those sediments. As indicated previously, riparian zones are effective at retaining sediments and thus effective at retaining particulate phosphorus.

Pesticides washed from agricultural fields are a growing problem in aquatic ecosystems. Little research has been completed to date on the effectiveness of riparian zones in retaining pesticides, but early data from France provide some encouragement. Gril *et al.* (1997) reported that grass buffer strips reduced concentrations of lindane by 72–100%, atrazine by 44–100%, isoproturon-IPU by 75–99% and diflufenican-DFF by 68–97%. Pesticides are stripped from runoff waters via adsorption on to soil particles. The pesticides can then be broken down by soil microbes to eliminate them completely from the system. Riparian zones assist in this process by slowing the flow rate of runoff waters and allowing additional time for sorption reactions to take place. The effectiveness of riparian zones at retaining other contaminants such as organic wastes and pathogens is still very unclear (Coyne *et al.*, 1995).

Wetland and flood plain processes

Wetlands are widely distributed in river basins and may or may not be in direct hydraulic connection to the surficial drainage system. This discussion is only concerned with wetlands that have some superficial hydraulic connection with the river system, and the term 'wetland' is used to refer to those areas that are saturated on an inter-annual time scale. The term 'flood plains' refers to depositional plains running along the margin of river and stream channels, which are flooded sporadically and most often seasonally. Each of these systems possesses ecohydrological conditions that are distinct from the river and stream channels, and each exhibits a certain capacity for buffering river and stream water against the ill effects of upstream contamination and flood events.

Small wetlands in the headwater portions of watersheds act as buffers against contaminated runoff in a manner much like that of riparian forests and grass strips. In fact, small fringing wetlands are commonly integrated into riparian zones and interact closely with riparian vegetation, soils and groundwaters. These wetlands collect fast-moving runoff waters and facilitate the deposition of entrained sediments. They also harbour aquatic vegetation and macrophytes that take up excess nutrients from runoff waters. While deposition of sediment has potential for long-term storage, nutrient uptake by vegetation is generally of short-term duration (Reddy *et al.*, 1999), especially when taken up by rapidly turning over species. Longer-term storage occurs when nutrients are taken up into wetland trees. Mitsch *et al.* (1979) reported, however, that phosphorus uptake by cypress trees amounted to only 10% of the total phosphorus uptake in an Illinois wetland. Hence it is likely that in other wetlands too only a small portion of nutrients are taken up into long-term storage pools. Headwater wetlands may also develop pockets of anoxia that support nitrate removal via denitrification (Groffman, 1994). On a river basin scale, Whigham *et al.* (1988) have suggested that sediment and nutrient retention in headwater areas should be proportional to the area covered by wetlands, but conclusive data to support this suggestion are still lacking.

Sediments and dissolved contaminants that penetrate or bypass headwater buffer zones move downstream and into larger sections of the river system. During high water phases of the river's hydrograph, some fraction of river water may move laterally out of the channel and into flood plains and riverine wetlands. Sediment and contaminant concentrations often increase during the early rising limb of river hydrographs. Thus it is likely that flood plains will receive significant pulses of these materials when they are initially inundated. The efficiency with which flood plains and riverine wetlands act as buffers depends on a number of hydraulic and biological variables. As with other ecosystem components presented in this section, flood plains and riverine wetlands slow the flow rate of river waters and thus decrease its sediment-carrying capacity and increase its interaction time with biota and sediment/soil surfaces. Sediment deposition is generally the most important long-term mechanism for retaining nutrients in river systems. In 17 riverine wetlands, Johnston (1991) reported rates of particulate nitrogen and phosphorus retention as high as 15 $g/m^2/year$ and 1.5 $g/m^2/year$, respectively. In a riverine wetland in Wisconsin, Klopatek (1978) reported nitrogen and phosphorus uptake rates by vegetation of 20.8 $g/m^2/year$ and 5.3 $g/m^2/year$, respectively. However, 74% and 62% of this nitrogen was returned to the water column through leaching and litterfall. An even larger percentage may have been returned through longer-term decay processes.

In-channel and hyporheic processes

In addition to buffer zones upstream and adjacent to river channels, rivers possess a certain capacity within their own channels to mitigate levels of sediments and contaminants that have come from upstream areas. The processes are similar to those discussed previously and include sediment deposition in areas of slow-moving water, plant uptake of excess nutrients, and sorption–decomposition of organic wastes. Exchange with the hyporheic zone is suggested to increase with increasing river size (Hill, 1997), provided geomorphological features (i.e. permeability and river bed topography) of the river are such that exchange is facilitated. Within the river bed sediments and hyporheic zone, denitrification and other anaerobic processes may be important.

Again the bulk of our current understanding relates to excess nutrient levels. In-channel aquatic vegetation has been reported to take up between 40 and 75% of nitrate input from upstream areas (Cooper and Cooke, 1984; Jansson *et al.*, 1994), but of course much of this nitrogen is later returned to the aquatic system. Periphyton and microorganisms can take up significant amounts of phosphorus (Newbold *et al.*, 1983), but rapid turnover rates in these biological pools leads to only short-term storage in any single location. With all of these biological pools, however, only a small percentage of organic material enters refractory pools and is stored long-term (Gachter and Meyer, 1993).

Whole-stream experimental methods applied in the south-eastern USA have demonstrated that phosphate uptake is significantly greater in streams with greater hyporheic flow (Mulholland *et al.*, 1997). Two streams investigated were similar in many physical characteristics but differed in their ratios of surface to subsurface water volumes. One stream contained a hyporheic zone 1.5 times the volume of the stream channel, while the other contained a hyporheic zone only 0.1 times the volume of the stream channel. Consequently, water residence times were considerably higher in the stream containing greater hyporheic flow. This greater residence time is thought to be the main factor facilitating greater mass uptake rates of phosphate. Phosphate was taken up in the hyporheic-influenced stream at a rate nearly twice that of the more channelized stream (Mulholland *et al.*, 1997). Hyporheic zone processes have been shown to be quite variable in terms of nitrogen retention. Studies by Triska *et al.* (1989) and Jones *et al.* (1995) have reported a net increase in stream nitrate concentrations as a result of hyporheic processes. While it is likely that denitrification consumed some fraction of nitrate in the subsurface of these systems, this loss was more than compensated for by subsequent nitrification of ammonium liberated from decomposing organic matter.

In a review of in-channel and hyporheic retention of nutrients in lowland agricultural streams, Hill (1997) cited examples of nitrate removal that ranged from 1 to 68% of annual nitrate loading. Summer low flow retention reached as high as 76% in one Canadian stream. Hill indicated that water residence time in the stream reach was the single most important factor in determining efficiency of nitrate retention. In the same review, Hill (1997) cited examples of phosphorus removal that ranged from 20 to 92% during summer low flow. Annual removal efficiencies were presented for only one study, and those ranged from 56–59% of total phosphate loading to <5% of total phosphorus loading. Because phosphorus is generally most abundant in particulate form, sediment entrainment during occasional storm events may quickly erase gains in phosphorus retention made during low water periods. Hill (1997) indicated, however, that both nitrogen and phosphorus retention increase with increasing water temperature.

The Amazon Basin

Studies of buffering processes in riparian zones, wetlands and river channels of the tropics are exceedingly rare. While limited, investigations in the Amazon basin of Brazil may provide the most comprehensive understanding currently available on buffering processes within a large tropical river basin. Research in the Amazon has touched on the effectiveness of riparian buffers, flood plain processes and in-channel microbial activities.

In a small catchment from the central Amazon basin, Williams *et al.* (1997) evaluated the effects of deforestation on a nearby stream. They found that NO_3^-, NH_4^+, Na^+, K^+, Ca^{2+}, Mg^{2+}, Cl^-, SO_4^{2-}, Al, Fe, Mn and dissolved organic carbon all increased in concentration in soil water filtering through a 'cut and burnt' experimental plot. They further found that overland flow had elevated concentrations of these same ions as well as PO_4^{3-}, total dissolved phosphorus, total dissolved nitrogen, dissolved inorganic carbon and dissolved silicon. Concentrations of several of these ions were observed to increase in stream water as well, but the authors noted a significant reduction in sediment loads and NO_3^- as surface and ground waters passed through the vegetated riparian zone. In fact, sediment was completely retained in the riparian zone, and it was only in areas without riparian buffers that stream sediment concentrations increased (Williams and Melack, 1997). Thus the maintenance of riparian buffer strips along rivers and streams of the region should provide an effective shield against excess sediment and nutrient runoff from agricultural fields and pastures. As these are the dominant forms of land use in the region, riparian buffer strips will serve as the most important regional tool for surface water control.

In the Amazon basin more than 90% of all sediment transported by the river originates in the Andes mountain range. Approximately 1400 Mt/year enters the river's mainstem, from which about 200 Mt/year, or 14%, are deposited on the river's flood plain and within its channel (Dunne et al., 1998). In the river water entering the flood plain, greater than 90% of suspended sediment was deposited, leading to increased water clarity and increased phytoplankton growth (Engle and Melack, 1993) (Fig. 11.2). Dramatic decreases in dissolved nutrients were also observed as river water moved across the flood plain. In addition to cleansing river flood waters, these flood plains provide habitat and food for many of the most important commercial and non-commercial fish species in the Amazon basin (Goulding et al., 1996). They are thus essential centres of protein production for the region's inhabitants.

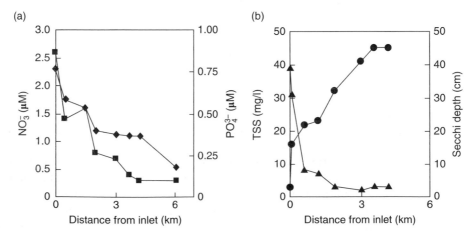

Fig. 11.2. Plots illustrating the deceasing levels of (a) nutrients (■, NO_3^-; ◆, PO_4^{3-}) and (b) sediments (▲, total suspended solids (TSS)) in river waters passing through a flood plain lake of the Amazon river. Increasing water clarity reported as Secchi depth (●) is also shown (b). (Adapted from Engle and Melack, 1993.)

In the mainstem river channel, available organic matter limits bacterial metabolism. Amon and Benner (1996) found microbial respiration rates to increase by a factor of 3, rising from 0.49 to 1.76 μM O_2/h, when glucose was added to the sample. Integrated across the width and depth of the river channel, these authors calculated a current consumption rate of 13.0 mmol C/m^2/h. Under conditions of greater carbon availability, this rate could rise to near 40 mmol C/m^2/h. Assuming an untapped consumption rate of 27 mmol C/m^2/h over the 2-week travel time of water along the Amazon mainstem, the river could potentially process an additional 200 grams of organic matter per square metre of river. Given the extremely large size of the mainstem river (3–5 km wide), this rate translates to nearly 1 Mt of organic matter consumed per metre of river length. Thus the Amazon mainstem has an enormous potential for processing organic wastes discharged into it, but it is not advisable to rely on this internal cleansing potential to control organic pollution.

Integrating Ecohydrological Principles into Tropical River Management

The suite of ecosystem components and processes described in this chapter constitute the most basic and intrinsic mechanisms for maintaining the environmental integrity of river systems. Riparian zones, wetlands, flood plains and in-channel/hyporheic zones effectively regulate water fluxes and quality, thereby maintaining appropriate habitat for aquatic organisms, supplying consistent quantities of clean water for human needs, and delivering needed nutrients and organic matter to unpolluted estuary zones. While intact, these ecosystem components significantly increase the river's tolerance of both natural and anthropogenic disturbances. These ecosystem components might be viewed as the river's immune system – its 'intrinsic purification systems' (IPSs).

But how can these natural riverine components be integrated into official water management actions and policies, and what function might they realistically serve in river management plans? To some extent IPSs have already been recognized in international agreements. The 1992 International Conference on Water and the Environment, in Dublin, Ireland, made clear references to the importance of land–water linkages and 'ecosystem integrity' in achieving water resource management goals. These references were amplified and further articulated in subsequent conferences such as the 1992 United Nations Conference on Environment and Development (the Earth Summit) in Rio de Janeiro, the 1994 Summit of the Americas in Miami, the 1995 Pan-American Conference on Health and Environment in Sustainable Development in Washington, DC and the 1996 Summit Conference on Sustainable Development in Santa Cruz, Bolivia.

In the midst of this new global focus on more holistic approaches to water management, the World Bank articulated its own new approach to guide lending for water resource projects (World Bank, 1993). Recognizing past failures brought by an over-centralized approach and insufficient account of environmental concerns, the Bank's new policy called for the preparation of comprehensive national strategies that incorporate policy reforms, institutional adaptation, capacity building, and environmental protection and restoration. Explicit attention was given in the policy to:

> incentives and programs. . .to improve land management practices and to restore, then protect, environmental resources in floodplains and wetlands [with the aim of] reducing pollution, soil erosion, waterlogging, and flood runoff. . .with non-structural measures that are less costly, yet no less effective. . . (p. 61)

The plan also referred to wetlands and riverine flood plains as:

> biophysical filters [that] safeguard biological diversity and conserve water resources (pp. 74–75)

Ecuador is one of many countries that have worked directly with the Bank to develop a national strategy that is in accordance with the new guidelines (CNRH, 1998). It is important to note that both the Bank policy and resulting national strategies call for a national prioritization of problems within each country and a global prioritization of the countries and regions with the most urgent needs across the globe.

While the importance of protecting riparian zones, wetlands and flood plains is now routinely acknowledged in the text of national and international water policies, the role of IPSs could be expanded such that they are employed explicitly as tools in water resource management. In urban and industrial areas, IPSs should be viewed as a crucial set of controls to fortify nearby aquatic systems and increase the likelihood that other, engineering-based controls will achieve water quality goals. Amplifying the chances that more technically complicated management activities will succeed is a wise strategy, which is why it is the guiding principle within the Ecohydrology Programme of the United Nations Educational, Scientific and Cultural Organization (Zalewski et al., 1997). In rural areas, IPSs may serve as a main means of controlling surface water quality in conjunction with simple drinking water treatment and waste treatment systems (e.g. solar latrines). Other simple engineering systems in rural areas may be designed to interact with IPSs. For example, sewage from small communities can be piped into wetlands designated for the purpose of cleaning the wastewater prior to discharge to an adjoining river. Of course, in this example the value of the wetland as an IPS must be weighed against other services (e.g. fish nursery habitat) that may be lost as a result of sewage inputs.

It is beyond the scope of this chapter to propose a detailed plan for the incorporation of IPSs into river management plans in the tropics. Specific plans are likely to vary greatly in their details, just as environmental, social and economic conditions vary between the individual river basins. A few guidelines that may be widely applicable are explained here:

1. IPSs should be explicitly identified in water policies and plans as tools to be used towards maintaining water quality and reducing flood hazards. Wherever possible, the services of IPSs should be quantified to justify, from an economic standpoint, the preservation of these ecosystem components. Other services of intact IPSs should also be considered in the calculation of economic value. These may include nursery habitat for juvenile commercial fish, habitat for plants and animals that are harvested in the wild, and touristic value. Other less tangible values such as non-commercial biodiversity and ecosystem integrity should also be considered.

2.   IPSs should not be viewed as independent and stand-alone solutions to water quality management problems. They should be viewed as a fundamental component of more multifaceted solutions, which should include additional systems to treat water for domestic supply and treat wastes prior to disposal. Wherever possible, the varied components should be interconnected to form a continuum of water quality management solutions. In urban and industrial situations, IPSs will play a smaller, but still important, role in management activities, while in rural situations IPSs may play a major role, even serving as the sole management tool outside individual supply and waste treatment systems in households.

3.   Natural riparian zones, wetlands and flood plains are the most economical and best adapted tools to employ in water quality management plans. They need not be purchased and they are already perfectly placed in the natural hydrological pathways of the basin. Natural IPSs also provide additional services and hold greater intangible values, as listed in item 1 above. However, in basins that have already been subjected to widespread anthropogenic disturbance, it may be necessary to rehabilitate IPSs and perhaps to even construct artificial buffer systems. Artificial buffers, when properly managed, may also be used in multiple ways such as recreation, low-intensity agroforestry in the case of riparian zones, and aquaculture in the case of wetlands and flood plains. In situations where contaminant loads are unusually high, such as near mines or petroleum exploration sites, it may also be prudent to construct artificial buffer systems to enhance the capabilities of IPSs.

4.   Special attention should be given to the maintenance of IPSs in headwater portions of tropical river basins. This is especially true in humid areas of high relief that are likely to be sensitive to flash flooding and mass wasting.

5.   The effective use of riparian zones, wetlands and flood plains in rural water quality management plans will require the complete and voluntary participation of local communities and landholders. Local people should play a leading role in developing plans to employ IPSs, such that they share a sense of ownership of the plan. This will not only increase the likelihood of compliance among the population; it will also foster a mechanism for community-driven enforcement when individuals do not comply. Long-term community participation will perhaps be best accomplished through the formation of basin committees composed of respected local community elders, leaders and normal individuals.

6.   IPSs should be incorporated into formal management plans, but ideally they will also be incorporated into cultural norms of commonsense resource management. In rural situations they are tools that nearly every landowner can make use of with no need for outside investment or expertise. If the utility of IPSs can be instilled into cultural norms, IPSs are likely to be employed in water resource management with or without formal management plans. In order to bring about this change in cultural perception, the multiple services of IPSs must be demonstrated to local communities through education programmes and example projects. Financing of these programmes in the tropics will likely require a degree of initial external financial support, but the financial benefits from protected water quality, diminished health problems and sustained aquatic food production would far outweigh the initial investment. This would serve as a perfect example of the power of education to bring about long-term and self-sustained financial and societal benefits.

Concluding Remarks

Abundant investigative evidence from ecohydrological and other research demonstrates that riparian zones, wetlands and flood plains act as efficient 'intrinsic purification systems' in aquatic environments. Although data are lacking in the tropics, it is apparent that these IPSs also provide essential services to natural tropical ecosystems and have potential for mitigating contamination from human activities. In the impoverished river basins of the tropics, IPSs have special importance in controlling surface water quality given that more costly, engineering-based controls are not economically feasible today, nor are likely to be feasible on a wide scale for the foreseeable future. As the World Bank has recommended, water problems should be prioritized within, and among, the nations of the developing world, and the bulk of international financial aid should go towards solving the most urgent problems. On a much wider scale, however, the water quality needs of countries within the large tropical basins will be best served by conserving and rehabilitating IPSs and explicitly incorporating these systems into water management plans that also call for appropriate technologies of treatment for water supply and liquid wastes. The time to implement these regional actions is now.

On a longer timescale, the effective use of IPSs in water quality management requires that we develop a more thorough and quantitative understanding of how these systems function, what their real capacities are for processing a wide range of contaminants, and their tolerances and limitations. We must develop techniques to better integrate these natural features into human resource systems. At the same time, however, we must understand how and to what extent their use as water quality buffers impinges upon the other services they provide to ecosystems, such as critical habitat to terrestrial and aquatic organisms and corridors for animal migrations. Thus, there is a profound need for continued research which is applied and objective-driven. This need for research should in no way diminish the need for quick action to preserve and rehabilitate IPSs today.

Water problems facing the Earth's great tropical river basins are multifaceted and complicated in ways not well understood in the temperate zones. Creative and region-specific solutions must be developed, and these solutions must, even more than in the North, involve local people. One benefit, however, of the slow pace of development in the tropics is that tropical river basins tend to preserve more of their original ecosystem components than large rivers in the temperate zones. As one aspect of the move towards creative solutions in this region, the intrinsic water purification capabilities of these rivers should be recognized and more importantly put into service.

12 Ecohydrological Analysis of Tropical River Basin Development Schemes in Africa

N. Pacini[1] AND D.M. Harper[2]

[1]Department of Ecology, University of Calabria, Arcavacata di Rende (Cosenza), Italy; [2]Department of Biology, University of Leicester, Leicester, UK

Introduction: The Scope of Ecohydrology within the Context of Tropical Catchments

Processes influencing ecological change in river basins differ significantly across continents as a result of differences in the natural resource base, geomorphological history and climatic patterns, in varying population densities, and related resource demands, and in catchment land uses and socio-economic trends. In this context, ecohydrology serves the purpose of integrating the physical and ecological setting and of highlighting the critical properties of river basins that are relevant to solve potential conflicts in multiple resource-use scenarios. As stressed in the previous chapter, ecohydrology acquires a specific relevance when applied to tropical river basins. An analysis of African river basins in ecohydrological terms requires the recognition of the geological, ecological and socio-political setting at the continental scale and within the main ecoregions that are found within the African continent.

In most low-income food-deficient (LIFD) countries, water is – at least seasonally – a scarce resource. Competition for water resources between stronger economic sectors tends to affect catchment productivity in terms of pastures, wildlife and fish and to undermine water quality; water stress is further magnified in arid and semi-arid lands where seasonal and inter-annual variations in rainfall are great. Fisheries in particular represent a valuable, often 'hidden' resource, the management of which requires the identification of the cumulative influence of physical, biotic and socio-economic components. Referring mainly to LIFD countries, the Food and Agriculture Organization of the United Nations maintains that keeping an open and fair competition for water resources among potential users represents a major challenge (FAO, 1999). Aquaculture and inland fisheries often fail to preserve fish stocks and high production levels due to unfair competition from subsidized sectors of the economy such as industry and agriculture. During the development of damming schemes, for example, too little weight is accorded to fishery stakeholders (Miranda, 2002). Developing countries with a rapidly expanding industrial sector, such as India, South Africa or Brazil, face similar challenges of combining economic growth with the protection of some of their most valuable natural assets. In these cases as well as in the case of LIFD countries, subsidies could become valid management instruments if directed to sustain policy interventions which, while apparently suboptimal under a short-term economic perspective, respond to environmental concerns and preserve the long-term sustainability of natural assets. The protection of extant fishery resources generally works out as the most beneficial form of management in terms of preservation of the wider ecological functions of riverine and flood plain habitats, as this is closely dependent on their overall environmental quality and resource renewal capacity.

The ecohydrological principle of a strong integration of cause and effects throughout the entire catchment implies that no single area should be managed independently. While the role of a basin management authority is to oversee the hierarchical distribution of responsibilities, it should be stressed that local-level management is important and necessary to maximize total harvest, ensure the reproducibility of renewable natural assets and reduce cumulative impacts.

Although it may appear self-intuitive, the harmonization of local versus catchment issues should be not be taken for granted; it often requires conflict resolution and the difficult setting of a hierarchy of priorities to be accepted by all stakeholders. Under multiple resource use, it may often be difficult to determine which sector is actually causing a given impact. Contrasts in tropical river basins emerge typically between upstream and downstream as well as between riparian and dry-range users, as illustrated in the case studies of the Senegal and Tana rivers (see below). The analysis of different resource-use options and the different scales of impact that need to be taken into consideration fall within the interest and scope of the implementation of the ecoregional approach.

Integrated river basin management tends to fail in developing countries in particular, where water-related conflicts emerge as a consequence of scarce water availability and high population densities, and local issues prevail over catchment perspectives; where river basins are cross-cut by ethnical, administrative or national boundaries; and where clearly defined hydrological and geographical boundaries are not evident due to landscape conformation or to the impact of large inter-basin transfer schemes.

An ecoregional approach for an ecohydrological perspective

With the purpose of establishing integrated land-use scenarios for better aquatic ecosystem management, the application of ecohydrology to river basins in the African continent recognizes the utility of the ecoregional concept. This was introduced and supported by the Consultative Group on International Agricultural

Research at the beginning of the 1990s, to enhance sustainable resource management in achieving poverty reduction, food security and environmental protection. More specifically, the ecoregional approach is considered adequate to compare resource-use potentials and actual management effects (i.e. status, pressure and response) as it is based on the integration of biophysical and socio-economic indicators.

An ecoregion can be defined as the land unit, determined by a set of biophysical and socio-economic characteristics, which is most appropriate for the analysis and resolution of a given management issue. It does not seem relevant to establish well-defined ecoregions throughout the globe unless this is done with the specific target of identifying specific environmental issues and well-defined processes. The ecoregional approach concept is particularly relevant for socio-economic scenarios characterized by a high influence of subsistence activities and a close linkage between the resource base and human livelihoods, as in most LIFD countries.

In a review of the freshwater biodiversity of Latin America and the Caribbean (LA/C), Olson *et al.* (1998) define an ecoregion as:

> a large area of water and land that contains a geographically distinct assemblage of natural communities that (a) share a large majority of their species and ecological dynamics, (b) share similar environmental conditions, (c) interact ecologically in ways that are critical for their long-term persistence.

Forty-two freshwater ecoregion complexes were identified for LA/C, within which 117 ecoregions were delineated based on biological distinctiveness (species richness, endemism, ecosystem diversity and the rarity of habitat types) and conservation status.

The crucial phase of this work stemmed from a 1995 biodiversity workshop, during which major biogeographers and natural science experts for LA/C convened to refine a preliminary ecoregional analysis generated by a World Wildlife Fund survey. The team felt that ecoregions tend to coincide with conservation management units as they define the areas that are driven by major ecosystem processes and pressures upon them. Ecoregions integrate quantitative information within the scale of governing processes and provide guidelines for river basin strategies. Within South America, ten ecoregions were attributed an outstanding priority status for their intrinsic biological value and the high level of threat. Ecoregional maps serve as a useful tool in the allocation of scarce resources for conservation management. Finally, ecoregional complexes were defined as 'larger biogeographic regions that encompass ecoregions sharing strong biogeographic affinities and large-scale ecological linkages'. Such a detailed analysis has not yet been carried out for the African continent. An interesting attempt of what could be likened to an ecoregional classification in Africa for the purpose of establishing agricultural potential is represented by the agroecological soil classification of Jaetzold and Schmidt (1983).

The ecoregional approach concept can be implemented at three main scales in ascending order:

- Ecological zones (such as agroecological zones) define the scale of local processes.

- Catchments define their cumulative impacts.
- Continental scales integrate the impacts of climate change.

While the ecohydrological approach is based on the recognition of catchments as unitary elements within the landscape, the understanding of relevant processes often requires the consideration of different scales. This is particularly true in Africa where very large rivers encompass 'ecoregions' such as Sahelian flood plains or gallery forests within a single catchment which share greater affinity with equivalent zones in other river catchments than within the same basin. The catchment remains, however, the preferred scale of operational application of ecohydrological solutions owing to the inherent unity of river ecosystems, governed by longitudinal flows such as water, sediment and nutrient transport and by natural connectivity processes, which establish catchments as single biogeographical entities.

Ecohydrological Approach at the Continental Scale

Past and present climatic conditions explain some of the main physical and biogeographical characteristics of African river basins. Africa was part of the first land mass that separated from Gondwanaland during the mid-Cretaceous, some 100 million years ago. The inter-tropical region of the continent represents a distinct and well-delimited zoogeographical entity. At a continental scale, its landform is more homogeneous than that of other tropical biomes such as South America, which is divided by the Andean chain, or South-East Asia, which includes large lowland rivers, peninsulas and islands. The early positioning of the continent about the equator means that one of the most ancient portions of the Earth's crust has been least affected by the recurrent formation and regression of glaciers. The consequences are that tropical African soils tend to be very deep, mature, leached, rich in metal oxides and weathered clays, and that they effectively prevent surface waters from coming into contact with the bedrock; a characteristic that explains the prevalently dilute stream water composition.

Deserts, large flood plain rivers and large ancient lakes are the most distinctive features of the contrasting scenario offered by the water resources of the African continent, underlying the need for more closely defined ecoregions. Seasonal patterns of rainfall and discharge are defined by movements of the Inter-tropical Convergence Zone (ITCZ) about the equator, under the influence of the monsoon winds. The strongly seasonal precipitation pattern is a distinctive feature of tropical biomes. Its major consequence is that it favours the formation of extensive, seasonally inundated flood plains. The intermittent discharge regime on one hand and the high radiation on the other imply that rivers that are relatively large during the floods may retreat into trickles or become separated isolated pools during the dry season and in years when floods fail. It is not surprising, therefore, that the most extensive permanent rivers are connected to large tropical lakes. Lake Tanganyika discharges into the Congo, Lake Victoria into the Nile and Lake Malawi into the Zambesi. Besides their current general role as bio-corridors, several of these rivers played a major role as biogeographical refugia during times of climatic stress during the last Pleistocene glaciation, such that the present distribution of animals and plants often reflects ancient hydrological networks.

As with the Nile, many African rivers such as the Senegal, the Niger and the Chari in the west, and the Jubba, the Tana, the Rufiji and the Okavango in the east, originate in well-watered highlands but then run through semi-arid savannahs for a significant portion of their lower course; a situation that induces the formation of gallery or riparian forests. Rivers are able to sustain life in regions that have a high precipitation/evaporation deficit. Gallery forests and their associated riverine wetlands act as important wildlife refuges and ecological corridors for the dispersal of animal and plant species (Medley and Hughes, 1996). The Flood Pulse Concept (Junk *et al.*, 1989) describes the functioning of tropical river basins as closely related to the repeated occurrence of extensive and long-lasting flood events. Floods are the main engines shaping the morphology of the landscape and provide a major trigger regulating reproductive cycles of animals and plants.

Stream water chemistry tends to be dilute. Chemical and physical denudation processes are enhanced by alternating cycles of drying and wetting. The prevailing high temperatures tend to bake the dry soil surface and then influence lixiviation in the soil profile at the onset of the rains. Temperature plays a major role, as the ionizing power of water and the development of deeply leached soil profiles rises with temperature. The prevalence of highly degraded, mature clays in tropical catchments implies that phosphorus, which is the main limiting nutrient in most terrestrial and inland aquatic ecosystems, becomes strongly sequestered within the soil matrix. This is a main limitation for agricultural production in African soils (Fardeau and Frossard, 1991). A further effect of high temperatures is that nutrient cycles are closely coupled with biological uptake and the presence of free limiting nutrients in the environment is extremely short.

Despite a continuous vegetation cover throughout the year, rivers tend carry huge amounts of silt. Sediment-laden, 'white-water' rivers are common in tropical lowlands. This effect is due to the combination of natural factors such as the presence of poorly structured soils and the frequent occurrence of highly erosive rains, as well as to the impact of deforestation and agriculture on steep slopes and in agroecologically (erosion-prone) marginal areas. High sediment loads impact in-stream habitats, reducing underwater light, macrophyte living space and affecting fish habitats. In tropical flood plains, of which the lower Nile basin before the Aswan Dam construction is cited as a prime example, the distribution across the landscape of flood plain silt during inundation renews the fertility of riparian areas and supports rural livelihoods.

Africa is inhabited largely by low-income communities whose life is based on subsistence activities. Survival is closely dependent on the amount and quality of basic renewable natural resources such as water, aquatic and terrestrial wildlife, forests, pastures and cropland. Rainfed agriculture is largely predominant while flood recession agriculture, which closely follows seasonal inundation cycles, is becoming progressively replaced by intensive irrigation schemes. In Africa, population increase, climate change and resource degradation imply that the per capita allocation of water is increasingly scarce. This contributes to tighten the dependence of socio-economic development on the protection of aquatic ecosystems.

As a response to this, high dams are now present in all the major river basins with the exception of the Congo which, in comparison to the rest of Africa, remains the last large, naturally pulsing system. River regulation has induced changes in other major rivers, such as the Nile, the Zambesi, the Senegal, the Volta and the Tana, which has demonstrated the need for designing appropriate ecohydrological solutions for catchment and for flood plain management.

Ecohydrological Approach at the Ecoregional Scale

This chapter examines African catchments within distinct ecoregions. Each of these is characterized by ecohydrological issues that need to be resolved to optimize the management of river ecosystems. Issues are not exclusive but assume a higher importance in some regions as opposed to others.

The northern ecoregion

The Arab states situated north of the Sahara constitute the northern African ecoregion. The great majority of this region is arid or very arid with precipitation sometimes well below the 200 mm isohyet (Wishart *et al.*, 2000). In the northwest, rivers are of short length, torrential in character and their flow is naturally interrupted during the dry season. Intensive agricultural exploitation reduces the baseflow period further. During the long dry season, most of the flow is made up of treated and untreated sewage effluents, which are nevertheless intensively reutilized for irrigating crops, posing a threat to human health. Irrigation systems in the whole of northern Africa are generally inefficient and lack adequate drainage, resulting in severe waterlogging and salinization. Demographic expansion, which started during the second half of the 20th century, and improvements of living standards contributed towards increasing water demand to a level that can be sustained only by exploiting fossil, non-renewable groundwater resources. Pumping schemes, inter-basin transfers and man-made rivers continue to expand throughout the region, creating new settlements and facilities which have an uncertain future. As a consequence of the increasingly critical condition of the groundwater supply, policy responses are increasingly directed towards demand management.

From a biogeographical point of view, the northern ecoregion can be further subdivided into: (i) a western sector, characterized by a strong Mediterranean and European influence owing to its proximity to the Iberian peninsula; and (ii) an eastern sector characterized by biogeographical ties with southern Asia due to the former continuous vegetation belt running through the Arab peninsula and with tropical Africa due to the Nile, which represents a major bio-corridor. Libya, situated at the crossroads of these major biogeographical influences, is characterized by a combined faunal composition.

The socio-economic situation is relatively more promising for Egypt, whose long history and development have been related to the presence of the Nile. The former traditional flood plain agriculture, relying on river spates and the distribution of fertile silt across the valley, has been replaced by modern agrotechnology combined with intensive artificial irrigation and fertilization. The lower Nile Valley has been profoundly transformed; it still produces more than 80% of the country's export earnings, but soil degradation as a consequence of the intensive use of chemical fertilizers and pesticides is posing an increasing threat to the overall resource.

Ecohydrological solutions for the northern African ecoregion need to address processes that regulate the torrential character of river flows, trying to retain water within channels and flood plains for an enhanced exchange with riparian zones. The constantly high temperatures can be usefully exploited for the creation of effluent treatment schemes based on evapotranspiration and phytotechnology, with a cost-effective reduction of human health hazards. Groundwater recharge needs to be carefully checked and managed according to carefully selected priorities and long-term targets.

The Sudano–Sahelian belt

This extensive and relatively uniform region, comprising much of West Africa apart from Cameroon and the Guinea highlands, stretches towards the Horn of Africa including Ethiopia, Eritrea, Somalia and northern Kenya. With the notable exception of the Nile, most rivers run almost parallel to the equator. This contributes to a strong latitudinal pattern of moisture, highlighted by the distribution of humid tropical forests, which are concentrated mainly within a broad vegetated strip between 10°N and 10°S. Large Sahelian rivers, such as the Senegal and the Niger, are characterized by a single rainy season causing extensive inundation. Catchments situated closer to the equator, such as the Tana, exhibit instead two unequal rainfall periods, as the ITCZ crosses the equator twice in a year (Pacini and Harper, 2000). The main West African Sahelian flood plains include the Senegal, the Niger (both the Ouémé inner delta and the Yaérés plains), the Hadejia-Jama'are flood plains, the Chari-Logone plains, and the Bahr Aouk and Bahr Salamat plains along the upper Chari. In the east are the extensive Sudd Swamps within the Nile basin and shorter flood plain rivers flowing off East African highlands, such as the Shebeli and the Jubba in Somalia; the Tana and the smaller Athi-Sabaki in Kenya; and finally the Ruaha/Rufiji in Tanzania, which is sometimes considered the southern frontier of the Sahelian ecoregion. However, due to biogeographical considerations these last river basins are assigned to the East African ecoregion.

Sahelian flood plains are relatively fertile, well-populated and biodiverse ecosystems, which typically flow through arid and semi-arid landscapes, acting as bio-corridors for the dispersion of animals and plants. Flood plain inundations are irregular in frequency and extent, but in many years can reach significant proportions of the river basins' area due to the prevailing low slope of Sahelian rivers. In some rivers, groundwater and tributary rivers' contributions enhance the flooding effect (Thompson, 1996; Pacini and Harper, 2000). The presence of sharply contrasting wet and dry seasons implies that Sahelian rivers are highly dynamic and means the vast regeneration areas are subject to intermittent

disturbance (Hughes, 1994). This characteristic is tied to aridity, which is positively related to the geomorphological effectiveness of extreme flood events
(Friedman and Lee, 2002). In this, as well as in the northern ecoregion, floods
are expected to assume a particular relevance for defining the structure of flood
plain habitats.

Despite the availability of sophisticated remote detection systems, there is
high uncertainty concerning the extension of some of the major African flood
plains. This is because of the naturally high proportion existing between the area
inundated at high water levels and the area inundated during baseflow. This factor can be as much as 480 for the Hadejia-Nguru wetlands in Nigeria (Hollis
et al., 1993). Estimates for the Upper Nile (Sudd) swamps, for example, vary
between 30,000 and >90,000 km^2, according to different observers cited by
Tockner and Stanford (2002).

In the Sudano–Sahelian belt, water is increasingly becoming a primary limiting factor of human development, and the great flood plains are rightly seen as
representing some of the most valuable natural resource assets of this ecoregion.
Ecohydrology is needed to support reservoir management and irrigation schemes
for achieving long-term sustainable options, keeping in mind the socio-economic
priorities of LIFD countries. Flood releases have been proposed in a number of
cases to reduce the negative environmental impacts of damming, but cannot be
considered a panacea applicable to every situation (Acreman et al., 2000). As
stressed by the Senegal and Tana case studies below, integrated flood plain management, maximizing the diversity and the overall value of flood plain products
(cereals, vegetables, cattle, wildlife, fish) and the sustainability of natural wildlife
assets, requires great detail in the analysis of catchment processes rather than the
design of ecotechnical measures at the dam or at the inflow of irrigation schemes.
Despite international catchment conservation initiatives, deforestation remains a
major concern and soil erosion a relevant problem, requiring careful, small-scale
interventions and intensive management. Much has been achieved by means of
agroforestry projects, which are generally well accepted by the local population
and remain the most effective solution.

East Africa

This ecoregion is distinguished by the presence of the Rift Valley, interspersed
with large ancient lakes and by its associated volcanic formations, providing
well-watered highlands from which a number of rivers flow towards the Indian
Ocean. East African rivers such as the Tana, the Athi, the Rufiji and the Zambesi
are shorter than the large Sahelian ones and are characterized by a steeper average slope, giving rise to less extensive flood plains but more favourable to hydropower development. The Rift Valley Lakes region is blessed by abundant water
resources; nevertheless even here (Uganda, western Kenya, western Tanzania)
basic human needs are poorly serviced and the sustainability of wetlands, springs
and streams is increasingly threatened by inappropriate exploitation initiatives.

The coastline of Kenya, and to a smaller extent the one of Tanzania, has undergone intensive development of tourist resorts and fast demographic expansion

also due to incoming flows of refugees. In Mombasa, the provision of swimming pools and facilities designed to accommodate foreign tourists can hardly be accommodated by the dwindling water resources and the lowering water table. Proposed water diversion schemes, from the Tana into the Athi and by pipeline directly from the Mzima Springs of Tsavo National Park, all have severe environmental consequences. Wildlife water resource uses are increasingly threatened as illustrated by proposed diversions that could severely affect the Serengeti ecosystem (Gereta *et al.*, 2002; Chapter 9). The increasing degradation of the highland habitats of the East African water towers – Mount Kenya and Mount Kilimanjaro – is possibly also having an impact on ecoregional climate changes.

In East Africa, the implementation of ecohydrology can rely on the experience and data gathered by a series of non-governmental organizations (NGOs) and internationally backed projects on soil erosion, agroforestry and river basin management.

The Congo basin

Relatively little is known about the status of this, one of the largest and most pristine tropical rivers of the planet. The inundated swamp forests of central and western Congo (right bank) still represent some of the least-visited areas on the planet. With its numerous tributaries, flowing from both sides of the equator, the Congo has enjoyed a relative stability throughout geological ages and represents today the most important biogeographical refuge on the continent (Pacini and Harper, 2000). Sediment transport estimates (Laraque and Olivry, 1996) tend to confirm its relatively well-preserved state, which is undoubtedly also related to the limited accessibility of this region due to long-term political instability. Although at the moment the Congo represents the least fragile river basin, maintenance of its intact ecological functions will require the application of ecohydrological principles in the near future.

Southern Africa

Large flood plains are present only on the eastern side of the southern African region; they include the Limpopo, the internal delta of the Okavango, the Barotse flood plains of the lower Zambesi, the Phongolo/Inkomati flood plains and the ones along the Berg river. The region includes highly specific biogeographical enclaves such as the southern coastal mountain chain, which represents a climatic refuge for cool-adapted fauna and for rare Pan-Gondwanian (Australo-African) lineages, and the Kalahari desert in the north. Climate is semi-arid to arid, with least precipitation in the north-west (Davies and Wishart, 2000). Overall rains are infrequent and unpredictable; in contrast to inter-tropical regions they exhibit little seasonal pattern. Basic human needs are far below being met and are constrained by scarce water resources, as illustrated by the proposed Okavango diversion schemes for the benefit of Windhoek township in Namibia (Davies and Wishart, 2000). In South Africa, demographic growth and rapid socio-economic changes call for a rapid policy response while the recent democratic water legislation (South Africa's National Water Act 36 of 1998) advocates the prime objective of ensuring basic resources for all, on the basis of equity, sustainability and efficiency. New, democratic inter-basin transfers and dams are on the horizon. In contrast to most of the rest of Africa, South Africa can rely on a well-established tradition of environmental assessment and experience in integrated river basin management. This experience provides an invaluable basis for applying the principles of ecohydrology.

Riparian habitats are increasingly encroached by human settlements. Fuelwood extraction from riparian areas represents a significant threat for water quality and for their persistence (Davies and Wishart, 2000). Ecohydrology needs to confront these issues with socio-economic alternatives for the definition of sustainable livelihood scenarios. Water harvesting techniques and protection of vital drinking water supplies may be closer in finding political consensus than large, centralized diversion schemes. Small-scale projects, carried out with the support of local communities, are efficient also because they contribute in education towards sustainable water resource management. However, the coordination of local initiatives into integrated catchment management remains a challenge. Recent legislation in Kenya (2002 Water Act) provides a framework for this, although hindered by a lack of supporting infrastructure. The Act establishes Water Basin Management Authorities for the first time and under their auspices encourages local water abstractors to form their own Water Users Associations.

The Ecohydrology of Two Selected River Basins: the Senegal and the Tana

The Senegal river, West Africa

Divided by the 15°N meridian, the Senegal river valley marks the southern frontier of the sub-Saharan Sahel in West Africa. The river originates from well-watered, sub-equatorial highlands but flows for most of its course (ca. 1800 km) through the semi-arid Sudano–Sahelian zone. The lower half of the catchment receives less than 500 mm rain; animals and plants depend on the floods for their survival as does the local economy. The Senegal catchment extends over 290,000 km^2. During wet years, before the construction of dams and embankments, as much as 250,000 km^2 could be inundated (Acreman and Howard, 1996) (Fig. 12.1). As is typical of rivers in arid and semi-arid lands, both inter-annual and inter-seasonal precipitation regimes exhibit very high variation; between the wet and the dry season the Senegal river discharge may change from 4000 m^3/s to less than 10 m^3/s. Flooding and flood recession create hydro-pedological conditions, which lead to fertile soils, due to the distribution of silt over the flood plain, and to high primary and secondary productivity. Traditional human activities in the Senegal Valley undertaken by around half a million people include rainfed, flood recession and irrigation agriculture, animal husbandry and fishing (Adams, 1992).

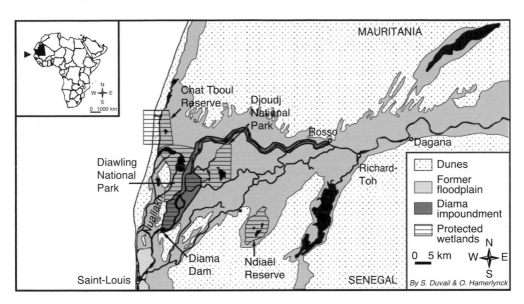

Fig. 12.1. The Senegal river basin. (After Hamerlynck and Duvail, 2003.)

Not far north, the Sahara landscape offers a strong contrast to the fertility of the Senegal Valley. Despite this, a series of unusual droughts starting at the beginning of the 1970s combined to cause a progressive exhaustion of ground-water reserves, a deeper introgression of the underground saline waters and a southward progression of the desert frontier (Olivry, 1987). In the following period, exacerbated demands for scarce resources and steeply rising population numbers offered arguments for drastic changes in river basin management. Technical development schemes arrived at a time when the environment had the least opportunity to compensate their negative impacts. River basin management drastically changed the character of human activities and caused social and environmental disruption, leading to repeated outbursts of inter-ethnic violence.

The large area liable to flooding (on average 17,970 km^2; Thompson, 1996) consists of a low-lying plain separated from the Atlantic Ocean by accumulated river sediments and interspersed with semi-permanent dunes and depressions, some of which are below sea level, and which used to provide the delta with a series of transient saline-to-freshwater habitats. Before development, between February and June, a salt wedge used to penetrate the river channel reaching sometimes more than 200 km inland. Every year, in early July, the river used to flood and the stagnant saline water was flushed out; by the beginning of August the river would be clear of salt down to the estuary in St. Louis, which was dominated by coastal mangroves (Grove, 1972). Immediately after river regulation, it was noticed how flood discharge started to fall off exponentially with an increasingly rapid decline in comparison to a historical average discharge decline coefficient measured in the valley (Olivry, 1987). This change in the basic hydrological pattern is highly relevant for flood recession agricultural practices and for the survival of wildlife in riparian areas.

Wildlife

The greatest interest of the international conservation community has concentrated on the extended coastal delta region. The last 100 km of river, before reaching the estuary, supported many flood plain lakes and wetlands of various sizes. Here the main channel divides into a large number of smaller courses, which then reunite before entering a series of brackish lagoons at St. Louis. The variety of waterbodies of different degrees of salinity and the presence of permanent water surfaces under arid climatic conditions gave this area a high conservation value. The Senegal Delta provides regular refuge for western Palaearctic bird species, which come to overwinter and are protected by a dozen reserves of various types including three Senegalese and one Mauritanian National Parks (Vincke and Thiaw, 1995). The Djoudj National Park in Senegal is important for aquatic birds (78 species) including some northern hemisphere taxa, which have become rare in Europe (Schwöppe, 1994). The Senegal Valley is the last watering point for migrant birds that undertake the journey along the Atlantic Flyway every year. Equally important is its role as a main breeding station for North-West African species, as other suitable large freshwater surfaces are distant (the Niger in Mali and Lake Chad). Two thousand kilometres of desert to the north represent a formidable barrier to the migration of animals and plants. For this reason the Djoudj National Park was listed as a wetland of international importance under the Ramsar Convention and as a UNESCO (United Nations Educational, Scientific and Cultural Organization) World Heritage Site.

Beside birds, the park and the Senegal delta region shelter a number of other vertebrates. Ancient documents show the presence up to the end of last century in the Senegal Delta of a very diverse wildlife, including large herbivores (elephants, giraffes, antelopes) and carnivores (leopards, lions, cheetahs, hunting dogs) today extinct from this region (Schwöppe, 1994). The Senegal Valley represents the relic of a former much more extensive and wetter ecosystem, which has been regressing over the past two centuries. Formerly widespread terrestrial vertebrates find refuge today in the Djoudj National Park. Among these there are two species of jackal, several wildcat species, several antelope species, and a few specimens of hippopotamus that were once common in the region. Extensive gallery forests (today much reduced) sheltered a varied fauna of primates; while the river and its estuary were inhabited by an endemic African sea-lion (*Trichecus senegalensis*).

The vegetation of the flooded parts of the lower catchment is characterized by the dominance of *Acacia nilotica* woodland interspersed with annual grasses, referred to as the Gonakié Forest. The forest is protected, but nevertheless it is experiencing progressive deterioration. Permanent grasses and bush vegetation are characteristic of the savannah belt outside the perimeter of the periodically inundated flood plain. The disappearance of the regular annual flood tends to prevent regeneration of the Gonakié Forest and is causing the progression of savannah vegetation. The situation is indicated by the appearance of a secondary climax dominated by *Acacia raddiana*, which is progressively replacing *A. nilotica* as desertification advances (Schwöppe, 1994).

Impacts of river regulation

Today the Senegal river discharge is controlled by two dams. Diama Dam was completed in 1985; it was constructed within the delta, on the major river channel as it exits from Diawling National Park (Mauritania). The presence of the reservoir created a barrier to the ingression of saline waters towards the inner catchment. Within 4 years the salinity of the Lac de Guiers, situated not far above the reservoir, decreased by a third, reaching 240 mg/l and allowed the use of the water for irrigation. Below the reservoir, however, salinity increased rapidly and the river became hypersaline, disrupting the formerly brackish aquatic ecotones. In 1990, a combination of low discharge levels and high tides flooded the deltaic pastures and killed off the remaining mangroves (Hamerlynck and Duvail, 2003).

The Diama Dam was developed in order to establish large-scale modern rice plantations by providing a higher head of water to be used for irrigation. Studies conducted by the Institute of Development Anthropology (IDA) warned, however, that due to the lack of competent labour in the area, the need for high investment and the risk of soil salinization under intensive irrigation practices, the development venture would incur major financial losses. As predicted, extensive areas of productive land were turned into desert after the building of the first raised river embankments (Roggeri, 1995). Farmers on the Mauritanian side were faced with the effects of increasing salinization without the benefits of efficient irrigation systems that would support the projected increase in yields. Of the expected 12 t/ha/year, only 3 t/ha/year were realized (Hamerlynck *et al.*, 1999).

The full consequences of these changes, underestimated at the time of dam and embankment construction, are yet to be assessed as the processes are still evolving. Some of the conservation alarms are as follows:

- Reduction of some aquatic plant communities that represent food and shelter for migrant water birds.
- Shrinking of the *A. nilotica* woodland due to drying, the increased salinity of the soils (42,000 ha) and permanent flooding (7600 ha).
- A 95% reduction of pastures (more than 200,000 ha) with consequent accelerated desertification caused by cattle.
- Shrinking of the coastal mangrove fringe (*Avicennia nitida*, *Rhizophora racemosa*).
- Blockage of the upstream migration of estuarine and marine fish, which used to move up to 200 km inland and constituted a valuable food resource in this upper area (Watt, 1981; Albaret and Diouf, 1994).
- A permanent *Typha* (*T. australis* and *T. domigensis*) reed curtain developed at the margin of the flood plain, impeding fishing operations in the reservoir shores and replacing part of the former flood plain pastures (Hamerlynck *et al.*, 1999).
- Native plants – such as *Sporobolus robustus* used for weaving, a water lily used as a substitute for cereals and a floating wild rice variety traditionally

farmed by flood plain farmers – have become seriously restricted to a narrow range, lost their economic relevance and have been put at risk of extinction (Hamerlynck *et al.*, 1999). Seedpods of *A. nilotica* used in hide tanning are now rare.

- Destruction of fish spawning areas and loss of an estimated 350,000 ha of fish habitat (Hollis, 1996).
- Continuous decrease of fish stocks in the estuary.
- Impact on the ecology of the National Parks in the delta where funds are being raised, with great difficulty, to reproduce artificial hydraulic conditions similar to those existing before damming (Beintema, 1995).

Fish are among the vertebrates that suffered most directly from the consequences of changes in the hydraulic regime. Fish assemblages of the large Sahelian rivers typically include large-bodied migrant species, characterized by high growth rates and capable of adapting to a harsh, variable environment. Due to the sharp environmental changes characterizing the Sahelian ecoregion, these assemblages tend to include relatively low species numbers when compared with the African lakes or large rivers (such as the Congo) characterized by more stable discharge conditions. Lateral migration is known to have a great importance for the rapid development of young stages, which find high temperatures and abundant food sources in the fertile, shallow floods. In the Sahel, fish productivity is closely related to the seasonally inundated flood plain surface (Laë, 1997). The building of dams and embankments prevented longitudinal and lateral fish migration and the retention of sediments within the dams decreased nutrient availability. Furthermore, the shorter flood duration reduced the time allowed for reproduction. The combination of these three factors may not have led to short-term species extinctions but undoubtedly severely affected fish stocks.

Beside riverine species, estuarine and coastal fishes that frequently exploited the lower flood plain wetland habitats became affected by the rising salinity levels in the delta after the closing of the Diama. Tropical coastal lagoons have a high spatio-temporal variability of ecological factors and this confers upon them the potential to harbour a great diversity of fish species. In particular, they serve as refuges for the reproduction of a large number of fishes and invertebrates including riverine, estuarine and marine species (Albaret and Diouf, 1994). Geological history, size and morphology, vegetation and climate, freshwater flows and sea water intrusions, and finally diverse human activities, all contribute to generate a complex array of multifunctional habitats in the estuarine zone. At the same time, coastal lagoons exhibit a degree of environmental stability unequalled by the river or the estuary itself, which is necessary for stenotopic species (habitat specialists) to evolve and speciate (Ribbink, 1994). In the Senegal delta, the salinity zone between 5 and 15% used to be the major reproduction zone for brackish water species. After the closing of the Diama, this zone virtually disappeared. Increased salinity also damaged fish stocks by causing the disappearance of sensitive crustacean species, which constitute important food resources for the fish (Albaret and Diouf, 1994). It was estimated that the initial overall effect of the regulation of the Senegal river cut the river's fish production by as much as two-thirds (Albaret, 1994).

Flood plain conversion into an irrigation scheme

Overall impacts on fish catches were produced directly by the Diama dam clo-sure and by the modification of the discharge regime. On a local scale moreover, deliberate flood plain regulation by the building of levees and abstraction schemes was planned to benefit irrigated plots for cash crop production. This form of development was sought despite expert warnings issued by local and international consultants concerning the fragility of the lower Senegal soils and the small surface that could safely have been developed into managed irrigation plots.

Specific concerns brought by irrigation development included:

• Early abandonment of some managed agricultural plots due to increased salinity. While floods wet and flush out salts from the land, irrigated plots are hydrologically semi-enclosed systems that lead to the formation of salt pans over the years (Hollis, 1994). In the early 1990s, groundwater conductivity in the delta reached around 100 mS/cm (about twice that of seawater; Schwöppe, 1994), confirming a trend towards rapid salinization.

• Invasion of grain-feeding birds (*Quelea quelea*) on irrigated surfaces due to the presence of permanent water surfaces and the absence of dry-season mortality. This obliged farmers to use avicides and created a potential threat to neighbouring bird sanctuaries.

• Water pollution due to the application of heavy doses of pesticides and chemical fertilizers, including episodes of cattle death (Schwöppe, 1994).

• Increased flood risks for human settlements and regions that had never previously flooded, due to raised artificial embankments built to protect irrigated plots (Hollis, 1996).

• Exceptionally high levels of urinary intestinal bilharzia.

• Increased occurrence of malaria and onchocerciasis.

Despite these immediate negative impacts caused by the closure of the Diama, a second, 65 m high dam was completed in 1988 at Manantali on Mali soil, 1250 km from the estuary, in the upper basin of the Bafing tributary. The result-ing 477 km^2 large reservoir controls more than 50% of the flow within the Senegal river basin. The discharge hydrograph downstream is affected by the dampening of flood peaks and by the increase of dry-season flows. The initial purpose of the project was:

• To produce hydroelectricity for the town of Bamako in Mali.

• To provide a head of water for extensive irrigation schemes.

• To allow navigation between Mali and the ocean during most of the year.

More than a decade from the start of the scheme the turbines had not yet been installed. Accessory engineering works for protection of the areas to be irrigated and of natural areas (small dams, dykes, canals) had not been completed. The delays were due to the large scale of the project and the conflicting interests of the three countries involved: Senegal, Mali and Mauritania.

From the point of view of sustainable natural resource management, the convergence of financial difficulties and political uncertainties in the manage-ment of this large trans-national and trans-ethnic development scheme has been somewhat fortunate. It soon became apparent that modern irrigation agriculture could not be profitably implanted in the basin in the short time proposed. Therefore, shortly after the Manantali dam closure, developers agreed to operate the reservoir in an artificial discharge mode, allowing controlled flooding for a period of 10 years. These partial flow releases had initially been designed for the benefit of pastures and flood recession agriculture on the left bank of the river with little concern for the fish and birds in Diawling National Park, Mauritania (Hollis, 1996). Water was finally released experimentally for the first time in 1994, to flood the Bell and Diawling basins through a sluice gate built in the embankment. Its primary effect was to sustain the aquatic and riparian vegetation of the proximal inundated zone, which is crucial for the protection of young stages of fish, birds and other vertebrates.

Controlled artificial flood releases from Manantali were initially planned for 10 years with the aim of minimizing the environmental impacts of the scheme, while allowing a reasonable amount of irrigated and flood recession agriculture. An evaluation of this strategy was made by Hollis (1996), who commented on the performance of artificial flood 'A' (as it was baptized). Even with the limited hydrological data available in such a large and poorly developed basin, Hollis managed to highlight essential hydrological relationships between discharge, flooded area, cultivated area and groundwater recharge. After an initial period of mutual mistrust, collaboration with the Organisation pour la Mise en Valeur du fleuve Senegal (OMVS), the international body for the management of the Senegal Valley, under sponsorship from the US Agency for International Development, allowed a better follow-up of hydrological investigations and ecological monitoring activities. During the first 7 years, artificial flood 'A' inundated at best 50,000 ha, with poor consequences for traditional agriculture, which used to cover far more extensive grounds. In some years, such as in 1994, ill-timed releases caused unnatural flood peaks that ended up damaging cultivated plots. Only during the last years of the proposed management, that is after the complete filling of Manantali reservoir, did the artificial release work in an efficient flood mode with promising results for traditional agriculturists who could rely on more regular floods and on the avoidance of damaging flood peaks. Throughout this period, other sectors, such as fisheries and cattle breeding, were largely at a disadvantage. The loss of natural pastures caused over-exploitation of those grazing areas left and produced, in turn, a loss of natural manure brought by cattle on to such traditionally managed plots. The contribution of cattle to the fertility of the Senegal Valley had been estimated to 350 kg/ha prior to development (Reizer, 1984; cited in Roggeri, 1995). Such natural manuring is considered highly significant for the achievement of high fish catches (Welcomme, 1985).

The IDA, backed by the Senegal government, applied a comprehensive cost–benefit analysis to demonstrate how a better-managed future flood release would prove a cost-efficient and environmentally sound solution (Hollis, 1996; Salem-Murdock, 1996). Their arguments were based on the high investment costs necessary to transform the socio-economic base of agricultural production and on the lack of competent labour (there had been much migration out of the rural areas to look for work elsewhere, as is typical of marginal rural areas). Despite this, the Senegal basin management authority (OMVS), with the agreement of representatives of Mali and Mauritania, concluded at the time that a 25% loss of hydropower due to flood releases could not be compensated by the agricultural income of the rural poor (Acreman and Howard, 1996), and the artificial release scheme risked being suspended.

While the environmental costs of dam closure were rapid in appearing, the full hydroelectric potential benefit of the Senegal River Scheme is yet to be completely realized. It is estimated to provide as much as 800 GW of power (Gassama, 1999). However, the net economic returns have yet to be properly assessed, due to high distribution costs. Early estimates are that the energy produced could hardly be profitably used beyond a 300 km range of the turbines; a poverty-stricken region that lacks basic infrastructure for industrial development (Mounier, 1986). Eventually, given the wide range of environmental impacts cited above that were not evaluated, the OMVS started an environmental impact mitigation and monitoring programme (PASIE) in 1997, which includes collaboration with local NGOs, immediate remediation projects and an improved monitoring network coordinated into a shared Environmental Observatory. An important role of the PASIE is the redefinition of the respective jurisdictions of OMVS (States and Haut-Commissariat), the local authorities, the principal contractor, private companies, SOGED and SOGEM (reservoir management companies) as well as the national electricity utilities.

Twenty-first century ecohydrology in the Senegal: challenge and opportunity
Following the first experiences with artificial floods, it appears that the future of the Senegal Valley is now closely dependent on the application of ecohydrological solutions within a framework of integrated river basin management. A complex flood routing software designed to estimate optimized floods for multi-purpose use has been proposed by the OMVS. Pushed by the recognition of unforeseen environmental impacts, the river basin management authority has evolved its mission to include consideration of a moderation of the ecological impacts of the Manantali and Diama dams. The World Bank agreed to finance turbine installation in 2001, if provision for flood releases was included in the project.

The current strategy is directed primarily at low flow augmentation and preliminary effects seem to be beneficial for traditional agriculture and fisheries. However, the restoration of riparian zones requires a careful observation of the processes occurring in the field. Extension work carried out in Diawling National Park in close collaboration with resident farmers, fishermen and mat weavers demonstrated the potential for an integrated sustainable local economy in the delta region, which proves far more successful than planned large-scale irrigation (Hamerlynck and Duvail, 2003). Sustaining traditional activities within a regulated river regime cannot be achieved without high capital expenditure. The building of dykes, sluices and artificial channels was necessary to achieve effective restoration measures on more valuable sections of the flood plain. Empirical tests and trial and error were successfully employed to establish a relationship between catchment hydrology and the phenology and reproduction of natural resources. These trials confirmed that the situation in the lower Senegal is ideal for implementation and it was demonstrated that reservoir operations could be tuned to mitigate negative impacts by sustaining the mangrove ecosystem, dryland farming and pastures and by reducing reed infestation (Hamerlynck *et al.*, 1999). Short-term, temporary flooding could be used to reduce the encroachment of unwanted invasive plant species. At the same time temporary drought could reduce human parasites and infestation caused by plants preferring permanently moist conditions such as *Typha* spp., particularly damaging for its tendency to overgrow ditches and sluice gates.

Flood releases constitute a major ecohydrological option; however, consensus over their operation remains a major issue owing to their cost in terms of both infrastructure and operation. Sophisticated hydrological surveys and models, careful economic evaluations and intensive field observation are needed to establish an optimized resource balance. Transparency and mature participatory consultations are essential, but difficult to establish given conflicting interests and the high number of interested parties. In Diawling National Park, this issue was addressed by the confrontation and progressive refinement of flood scenarios and seasonal discharge calendars linked to sustainable livelihood activities (Fig. 12.2). Both the timing and duration of flooding were crucial for obtaining the required result.

Releases from Phongolapoort Dam, on the Phongolo river in South Africa, were defined by a democratic process involving consultations with local farmers, fishermen and other stakeholders (Acreman *et al.*, 2000) after a period during which flood releases had caused more damage than good. This case study stressed that the participatory flood management approach is an important exercise which generates an active involvement of the resident population, but should be elicited and organized with appropriate times and forms. This seems to be a major challenge in the Senegal due to the basin size, the contrasting interests of three governments, and the large number of tribal groups and stakeholders who lack the means and organization for conveying their active participation to the planning process (Adams, 1999). After a decade of large-scale regulated catchment operation, the Senegal case study provides a fruitful opportunity for deriving a number of lessons and for comparative assessment exercises.

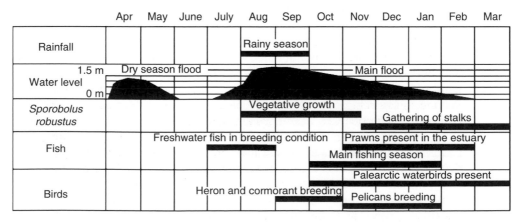

Fig. 12.2.   Seasonal succession of discharge events and local resource harvesting activities in Diawling National Park. (After Hamerlynck and Duvail, 2003.)

The Tana river, Kenya

Although smaller and entirely within a single country, the River Tana offers several similarities with the better-studied Senegal basin. It rises from the well-watered slopes of the Nyandarua and Mount Kenya ranges, two of the five 'water towers' of the country, and flows towards the Indian Ocean, describing a wide arc across the dry eastern province, parallel to the border with Somalia. The Tana is the largest river in Kenya, and one of the three major flood plains in East Africa together with the Kilombero and the Rufiji. It had been associated with a wild, untamed country, exclusive domain of wildlife. The Tana river gallery forest grows along a >3000 km^2 temporarily inundated flood plain and delta, running through an otherwise semi-arid region. During the wet season it becomes impassable and still represents a frontier of administrative control and a major ethnic divide between Bantu and Somali tribes today. The flood plain itself is inhabited by more than half a million people, including indigenous tribes, which provides, once again, a scenario common in most Sahelian flood plains: the contrast between the pastoralist Orma and Wardei (ethnically related to the Somali), who disperse during the flood and return to the river during drought using traditional access routes known locally as *malkas*, and the agriculturist Pokomo and Malekote, growing a mixture of green vegetables and fruits on river terraces. Besides traditional residents, the current local community is influenced by ethnic Kenyan newcomers brought in to run two irrigation schemes on the lower Tana at Bura and Hola (Pacini *et al.*, 1998). New job opportunities, such as trading to supply the new community, break down the traditional role of resident tribes.

Since a first scientific survey conducted in the early 1970s (Andrews *et al.*, 1975), several studies have been conducted on the vegetation and conservation aspects of the flood plain, concentrating in particular on the Tana River National Primate Reserve (three endemic subspecies of monkey). A more careful assessment of the existing natural resources and of the livelihood of resident people was only carried out recently, when a number of critical issues related to hydrological requirements for the sustainability of the riverine corridor were identified by a team of consultants and discussed during field work with the local community (JICA, 1997). This report highlights some past mistakes of development in the catchment and suggests greater attention should be paid to impacts and benefits experienced by riparian residents and to the overall response of natural resources (pastures, agriculture, fisheries) to flooding. The most valuable components of the downstream environment were identified as flood plain grasslands, riverine forests and protected areas, together with their traditional agricultural practices. The flood plain grasslands are particularly valuable as they consist of highly palatable fodder varieties and provide dry-season grazing for a very wide area. Their productivity can be sustained only by a bi-annual flood regime. Traditional rainfed and flood recession agriculture is poorly quantified and little considered, as much agriculturally valuable flood plain was developed into irrigated rice schemes such as the ones of Bura and Hola, while some other areas may have disappeared due to flood plain shrinkage. River fisheries represent probably only a minor component of the local economy and have been evaluated as corresponding to around 4% of the overall flood plain economy. Fishery decline is also due to the increasing marginalization of flood plain wetlands, which become permanently separated from the main channel and cannot serve as breeding grounds for repopulating river stocks. Fisheries play a relatively modest role within the channel but are more significant within the large river delta.

Development history

As with many other flood plain rivers, the most valuable part of the catchment is the lower flood plain and delta, where slope levels off and the inundated area is largest. Towards its mouth the river reaches a >1000 km^2 delta, partly regulated by an early artificial outlet channel built by the Omani administration of Malindi during the 16th century. The Tana has never been an important channel for communication and no major settlements exist on its banks. In the latter half of the 20th century, however, the mid-upper reaches of the river became the location for a 'development' project including dams for hydropower generation coexisting with downstream intensive rice irrigation schemes (Pacini *et al.*, 1998). In the beginning of the 21st century, fewer than 50% of urban residences in Kenya (about 5 million people) are provided with electric power; nearly 75% of this is produced by the Tana Seven Forks Development Scheme with five existing reservoirs. (The name 'seven forks' comes from the identification of seven tributaries to the main river as suitable for a hydroelectric dam inundating each confluence.) At the same time, about 80% of rural households (nearly 35 million people) rely on fuelwood and biomass as their sole energy source. Small alternative energy sources within the country include a geothermal plant with an estimated potential of 2000 MW (about 10% of national power). Heavy reliance on hydropower caused significant distress during exceptional droughts of 1998 and 1999, when nationwide crop failure, famine and lack of water supply were accompanied by power rationing.

River regulation, introduced during the 1960s with the first, now partly silted-up reservoir (Kindaruma), brought some first unexpected negative impacts that were largely ignored. Since 1989, when the last dam was closed, floods have decreased dramatically (apart from an infrequent event in 1997 associated with El Niño) and this contributed to greater saline introgression in the lower basin, enhancing risks of soil salinization. Environmental impacts of damming the Tana are multiple, complex and not well quantified. The Seven Forks Development Scheme may have doomed what is increasingly seen as a biogeographical hotspot within East Africa, with endemic trees, birds and primates and perhaps much unknown to science. The composition of the Tana river forest is strictly dependent on floods as natural moisture in the region would not support such extensive growth. Hughes (1985, 1988, 1990) showed how irreversible changes to the flood plain forest composition are due to the prevention of low-frequency, extreme floods which actually mine the regeneration potential of the forest. Stream water chemistry has changed and the very significant reduction of particulate load has modified soil formation processes and geomorphological dynamics within the lower river course.

Besides direct environmental impacts, drought caused by dam operations produced conflicts between farmers and herders and induced wildlife to invade agricultural crops. The operational requirements of power generation occasionally produced destructive artificial floods and damage to harvests. The escalation of violence between resident tribes since 2000, as a consequence of the previous, particularly dry year, resulting in numerous casualties and loss of property, represents a direct outcome of river regulation that could have been predicted but that had not been accounted for. Flood plain shrinkage and violence produced a substantial number of internal refugees who then resettled in other areas of the flood plains, causing further instability. To prevent the outburst of extensive tribal clashes, a Tana River Peace, Development and Reconciliation Committee has recently been founded with the support of Oxfam (UK). Long-term unrest and war in neighbouring Somalia has contributed an influx of newcomers and unfortunate availability of cheap firearms. Current local political conditions do not welcome further hydrological changes in the flood plain that could increasingly undermine the basis of resident rural livelihoods.

A new dam on the Tana

Feasibility studies for the last two sites of Mutonga and Grand Falls are advanced and at least one reservoir may be built in the near future; it will be situated below the existing five dams and it may be the only one provided with a flood release facility. The planned Grand Falls Dam is expected to cost nearly US$1 billion; it would submerge a surface in excess of 100 km^2, displace over 5000 people and produce up to 200 MW at full capacity (Anon., 2004). In terms of hydropower it could rank therefore among the most productive plants on the River Tana. Its size should not be as great as foreseen in some of the preliminary plans. It is estimated that events with a return frequency of >10 years may well exceed reservoir capacity (Acreman et al., 2000); given this condition, the impounding period is not expected to cause severe adverse impacts on ecosystems downstream. Conditions at the Grand Falls dam site to include flood release facilities are considered favourable (Acreman et al., 2000), proposals to include sediment flushing as an option to increase reservoir lifetime and favour the distribution of fertile sediments to downstream habitats appear less viable. In the mind of developers, flood release gates should produce artificial floods of a size that would replicate the 'normal' flooding pattern in the downstream reaches of the Tana (JICA, 1997). According to this last report, the proposed artificial annual flood would reach a peak flow of the order of 785 m^3/s at Garissa, corresponding to an estimated median flood calculated over the years 1963–1993. This regime would ensure a regular overflow of the banks and it would correspond to natural events with a return period of 2 years. The results of intensive field studies carried out in the Tana riverine forest by Hughes (1994) suggest that a flood of similar characteristics is close to the minimum requirements for supporting the growth of evergreen forest species which are an important component of the gallery forest. The minimum flow required for evergreen species corresponds to an elevation at the Garissa gauge of approximately 4.1 m and has an estimated return time just over 2 years.

While great effort has been made to estimate the required maximum peak flood elevation and duration, bi-annual flood events, which mimic the natural flood pattern, are likely to be necessary for avoiding serious ecosystem degradation. The equatorial position of the Tana catchment implies a distinct bi-annual flooding pattern as the ITCZ crosses the equator twice a year. The second flood peak takes place around October and is on average significantly lower than the main March–May flood. Considering the ecohydrology of the catchment as a whole, concerns expressed in relation to Grand Falls come at a relatively late stage, when most of the damage to the flood plain has already been done by building the five upper reservoirs and in particular the first one in the series, Masinga Dam (1560×10^6 m^3 volume and 125 km^2 surface). Masinga retains up to 90% of the nutrient-rich sediments carried from the fertile volcanic slopes and dampens the main volume of the flood peaks directed to the flood plain (Pacini *et al.*, 1998). Most of the phosphorus passing through the reservoir is extracted from suspension and from solution, and ends up locked into reservoir bottom sediments (Pacini, 1994). Grand Falls Reservoir is expected to trap some 60% of the sediments transported by the Mutonga and the Kathita tributaries (JICA, 1997), which constitute the main sediment loads reaching the Tana flood plain since the Masinga Dam was closed in 1981.

Flood releases represent a further level of engineering control over hydrological unpredictability, specifically designed to moderate the impact of earlier hydropower schemes. They would reduce the impact of Grand Falls and this may be of relevance for the preservation of what remains of the Tana flood plain, which is slowly undergoing progressive shrinking, loss of fauna and soil salinization. Recent investigations support the view that dam impacts can be significantly alleviated by environmental flow releases and the provision of a bi-annual flood (IUCN, 2003). It remains highly questionable, however, whether the construction of a new dam including a flood release option could actually reduce current environmental impacts resulting from the operation of present dams, by means of an improved water management that would support the needs of current flood plain land uses, as argued in the JICA (1997) report.

Final remarks

Human-induced modifications, both demographic increase and river regulation, as well as longer-term climatic changes within the Sudano–Sahel ecoregion, act in synergy to produce declining inter-annual discharge regimes and a greater 'drought' effect. Studies conducted by the IDA in Senegal indicate that under such natural hydrological conditions, the control of basin hydrology can be only partially achieved. While there are benefits to be gained from flood control and baseflow augmentation during droughts, the strict imposition of management targets during suboptimal years is economically counterproductive. Non-linear relationships with flow imply that the 100% achievement of a given target may increase already high social and ecological costs to other river basin functions. At the same time, the maintenance of a flexible multi-purpose management scheme allows a greater spreading of risk and higher long-term returns.

Under its artificial release scheme, the Senegal became a large-scale river basin management experiment. In its basin, a number of ecological processes and dependent human activities are still being described in increasingly greater detail while observing the effects of management in landscape units under different flooding regimes. For each of them, a direct relationship to the amount and pattern of discharge was constructed in a learn-by-doing fashion and by field monitoring. Reliable relationships require comparison between different years. Hamerlynck and Duvail (2003) warn that such an approach is necessary and should be continued, to achieve a reliable, long-term data set that will allow the establishment of well-defined relationships. They have proved in the Mauritanian delta that such a strategy is effective in explaining immediate ecosystem responses. The artificial flooding scheme of the right bank of the Senegal Delta represents one of the very few projects in which the benefits of artificial flooding are assessed and documented. Things become more complex when the entire basin below Manantali is considered. What remains to be seen in the coming years is the direction of slow, long-term change in the basin. Slow processes such as soil and groundwater salinization, groundwater level change, decrease in biotic integrity and the progressive impact of nuisance species are difficult to predict and to relate to current decision making.

It is likely that, as in many other parts of the world under flood plain regulation, the basins of the Senegal and the Tana are undergoing changes that are not yet entirely reflected in the structure and composition of the respective ecosystems. Ecosystem structure may not be sustainable under current management and its present standing may not reflect its future evolution. In the Senegal this has been demonstrated by the drought 'memory effect' shown by baseflow discharge which, during exceptional years of high discharge such as 1985 and 1986, rose by less than would have been predicted by hydraulic models (Olivry, 1987). A significant rainfall increase produced only a moderate hydrological response due to the long-term groundwater deficit that needs to be replenished before a true amelioration can be observed. In recent years, extensive water abstraction for irrigation meant that during particularly dry years the lower Senegal ceased to flow (Tockner and Standford, 2002). The consequences of natural and human-induced droughts may last over several years or even bring long-term changes. In the Tana, lack of regeneration of one of the key tree species within the gallery forest, due to the absence of low-frequency floods (Hughes, 1985), indicates that current forest composition is the product of past, non-repeatable hydrological regimes, and is therefore undergoing a slow but irreversible change.

The whole Sahelo–Sudanian zone is threatened by an irregular discharge regime which undermines the recovery of fragile economies. This implies that water management is strategically far more relevant in these water-poor countries and, from this viewpoint, freshwater ecosystems represent what are potentially some of the most valuable economic resources. During times of water scarcity, such as in 1996 in the Senegal, flood plain areas become important environmental refuges and the sole resource able to sustain areas under food production deficit. In Kenya, water management regulates power and water availability for the sustainability of both rural livelihoods and the developing modern economy. The El Niño floods of 1997–1998 affected much of the country, including the Tana river, causing an economic loss of some 10% of the gross domestic product (Mogaka *et al.*, 2002). Still worse, the subsequent La Niña drought lasted for 2 years; its impact was comparable to 16% of the gross domestic product for both years. All sectors of the economy were affected including agriculture, tourism and industry. Floods and droughts are inherently unaccountable and are part of the ecohydrology of arid and semi-arid lands. These natural constraints call for a careful water management policy with a particular concern for the persistence of renewable resource assets. The application of ecohydrological principles in the management of Sahelian rivers should be followed, bearing in mind the constraint of a relatively scarce knowledge base. In this context, an important recommendation stemming from the Senegal case study is that river basin management has to address local livelihood options and follow primarily the path of supporting extant traditional economies before venturing into risky large-scale conversions based on tentative assumptions such as the suggested implementation of irrigated rice schemes on saline soils.

The Tana and the Senegal case studies reflect some of the challenges posed by Afrotropical river systems for the basic principles of ecohydrology. These include:

- Integrated catchment management in very large river systems, encompassing several international and inter-ethnic boundaries.
- Management of rivers through enhancement of sustainable livelihoods.
- Sustainable management of riparian ecosystems in rivers characterized by great seasonal variations and great inter-annual unpredictability of both flow and the extension of seasonally inundated flood plains.
- Preservation of the ecological role of riparian ecosystems in situations where these resources become fundamental for the survival of terrestrial wildlife, cattle and traditional economies.
- Deriving management models for large catchments based on poor hydrological and biological data.
- Moderate the social and environmental impact of some of the largest dams in the world and provide guidelines for the management of 'ecologically friendly' discharge schemes.
- Manage river sediment throughout their origin, transport and fate, from eroded topsoils to silting reservoirs and eroding coastal deltas.
- Understand nutrient cycles, in particular phosphorus, and optimize the natural fertilization of riparian ecosystems by spates.
- Implement cost-effective, small-scale structures of waste abatement (by means of phytotechnology) to reduce pollution discharges into tropical rivers and prevent the need for unaffordable sewerage and sewage treatment facilities.

13 Ecohydrological Management of Impounded Large Rivers in the Former Soviet Union

B. Fashchevsky[1], V. Timchenko[2] and O. Oksiyuk[2]

[1]*International Sakharov Environmental University, Minsk, Republic of Belarus;* [2]*Institute of Hydrobiology, Ukrainian Academy of Sciences, Kyiv, Ukraine*

Introduction

Mankind has profoundly changed the hydrological regime of waterbodies in the world. Links between the hydrological regime, flood plain development and the structure of ecosystems, including key biotic components such as aquatic plants, invertebrates, fish and water birds, are now appreciated in order to justify the sustainable use of earlier engineering works, or 'hydro-technical constructions', in river basins for the benefit of human societies. This chapter examines these relationships in large rivers of the former Soviet Union in western Asia, where highly managed hydrology has a strong impact on aquatic living processes, by controlling admissible levels of water resource depletion and pollution in river systems and by setting conditions for the recreational use of waterbodies. The main hydrological properties of interest in a river basin are its water regime (water levels, discharges, velocities), its temperature regime, its suspended sediment transport pattern and the transport of dissolved substances.

Hydrological Properties

Water regime

River flow is of crucial ecological importance in the maintenance of the flood plain and its aquatic and riparian ecosystems. Fundamental elements of a river regime in Eurasia are: (i) snowmelt floods; (ii) rain floods; and (iii) winter/summer droughts. Snowmelt floods are characterized by a steady rise of water levels and discharges occurring *every year* during the same season; usually they last for a prolonged period. In different climatic zones, such floods may occur in different seasons. In northern river basins, in regions such as Chukotka and in high mountain ranges (Altai, Pamir), floods occur in summer. In central and eastern Europe

(Germany, Poland, Belarus, Ukraine, Baltic countries, Russia), floods occur in spring. Rain floods are characterized by a relatively short and rapid water level rise caused by heavy rains. Baseflow occurs when rivers and lakes are mainly fed by groundwater and is characterized by stable water levels and discharges; a phenomenon that is well known in karst geologies.

Water motion has three components: (i) lateral motion due to gravity; (ii) lateral motion caused by wind (water mass movement from one portion of a waterbody to another); and (iii) oscillatory motion of a liquid with a free surface, accompanied by the deviation of the surface from its equilibrium position. Natural streams are characterized by both laminar and turbulent flows. The predominant type of current is turbulent flow, the main properties of which are pulsing velocities and pressure. The essential result of irregular turbulent movements is the transfer of quantities of water mass, heat, suspended solid particles and solutes.

Water movement affects the biota through the force that the motion exerts. Many species have adapted to water mass movements; among these the guild of migrating fishes moving to Siberian rivers for spawning. *Osmerus mordax* enters rivers with flow velocity not greater than 0.6 m/s. When the ice breaks up, spawning migrations of ciscoes (*Coregonus* spp.) begin at flow velocities up to 2 m/s. Later, when flow velocities decrease to 0.2 m/s the Arctic cisco (*Coregonus autumnalis*) moves downstream. In late August, towards the end of rain flooding, muksun (*Coregonus muksun*) enters rivers with flow velocities of 0.9–1.2 m/s. The last fish that moves to the spawning ground is the long-jawed cisco (*Coregonus alpenae*). This species appears in September–October, when flow velocity is no more than 0.5 m/s. Each of these species travels for spawning at a velocity to which they have become adapted by their specific hydrodynamic properties (Pyrozhnikov, 1932).

Water temperature

The water temperature pattern in rivers and lakes is generated by the combined effect of solar radiation, evaporation, heat transfer, disturbance and turbulent mixing. Temperature is one of the most important ecological factors influencing the physical, chemical, biochemical and biological processes that define ecosystem conditions. Water temperature is an essential factor for fish spawning, but populations of the same species living in different parts of the world exhibit different temperature optima. Thus, the spawning temperature of Pacific salmon living in the eastern rivers of Kamchatka is between 0 and 4°C, while for the western coast salmon it is 6–7°C (Kyakk and Lartsina, 1974). Simultaneous effects of temperature and other factors occur; as temperature rises, the toxic effects of pollutants increase in parallel to physiological oxygen demand and increasing food requirements. Lukyanenko (1987) showed that if temperature increased by 10°C, average fish survival in toxic solutions of various metal salts was shortened by 50%. Similarly, the survival of carp in a dilute phenol solution shortened twofold as temperature doubled from 10 to 20°C.

The reduction of air temperature below 0°C prompts the overcooling of rivers and lakes. At such air temperatures, velocity characteristics and mixing conditions change; 'frazil' ice crystals are formed at the surface and at the bottom, prompting the build-up of ice cover. Analysing the hydrodynamic conditions of the freezing-up and of the ice transit within rivers, Berg (1974) found that the linear dimension of ice-cover formations is proportional to the square of the surface velocity of ice movement. However, in mountainous rivers, surface velocities are several times higher than in rivers flowing across shallow plains. The linear dimension of ice-cover formations, under given weather conditions, can be considered proportional to square velocities in lowland rivers, but this relationship may not extend to mountainous river reaches.

During the freezing period, depending on lake size, there can be homo-thermia (in shallow and usually lowland lakes) or reverse thermal stratification, when water temperature is higher at the bottom than near the ice surface. Snow cover on the ice surface is a very important ecological factor. When snow is lacking, sunlight rays penetrate the crystalline ice surface freely and sustain photosynthetic processes, replenishing oxygen reserves and creating favourable environmental conditions. As the snow layer accumulates on top of the ice surface, sun rays are hindered, photosynthesis is stopped, and oxygen tends to become depleted (Fashchevsky, 1966).

Sediment flow

Gravity forces water to flow downhill, along sloping surfaces and in channels. In so doing, work is produced, which is then used to overcome bed friction, to create erosion and the suspension and transfer of particles to sediment sinks downstream. Natural erosion is a relatively slow process having minor effects on the ecological functioning of aquatic systems. The erosion and export of 10 kg forest soil/ha corresponds to 1000–5000 kg soil/ha created in the alluvial plain, which compensates for the erosion upstream. Anthropogenic erosion is a rapid process however, especially on mountain slopes, causing significant harmful effects on aquatic ecosystems.

Many species have adapted to the solid flow regime. Fish with a dense filtering apparatus do not migrate during peak floods when water is turbid, but swim upstream for spawning at the end of flooding, when an essential part of the sediments has been carried away. Permanent inhabitants of turbid waters have small eyes and mucous secretions to protect gills.

High levels of suspended sediments worsen the conditions of natural reproduction of salmon. Siberian salmon avoid spawning in waterbodies where suspended particle concentrations exceed 220 mg/l and lay their eggs in sections characterized by turbidity less than 20 mg/l (Ruhlov, 1973). Beyond a given turbidity threshold, over-silting of spawning grounds may occur. Under conditions of a high concentration of fine fractions (<2 mm) depositing over spawning grounds, 26% of sturgeon and 22% of autumn salmon reduced their reproduction by 40–50% (Ruhlov, 1973).

Soil composition and slope determine the mobility of the material covering the river bed. In mountainous areas, erosion results from the scouring effects of moving glaciers. Mineral debris produced by moraines has an abrasive effect.

This fine-dispersed material enters the river giving it a whitish or milky colour, depending on input volume and concentration. In the River Katun (Vozhzen-nikova, 1958), phytoplankton does not develop because of the great turbidity generated by the activity of upstream glaciers. Mineral sediments settle over the river bed biofilm, covering mosses and suppressing the photosynthetic activity of benthic microorganisms. In tributaries of the Katun river with no glaciers in their upper catchment, the water is transparent and autochthonous producers are represented by a variety of diatoms and green algae.

Table 13.1 illustrates the distribution of the bed material of the River Issyk from the inlet to the outlet of the valley. The mobility of river bed material determines characteristic properties of extreme ecological conditions in mountain rivers in comparison to slow-flowing and stagnant waters. Suspended sediments carry with them a great amount of nutrients that later accumulate in flood plains and in channels. Table 13.2 shows the area yield of suspended sediment-bound nutrients in some catchments in Central Asia. Differences in the annual suspended sediments and particle-bound nutrients load are due to physical/geological properties of the underlying rocks and the erosion capacity of stream flow.

Mud–stone torrents are formed in mountain basins under conditions of high flow (e.g. high precipitation, glacier and snow melt, outbreaks of moraine lakes)

Table 13.1. Distribution of bed material in the River Issyk. (Data from K. Brodsky, personal communication.)

Distance from the inlet (km)	Material on the bed of the river channel
0.75	Glacier ice
2.75	Large boulders, rock fragments
25	Coarse pebbles, rocks, boulders
27	Average-sized and fine pebbles
35	Fine pebbles, gravel, sand, silt

Table 13.2. Suspended sediment areal yield and nutrient content in some central Asian catchments. (Data from various sources.)

Catchment	Average watershed height (m)	Suspended sediments load (t/km^2/year)	N (kg/km^2/year)	P (kg/km^2/year)	K (kg/km^2/year)	Organic matter (kg/km^2/year)
Hunt-Horog	4,420	40	35	124	10	300
Yazgulen-Matraun	3,920	477	386	715	124	3,340
Vantch-Vantch	3,790	1,297	1,037	2,883	220	9,080
Zeravshvan-Fandoria	3,330	1,002	500	2,204	200	11,000
Vakhsh-Tutkaul	3,433	2,981	2,300	3,179	685	25,640
Zeravshvan-Dupuli	3,000	444	90	977	100	5,000
Chu-Kochkora	2,840	22	10	57	20	100
Talas-Budenovskoye	2,680	8	6	10	2	54

and availability of rock destruction products. Sudden high flows are a character-istic property of mud–stone torrents; they are of short duration (minutes, hours) and convey high suspended solids concentrations (10–80% of yearly load). Three torrent typologies can be distinguished: (i) muddy; (ii) mud with stone; and (iii) water with stone. Torrential flows lead to significant damage to channel spawning grounds, riparian communities, highways, roads and farmlands. Flood control is of high social importance. Much work is being done to prevent flood damage, such as the construction of mud–stone dams and other hydraulic struc-tures and the afforestation of steeper slopes. A striking example is the construc-tion of mud–stone dams protecting the town of Alma-Ata (Kovalenko and Fashchevsky, 1986). The first stage of the construction of the 110 m high dam was completed in 1973. The capacity of the mud–stone reservoir was 6.2×10^6 m^3. On 15 July 1973, the greatest mud–stone torrent for 100 years occurred on the Malaya-Almaatinka river. Its velocity was 10 m/s, its height 20 m and the total amount of the rock fragments carried downstream was about 4×10^6 m^3. The mud–stone reservoir filled up and saved Alma-Ata from destruction. To protect the dam against damage during a future torrent, the reservoir was enlarged by 40 m in height and its capacity was increased to 12×10^6 m^3. In addition, some activities were carried out upstream, on Mount Almaatinka, to hinder the pro-cesses of torrent formation. Moraine lakes were released, special retention ponds were built and measures were taken for the restoration of forests and grasslands on mountain slopes.

Flood plains

The flood plain includes parts of the lower river valley and surrounding wetlands inundated during high water periods. Inundation usually takes place in spring–summer (snowmelt) and in summer–autumn (rain). In some years, flood plains can be inundated in winter as well, as a result of heavy early snowmelt leading to the formation of ice jams. Floods play a vital ecological role in the life of aquatic and sub-aquatic ecosystems by ensuring the productivity of meadows and of flood plain forests. One of the early founders of modern flood plain science, Yelenevsky (1936), wrote that floodplains are like huge 'banks' where great potential riches have been stored.

Within flood plains, river margins are characterized by a specific micro-climate resulting from the warming effect of stream water. On large rivers running from south to north, such as the Pechora, the Ob, the Yenisei and the Lena, val-ley slopes situated in the tundra climatic belt are characterized by tundra vegeta-tion, whereas flood plains, at the same latitude, are covered in grasslands typical of more temperate climatic zones. In the River Anadyr flood plain, Shrag (1969) recorded poplars from southern latitudes, whereas within the surrounding water-shed the vegetation was represented by moss and bushes, where birches grew no higher than 50 cm. Spring to summer flood plain inundation with a thin (compared with the main channel) and weakly drained layer of standing water, leads to rapid heating by advective heat and direct radiation. Photosynthesis is intense due to significant heating of shallow water areas. Phytoplankton, followed by zooplankton, develop rapidly. As a consequence, food for fish, water birds and mammals is tenfold greater than within the main channel (Tables 13.3, 13.4 and 13.5).

To maintain the vital functions of the natural flood plain ecosystem, regular inundation is required to irrigate and fertilize the flood plain with suspended particulates and dissolved nutrients. The regular inundation of the flood plain is described with the Russian term: *poemnost*. This word refers to a very important hydrological characteristic of a river regime, related to the quality and productivity of entire flood plains. An example of the influence of *poemnost* on the hay harvest can be appreciated by a comparison of annual yields in the Pechora and Vychegda river basins. The middle and lower Pechora flood plain becomes inundated every year and for a longer period than the Vychegda flood plain. As a result, the amount and quality of hay in natural meadows in the Pechora valley are 1.5–2 times greater than those of the Vychegda flood plain.

Table 13.3. Zoobenthos numbers (specimen/m^2) and biomass in the Vah catchment (Adapted from Fashchevsky, 1996).

Waterbody	Common zoobenthos		Chironomidae	
	(specimens/m^2)	(g/m^2)	(specimens/m^2)	(g/m^2)
River waters				
Channel of Vah	1038	5.03	229	0.40
Tributaries	520	1.99	312	1.13
Additional ponds				
Anabranches	1544	4.51	194	0.33
Backwaters	3630	23.33	720	2.85
Flood plain ponds				
Meander cut-off	7540	79.14	31.60	14.7
Lakes of low flood plain	3705	19.92	14.79	4.61
Lakes of high flood plain	1139	3.72	588	2.60
Outer flood plain lakes				
Reserved	1480	3.02	600	0.63
Flowing	1053	1.27	407	0.30

Table 13.4. Waterfowl numbers in different natural habitats in 10 km^2 plots (Adapted from Fashchevsky, 1996).

	Outer flood plain		Flood plain		
	Forest	Marsh	Forest meadows and temporary ponds	Flat bog	Ponds
Northern Taiga	46	238	782	354	510
Southern Taiga	30	667	318		110

Table 13.5. Zooplankton and benthic invertebrates in different catchments of Komy Republic. (Adapted from Solovkina, 1975.)

Part of river valley	Name of region			
	Verhnepechorsky	Spednepechorsky	Nizhnepechorsky	Usinsky
	Zooplankton (specimens/m³)			
River channel	2,600	3,000	–	4,000
Flood plain	38,000	208,000	274,000	56,000
	Benthic invertebrates (specimens/m²)			
River channel	8,000	16,000	1,000	7,000
Flood plain	20,000	41,000	3,000	8,000

One measure of the ecological and economic value of a river system is the flood plain development coefficient (Fashchevsky, 1986), defined as the ratio between the average width of the inundated surface at the highest water level (1% frequency) and the width of the main channel. The coefficient of flood plain development is calculated as a mean estimated value from the headwaters to the river mouth, such that:

$$K_\text{p} = \frac{K_\text{p1}l_1 + K_\text{p2}l_2 + \ldots + K_\text{pn}l_n}{L} \qquad (13.1)$$

where

K_p = mean estimated value of the coefficient of flood plain development along the river,
K_p1, K_p2, K_pn = coefficients of flood plain development in separate reaches,
l_1, l_2, l_n = lengths of river reaches between characteristic geomorphological river features,
L = river length from source to mouth (km).

The coefficient of flood plain development for separate river reaches is calculated as:

$$K_\text{fd} = \frac{B_0}{B_\text{b}} \qquad (13.2)$$

where

B_0 = width of the inundated flood plain at a given site during high water levels at 1% of frequency (determined by large-scale topographic maps) (m),
B_b = mean channel width, determined from observations based on the relationship $B = f(H)$, where B and H are measured values of channel width and water level.

Table 13.6. Estimates of the natural potential of the Ob and Yenisei catchments.

River	Watershed area ($\times 10^3$ km²)	Precipitation (mm)	Wildfowl Flow (km³)	Coefficient of flood plain development	Fish catch (kg/km³)	Waterfowl reproduction (unit/km³)	Hay harvest (t/km³)
Ob	2,900	543	400	12.2	70,640	6,522	20,000
Yenisei	2,580	560	632	1.9	7,246	497	633

The comparison of the natural potential of the two large rivers, the Ob and Yenisei, located in similar climatic regions (Table 13.6), shows that the Yenisei, having a 30% larger flow than the Ob, yields smaller fish catches both in absolute terms and as fish catches per cubic km of annual flow, even though its waters are only moderately polluted. The mean flood plain development coefficient of the Ob is six times greater than that of the Yenisei. Water bird numbers along the Ob are 13 times higher than those within the Yenisei flood plain. The potential reserves of hay in the Ob flood plain are 20 times greater in absolute terms and 32 times greater when related to flow. The great development of the Ob flood plain is the determining factor of the great natural and economic importance of the Ob basin. The analysis of typical cross-sectional profiles of the Ob and Yenisei, indicating the flood plain development in these river basins, reveals that the Yenisei flood plain is not significantly inundated at the 50% frequency level, and therefore it cannot be as productive. On the Ob, a 50% frequency flood covers most of the flood plain.

Fish productivity is closely connected with flood plain development. A connection could be established between gross and specific fish productivity and the width of the Ob river flood plain. The Ob basin suffered oxygen deficiency for many years, causing fish mortality. Fish productivity could rise by 1.5–2 times if these adverse effects could be eliminated. In the lower section of the river, fish productivity is reduced to 10 t/km² at a flood plain width of 50–60 km, as a consequence of a reduction in food organisms related to the adverse ecohydrological conditions, i.e. low temperature, increased depth and increased current velocity.

Impounded rivers

Many large rivers in Eurasia (as well as elsewhere, such as the Columbia river, USA) are characterized by extensive impounded sections in their lower reaches. The hydrological regime of these rivers is regulated by the releases from their upstream hydroelectric power station reservoirs (HEPSR). Management of regulated hydrological regimes can be used to optimize the status and water quality of heavily modified waterbodies such as these. The overall discharge regime determines the functioning of the main channel ecosystem while diurnal fluctuations in water levels, caused by the peak operational regime of the HEPSR, maintain the health of the secondary network (side-arms, flood plain lakes, ox-bows and other wetlands) and the feedback effect of these on the water quality in the main channel.

Uneven daily HEPSR releases regulate water movement in the main channel and cause 'trouble waves' in downstream pools. Such trouble waves are formed as a reaction to sudden water releases. Trouble waves, as a signal of such events, move downstream with velocities that exceed current velocities by tenfold or higher. These waves create gradients between the main channel and the tributary network. The release volume can be measured, but an evaluation of water exchange requires the solution of a number of hydrological problems. One of them is the estimated transformation of the trouble waves as they move downstream from the HEPSR along the impounded section.

The change in the amplitude of water level fluctuations along the section can be described by the following exponential equation:

$$\Delta H_L = \Delta H_{HEPSR} \exp(-aL) \tag{13.3}$$

where

ΔH_L = water level fluctuation amplitude at distance L downstream from the HEPSR,
ΔH_{HEPSR} = water level fluctuation amplitude in the downstream pool,
L = distance from the HEPSR,
a = empirical coefficient in the range 0.02–0.03.

The calculation of water level fluctuations at any subsection provides an evaluation of water exchange between the main channel and the tributary network. In different subsystems calculation methods may be different, however. For determining the inflow (or outflow) into (or from) flood plain waterbodies, it is necessary to take into account both the morphology of these waterbodies and the morphology of the river branches that connect them to the main channel. For example, in the case of the Dnieper river section downstream from the Kakhovka HEPSR, the daily water exchange between flood plain lakes and the main channel (W_l) can be calculated by the following equation (Timchenko, 1996):

$$W_l = 0.0066 V_l h_l^{-1} - (ln^2 b^{-2} h_b^{-3.33})^{-0.205} \Delta H_L \tag{13.4}$$

where

V_l, h_l = volume and average depth of the lake,
l, b, h_b, n = length, breadth, depth and roughness coefficient of the river branch.

The volume of water entering the flood plain is evaluated by comparing the amplitude of water level fluctuation within the main channel observed at low water level, which does not prompt inundation of the flood plain, with the amplitude of water level fluctuation at higher water level, corresponding to flood plain inundation (Timchenko, 1990).

Water quality

Dissolved oxygen concentration and biochemical oxygen demand are integrated indices of ecosystem status and water quality. They reflect the structure

and functioning of aquatic ecosystems as a whole, and stress the dependence of
organic matter production and metabolic processes on hydrological parameters
and abiotic components of ecosystems. The dynamics of variables such as dissolved oxygen and organic matter (expressed as the total bacterial oxygen
demand, BOD_{tot}) in impounded river sections are dependent on water regime
parameters, which can be expressed by means of the following equations. For
the dissolved oxygen concentration:

$$C_n = C_0 + \sum_{i=1}^{n} \frac{W_{tr,i}}{W_i}(C_{tr,i} - C_{i-1})$$

$$+ \sum_{i=1}^{n} \frac{\tau_i W_{r,i}}{W_i}(A_{r,i} - R_{r,i} + At_{r,i} - G_{r,i}) \qquad (13.5)$$

$$+ \sum_{i=1}^{n} \frac{\tau_i W_{ap,i}}{W_i}(A_{ap,i} - R_{ap,i} + At_{ap,i} - G_{ap,i})$$

and for the BOD_{tot}:

$$C_n = C_0 + \sum_{i=1}^{n} \frac{W_{tr,i}}{W_i}(C_{tr,i} - C_{i-1})$$

$$+ \sum_{i=1}^{n} \frac{\tau_i W_{r,i}}{W_i}(A_{r,i} - R_{r,i} + F_{r,i}) \qquad (13.6)$$

$$+ \sum_{i=1}^{n} \frac{\tau_i W_{ap,i}}{W_i}(A_{ap,i} - R_{ap,i} + F_{ap,i})$$

where

C_0 = dissolved oxygen concentration in the water entering the impounded
section of the river through the HEPSR (mg O_2/dm³),
C_n = dissolved oxygen concentration at the outlet of the section (mg O_2/
dm³),
$C_{tr,i}$ = dissolved oxygen concentration in the tributary flowing at subsection
i (mg O_2/dm³),
n = number of all subsections,
W_i = volume of flow at section i, the sum of the flow volume in the
HEPSR downstream pool (W_{HEPSR}) and flow volumes of the tributaries
($W_{tr,i}$) entering before subsection i (m³/day),
$W_{ap,i}$ = water exchange between the main channel and tributary network (m³/day),
$W_{r,i}$ = that part of the flow passing through the main channel, equal to
$W_i - W_{ap,i}$ (m³/day),
A_i = gross primary production by phytoplankton, phytobenthos, periphyton
and macrophytes at subsection i (mg O_2/dm³/day),
R_i = oxygen demand consumed by organic matter catabolism (respiration)
in the plankton, benthos, periphyton and macrophyte communities at
subsection i (mg O_2/dm³/day),

At_i = atmospheric aeration at section i (mg O_2/dm³/day),
G_i = chemical oxygen demand at section i (mg O_2/dm³/day),
F_i = outer organic matter load at section i (mg O_2/dm³/day),
τ_i = lag (residence) time of water at section i (days).

The gross primary production A (mg O_2/dm³/day) is estimated on the basis of specific algal oxygen production capacity a (mg O_2/dm³/day) and algal biomass B (mg/dm³), such that:

$$A = aB \qquad (13.7)$$

These parameters are modified by different environmental conditions. In impounded river sections they are largely dependent on flow (Romanenko *et al.*, 1990). Specific production capacity values are inversely related to algal biomass, while algal biomass is inversely related to flow volume and hence to HEPSR releases (Fig. 13.1).

Biological organic matter decomposition R (mg O_2/dm³/day) is estimated as the oxygen demand used up in community respiration. This is calculated as a product of specific oxygen demand r (mg O_2/dm³/day) and biomass B (mg/dm³):

$$R = rB \qquad (13.8)$$

Bacterial oxygen demand BOD_{tot} (mg O_2/dm³/day) depends on the concentration of organic matter and is calculated by means of a bacterial decay coefficient (k_b):

$$R = k_b\,BOD_{tot} \qquad (13.9)$$

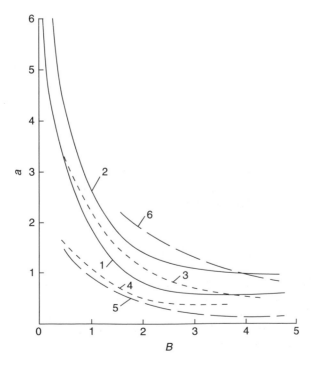

Fig. 13.1. Specific production capacity of algae at optimal depth (a, mg O_2/mg/day) as a function of biomass (B). Phytoplankton (B, mg/dm³): 1, in the main river channel; 2, in the append-age network of the Kaniv Reservoir's Kyiv section. Microphytobenthos (B, mg/cm²): 3, in the main channel; 4, in the flood plain waterbodies of the Dnieper river mouth zone. Phytoperiphyton (B, mg/g macrophyte biomass): 5, on the emergent; 6, on the submersed macrophytes at the Kaniv Reservoir's Kyiv section.

where k_b (mg $O_2/dm^3/day$) $= -R_1/BOD_{tot}$. R_1 is the daily bacterial respiration (mg $O_2/dm^3/day$).

The atmospheric aeration At_i (mg $O_2/dm^3/day$) is estimated from:

$$At_i = (C_s - C_{i-1})(1 - 10^{-k_2\tau})$$ (13.10)

where

C_s = saturated oxygen concentration (mg O_2/dm^3),
C_{i-1} = oxygen concentration at the upper line of subsection i (mg O_2/dm^3),
k_2 = atmospheric aeration coefficient (1/days) related to current velocity (u), depth (h), wind velocity (w) and temperature (T) such that

$$k_2 = 1.024^{T-20}(0.067wh^{-1} + 1.6u^{0.5}h^{-1.5})$$ (13.11)

The chemical oxygen demand is measured experimentally.

These equations have been applied to the impounded sections of the Dnieper river downstream of the Kyiv HEPSR at the Kaniv Reservoir's Kyiv section (Timchenko and Oksiyuk, 2002) and downstream of the Kakhovka HEPSR in the Dnieper river mouth zone (Timchenko *et al.*, 2000). At the Kaniv Reservoir's Kyiv section (Fig. 13.2), during the summer low-water period, the status of the ecosystem and the water quality deteriorated as a result of the dissolved oxygen deficit. Within the river section dependent on the Kyiv HEPSR releases (350, 650, 950 and 1250 m³/s) dissolved oxygen dynamics and daily water level fluctuations (at Kyiv HEPSR downstream pool: depth 0.5, 1.0, 1.5 and 2.0 m) were modelled by means of equation (13.10) (Fig. 13.3).

Controls of ecosystem status and water quality at the Kaniv Reservoir's Kyiv section during the summer low-water period are as follows:

1. Increase of the Kyiv HEPSR releases.
2. Provision of daily water level fluctuations for intensifying water exchange between the main channel and the tributary network.

A satisfactory dissolved oxygen concentration (above 70% saturation) is provided by Kyiv HEPSR releases of 950 m³/s and above. At middle (650 m³/s) and low (350 m³/s) releases, the dissolved oxygen concentration falls to 40–50%

Fig. 13.2. Location of the study areas: 1, the Kaniv Reservoir's Kyiv section; 2. the Dnieper river mouth (HEPS, hydroelectric power station).

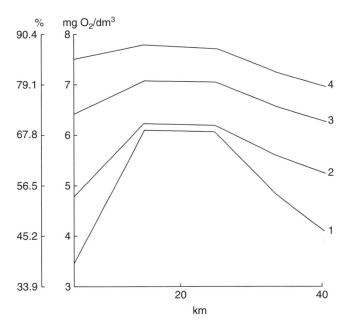

Fig. 13.3. Dissolved oxygen concentration dynamics in summer at the Kaniv Reservoir's Kyiv section following changes in releases of the Kyiv hydroelectric power station reservoir (m^3/s): 1, 350; 2, 650; 3, 950; 4, 1250 (water level variability in the hydroelectric power station downstream pool: 1 m).

saturation, as a consequence of which the ecological and sanitary conditions at the Kyiv section deteriorate. A positive effect of the tributary network on dissolved oxygen concentrations in the main channel is noticeable at low releases, such as between 350 and 650 m^3/s. At both low and high release volumes, water exchange between the main channel and the tributary network, dominated by the Kyiv HEPSR peak operating regime, plays a very important role for ecosystem health within waterbodies of the tributary network.

At the Dnieper river reservoirs and at the mouth of the Dnieper, the HEPSR releases necessary to support the functioning of aquatic ecosystems and to conform with water quality standards during summer baseflow were determined. Organic matter production and destruction were estimated as changes in BOD_{tot} (ΔC_{BOD}) in the whole system, and BOD_{tot} dynamics along the Dnieper downstream of the HEPSR were calculated using equation (13.11).

Releases from the Kakhovka HEPSR determine flow volume and act as major controls for the status of the Dnieper delta ecosystem (Fig. 13.4). In the main channel, at low releases (average daily discharge <470 m^3/s), an increase in the concentration of labile organic matter occurs (BOD_{tot} >13 g O_2/m^3), while at high releases self-purification processes predominate, with an average decrease of BOD_{tot} of some 4.2 g O_2/m^3.

During summer, at average daily releases of less than 1500 m^3/s, deltaic waterbodies are characterized by a water exchange period of 5–12 days. Under such

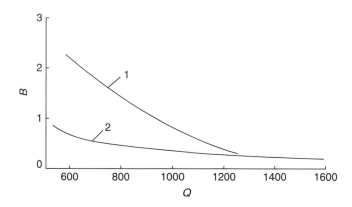

Fig. 13.4. The dependence of phytoplankton biomass (B, mg/dm^3) on releases of the hydroelectric power station reservoir (Q, m^3/s): 1, at the Kaniv Reservoir's Kyiv section; 2, in the Dnieper river mouth zone.

conditions the production/catabolism balance is positive and results in a high bioproduction potential. During this period, at any release rate, flood plain waterbodies supply organic matter to the channel network. A large proportion of this organic matter (up to 5 g O_2/m^3 and more) is produced at releases below 650 m^3/s. Wetlands possess a significant self-purification potential at any release rate; however their influence on ecosystem status and water quality in the main channel is small.

Overall, at an average daily discharge of 470 m^3/s, the production and destruction processes in the main channel at the delta are balanced. This organic matter equilibrium flow, that could be called 'ecosystem release', represents a characteristic discharge level that depends on internal ecosystem processes (Fig. 13.5). The total Kakhovka HEPSR release, necessary for the neutralization of the organic matter that is contributed both by the ecosystem itself and by the pollution that enters the main channel from outside in the form of anthropogenic load, can be termed instead an 'ecological release'. This is not a constant amount but includes the 'ecosystem release' together with a 'sanitary addition', ensuring the system's self-purification necessary to counterbalance anthropogenic pollution. The ecological release is always greater than the ecosystem one. In our case, an anthropogenic load of 43 t/day would require an ecological release of about 530 m^3/s.

The Kakhovka HEPSR releases support the functioning of the Dnieper deltaic wetlands (Fig. 13.6) from 470 to 1500 m^3/s. At higher releases, primary and secondary production are reduced and fisheries face financial losses.

Managing quality through dilution of waste waters and self-purification

Modern treatment plants can remove approximately 90–95% of organic matter and 10–40% of inorganic matter from waste water. In order to maintain water quality standards, it is often necessary to dilute treated effluents up to 10–50 times.

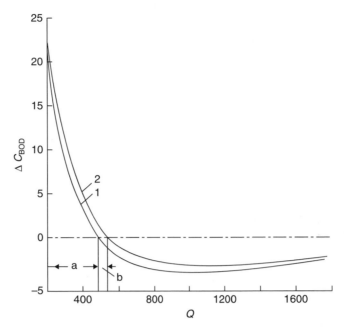

Fig. 13.5. Determination of release needs (Q, m³/s) from the Kakhovka hydro-electric power station reservoir by the organic matter concentration changes (ΔC_{BOD}, g O_2/m²) in the Dnieper mouth (subsection 3): a, ecosystem release; b, sanitary addition: a+b, ecological release; 1, with anthropogenic load; 2, without anthropogenic load.

Efforts to enhance treatment to 98–99% lead to an increase in electrical power consumption by as much as two- to tenfold. Dilution and self-purification processes in the receiving waters can reduce this consumption and are thus of economic as well as of ecological importance. The role of ecotonal riparian habitats in self-purification can be assessed from the Dnieper case study (Figs 13.7 and 13.8).

Dilution, solution and mixing are three relevant hydrodynamic factors involved in self-purification. Turbulent river regimes provide proper mixing of pollutants and decrease the concentration of their bioavailable fractions. Common waste-water dilution models include the Frolov–Rodzilles model for rivers and the Ruffel model for lakes and reservoirs (Lapshev, 1977).

The intensity of dilution processes is affected by multiple dilution, which can be expressed by the relationship:

$$h = \frac{\gamma Q_f + Q_w}{Q_w} \qquad (13.12)$$

where

h = multiple dilution coefficient,
Q_w = waste water discharge (m³/s),

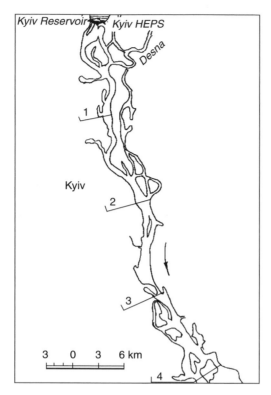

Fig. 13.6. The Kaniv Reservoir's Kyiv section, showing subsections 1–4 (HEPS, hydroelectric power station).

Q_f = natural discharge (m³/s),
γ = dimensionless mixing coefficient.

The mixing coefficient is calculated by the formula:

$$\gamma = \frac{1 - \exp(-a\sqrt[3]{L})}{1 + (Q_t / Q_w)\exp(-a\sqrt[3]{L})} \tag{13.13}$$

where

L = distance between the section under consideration and the waste water outlet,
a = coefficient accounting for hydraulic conditions, such that

$$a = \xi k_t \sqrt[3]{\frac{D}{Q_w}} \tag{13.14}$$

In the above:

ξ = coefficient accounting for the condition of the waste water inlet (for a bank inlet $\xi = 1$, for a channel inlet $\xi = 1.5$),
k_t = coefficient of tortuosity, determined as the relationship between the complete channel length from the inlet until the section under consideration (l) and the distance between these sections along a straight line (l_n),

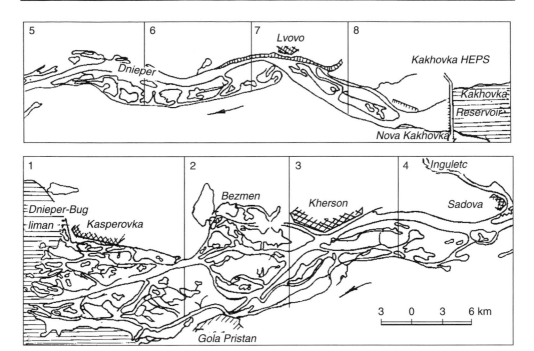

Fig. 13.7. The Dnieper mouth, subsections 1–8 (HEPS, hydroelectric power station).

D = turbulent diffusion coefficient estimated by

$$D = \frac{gVH}{37h_r C^2} \tag{13.15}$$

where

g = acceleration due to gravity (m/s^2),
V = mean velocity of the watercourse (m/s),
H = mean stream depth (m),
h_r = roughness coefficient of the river bed determined either by the use of classical tables (such as those of Srybny) or determined experimentally,
C = Chezy coefficient (m$^{1/2}$/s).

Research carried out at reservoirs and lakes show that their conditions of waste water mixing differ greatly from those of rivers. This is due to weak currents, which tend to slow down the transport of diluted waste water at waste water inlets and to episodic wind-driven currents which tend to carry these flows away in any direction. At first, the degree of pollution decreases rapidly but a complete self-purification takes place at very large distances from waste water inlets.

The waste water dilution estimation method suggested by Ruffel (Lapshev, 1977) for lakes and reservoirs was experimentally tested. The initial multiple dilution is estimated according to the location of the waste water inlet.

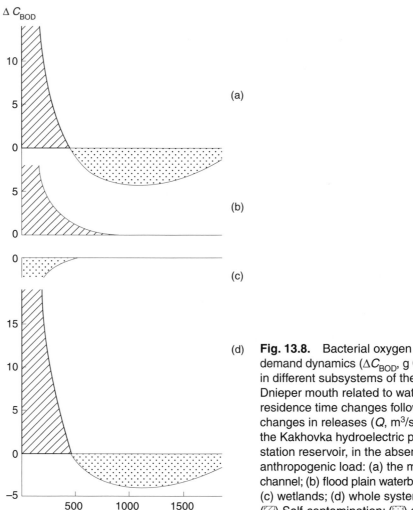

(a)

(b)

(c)

(d) **Fig. 13.8.** Bacterial oxygen demand dynamics (ΔC_{BOD}, g O_2/m^3) in different subsystems of the Dnieper mouth related to water residence time changes following changes in releases (Q, m^3/s) from the Kakhovka hydroelectric power station reservoir, in the absence of anthropogenic load: (a) the main channel; (b) flood plain waterbodies; (c) wetlands; (d) whole system. (▨) Self-contamination; (▦) self-purification.

1. When the inlet is in the upper third of the depth or close to shallow banks:

$$n_H = \frac{Q_w + 0.0118H^2}{Q_w + 0.000435H^2} \qquad (13.16)$$

2. When the inlet is in the lower third of the water column, a modified relationship is applied:

$$n_H = \frac{Q_w + 0.0087H^2}{Q_w + 0.000475H^2} \qquad (13.17)$$

where

Q = waste water discharge (m³/s),
H = mean depth in the bank region (m),
n_H = initial multiple dilution.

The complete multiple dilutions can also be calculated differentially.

1. For the outlet at the bank:

$$n_{com} = 1 + 0.412 \left(\frac{L}{6.53H^{1.57}} \right)^{[0.627+(0.000031L/H^{1.167})]} \tag{13.18}$$

2. For an inlet far away from the bank:

$$n_{com} = 1.85 + 2.32 \left(\frac{L}{4.41H^{1.167}} \right)^{[0.41+(0.0014L/H^{1.167})]} \tag{13.19}$$

In the above, n_{com} is the complete multiple dilution at distance L (m).

It is known that conservative substances in natural waters remain constant for a long period. These usually include the major ions (Cl^-, SO_4^{2-}, Mg^{2+}, Na^+, K^+) and some minor ones, present at low concentration (Sr^+, Li^+).

The self-purification capacity C_{sp} of separate sites is expressed as a percentage of pollution decrease against the initial concentration:

$$C_{sp} = \frac{C_{in} - C_f}{C_{in}} \times 100 \tag{13.20}$$

where

C_{in}, C_f = concentrations of pollutants in the initial and in the final section (mg/l)

and the rate of total self-purification processes in stream water or in lakes is estimated by the relationship

$$K = \frac{1}{\tau} \ln \frac{C_{in}}{C_f} \tag{13.21}$$

where

K = self-purification coefficient or rate of pollution decrease (days),
τ = empirical constant,
other symbols are the same as in equations (13.16) and (13.17).

Self-purification coefficients are determined by the following processes occurring in rivers and lakes:

1. Processes resulting from biochemical conversions within the water column, within suspended sediments and in bottom deposits.
2. Processes resulting in chemical oxidation by the consumption of dissolved oxygen and by photochemical oxidation.

3. Processes resulting from sorption and desorption, particle settling and coagulation.

4. Biological effects due to trophic chain dynamics involving bacteria, vegetation and animal aquatic communities.

The intensity of each process is characterized by self-purification coefficients related to factors such as current velocity, water and air temperature, solar radiation and oxygen content, and microorganisms and invertebrates, aquatic and riparian vegetation.

Oxygen concentrations increase in rivers and lakes due to aeration enhancement, which results from natural factors such as wind turbulence, high water velocities during floods, channel slope and roughness; human activities such as aeration plants and reservoir releases; and photosynthesis. These lead to the oxidation of organic and inorganic pollutants and to the renewal of ecological conditions.

The kinetics of the mechanisms behind the oxidation of organic compounds in rivers and lakes is specified by the chemical structure of the substances and by state conditions, such as pH, water temperature and ultraviolet radiation. Organic matter disintegration occurs mainly under the influence of microorganisms. At first, the carbon skeleton of organic compounds undergoes transformation with the release of carbon dioxide and water; subsequently nitrogen is oxidized to nitrates and nitrites.

Changes in physical and chemical properties of water cause changes in vegetation and in animal communities. Many fish species cannot persist under conditions of high suspended sediment concentrations, which favour the development of pathogenic organisms while settling through the water column. Suspended matter accumulates in motionless, oxygen-free zones, where the decaying organic matter has toxic effects on the biota. Self-purification processes in rivers polluted by suspended matter are much slower than in rivers affected by dissolved pollution loads. Self-purification from suspended particles depends on the ratio of mineral to organic compounds within the suspension. It is related to the decrease of water turbidity and colouring, and to the renewal of the conditions for the development of phyto- and zooplankton.

In addition to the chemical factors of water pollution produced by industrial and agricultural discharges, bacterial pollution also contributes by affecting the ecological status of rivers and lakes. As a result, streams may accumulate pathogenic bacteria and viruses causing dangerous diseases such as cholera, dysentery and typhus.

The reduction of bacterial contamination of waterbodies is determined by the death of microorganisms due to effects of physical, chemical and biological factors. It should be considered that each type of bacteria is characterized by specific survival characteristics. According to Cherkinsky (1971), in flood plain rivers bacteria decrease by over 90% within 60–70 h, but this is five to ten times faster in mountain rivers. The self-purification capacity of water sources should be adequately considered when choosing a water intake location.

Rtshiladze (1951) found that the percentage of total dead bacteria indicated that self-purification processes of the River Cura downstream of Tbilisi occurred in a similar fashion in winter, summer and autumn. They terminated at a distance of 38 km from the most polluted region, within 9.5 h. During spring floods, flow velocities increased and self-purification terminated at the same distance but within 5 h. Up to 98–99% of intestinal bacilli died during summer and autumn. Koltunova (1951) showed that self-purification processes are inhibited by copper at concentrations higher than 0.14 mg/l, under the same conditions. Consequently, biochemical oxygen demand was depressed and plankton development was feeble.

Biological pollution in rivers and lakes originates from organisms and from decay products associated with the production of organic matter and its subsequent death. A first contribution of organic matter is due to the leaching of surface soils during rains and snowmelt; a second is derived from primary producers such as plankton and benthic algae and higher aquatic vegetation; and a third from the mass development of microorganisms and their subsequent death. The majority of aquatic organisms are capable of purifying rivers and lakes of pollutants, by means of biogenic substances, through the formation of heavy metal salts and the complexion of phenols, pesticides and oil products. Higher aquatic vegetation plays a great role in these processes, as it accumulates pollutants, protects the riparian fringe of rivers, lakes and reservoirs from wave effects and bank erosion, improves the oxygen regime through photosynthesis and assists the sedimentation of the suspended load. According to Smirnova (1986), higher aquatic vegetation accumulates a great number of biogenic elements, heavy metals and pesticides, purifying the waters from these pollutants in the mouths of the Danube, Dnieper and Dniester (Table 13.7).

Table 13.7 shows that the aquatic plants of the Danube contain the lowest quantity of DDT and γ-BHC, due to the high turbidity of the Danube waters. Suspended particles absorb organochlorine substances and render them less accessible for absorption by plants. Reeds accumulate heavy metals mainly through the root system, transporting them into bottom deposits. Pesticides, metals and organic contaminants accumulate in macrophytes and can be transferred to humans through the trophic chain. Fish consuming contaminated water plants can show symptoms of heavy metal poisoning. Biological self-purification processes were enhanced by the introduction of planktonic algae into ponds designed for the treatment of waste water coming from a tuberculosis sanatorium (Telitchenko, 1972). The ponds were loaded with several million algal cells per litre of waste water. The algal polynoculate (AP) consisted of Cyanobacteria, Chlorophyceae and Bacilliophyceae. The introduction of AP caused extensive blooming in the ponds, which resulted in the mineralization and disinfection of waste waters. Experimental data showed that, 12–14 days after inoculation, tuberculosis mycobacterium disappeared from the experimental ponds, while in test ponds without AP, their quantity decreased only by 1.5–3 times in 3 days. Long-term exposure was accompanied by the pollution of waterbodies with dying phytoplankton until that had to be removed from the medium and used for fertilizing crops.

Alimov (1989) showed that the filter-feeding zooplankton of Red Lake were able to remove approximately 7110 t of suspensions within 6 months, 4060 t of which were accumulated by them and 2030–3050 t were returned to cycling. Kruchkova (1972) showed the great efficiency of zooplankton as an agent of self-purification. Experiments showed that inoculations into waste-water treatment ponds reduce the number of bacteria by 16–3000 times.

Table 13.7. Accumulation of biogenic elements, heavy metals and pesticides in reeds and pondweed (mg/g dry substance). (Adapted from Smirnova, 1986.)

Sampling site	N	P	K	Mn	Fe	Pb	DDT	DDE	γ-BHC
Danube[a]	1.86/0.91	0.18/0.17	2.89/2.03	99/299	240/1591	2.4/4.4	1.0/0.3	0.59/0.25	0.69/8.24
Dnieper[a]	1.59/0.86	1.19/0.12	0.82/0.86	257/400	750/4500	0/0	0/0	1.3/2.8	60/99
Dniester[a]	1.32/0.83	0.14/0.24	1.84/1.47	0.13/1815	309/5811	29.6/49.5	0/1.5	2.0/3.5	0/36
Danube[b]	2.70	0.44	2.75	469	2000	16.0	0	0	1.5
Dnieper[b]	1.56	0.33	3.35	937	225	4.2	2.77	8.35	6.4

[a]Common reed; [b]perfoliate pondweed. The numerator indicates the content of pollution in elevated bodies; the denominator the content of pollution in underground bodies.

The most dangerous river and lake pollutants are man-made radioactive isotopes derived from nuclear tests, accidents of atomic power stations, mining and processing of radioactive materials. Radioactive pollutants enter waterbodies in dissolved or in colloidal state, as part of precipitated salts and in sorbed form on the surface of suspended loads. Water radioactivity and its harmful effects on organisms are determined by nuclear processes (α, β and γ radiation), which are very difficult to modify by physical and chemical methods. Timofeeva-Resovskaya (1963) showed that the mean distribution of radioactivity among water, soil and biomass (27:45:28) differed widely in relation to the mass of the components considered (85:14:0.1, respectively). According to the character of radionuclide distribution among water, soil and biomass, contaminants can be divided into four categories:

1. Hydrotons, which remain predominantly in the water phase ($>70\%$), such as sulfur-35, chromium-51 and germanium-71.
2. Pedotrons, which concentrate in soils ($>75\%$), such as caesium-137, yttrium-91, iron-59, zinc-65 and zirconium-95.
3. Biotrons, which accumulate in biomass ($>50\%$), such as mercury-203, phosphorus-62, cadmium-115 and cerium-144.
4. Equitrons, which distribute uniformly among water soil and biomass, such as cobalt-60, strontium-90, rubidium-86, ruthenium-136 and iodine-131.

Aquatic vegetation and animal organisms living in lakes accumulate a great number of radionuclides at low concentration from the water column. Settling ponds for treating waters before their outfall to the hydrological network are used when radioactive pollution is not heavy. When significant concentrations of persistent radionuclides are present in waste, they have to be deactivated and subsequently buried. Methods for concentrating radioactive pollutants include vaporization, bio-filters, ion-exchange resins and coagulation. The burial of the concentrates is carried out by building concrete, clay and other screens. The more natural is a river, the more intensive are the self-purification processes from radionuclides (Fashchevsky, 1996). This self-purification from radionuclides depends on many factors, including physical and chemical water characteristics, the concentration of suspended loads, the presence of aquatic organisms and flow velocity. Higher aquatic plants and algae accumulate radionuclides and reduce their concentration in the aquatic medium.

'Ecological' discharges

Ecological status has to preserve not only economically profitable organisms (fish, mammals, birds, vegetation) but also the sustainability of organic matter cycling by the biota. Such conditions can be obtained by providing the necessary hydrological characteristics, such as water levels and discharges, flow velocity, depth, turbidity, gas regime and water temperature. It is necessary to distinguish between the quantitative depletion of water resources caused by natural processes (deficit precipitation, large evaporation with high temperature) and anthropogenic impacts, which often damage interrelationships between the living and non-living components of nature and cause the degradation of aquatic ecosystems.

In regulated rivers flood peaks are cut off, but discharges, water levels and current velocity are increased during periods of low flow, influencing vital cycles of aquatic flora and fauna. The preservation of aquatic ecosystems requires the limitation of the admissible degree of regulation so as to grant ecological flows (Fashchevsky, 1989). The quantitative determination of ecological flows stems from the interrelationship between elements of the hydrological regime and the dynamics of the ecosystem within the flood plain and the channel.

The term 'ecological flow' implies a comprehensive approach, taking into consideration all phases of the discharge regime. Terms such as minimal, required or admissible cannot be used to describe ecological flows. Ecological flows generated by snowmelt and rain floods in a year of 25% frequency are considered to be equal to the natural flow of 50% frequency! The consideration of ecological flows involves the inclusion of the whole natural complex of river systems: fish, amphibians, reptiles, birds, mammals, and their habitats such as flood plain meadows and forests, on top of agricultural and fisheries concerns. If we consider the harvest of flood plain meadows in terms of hay stocks (or supply), fish catches and mammal reproduction in the light of hydrological characteristics, it appears that decreases in biological productivity are observed during both very low flows and very large ones. Long-term averages indicate that it is during average discharge years that biological productivity tends to reach its maximum value.

Irreversible changes have not occurred during the last 1000 years (repetition once in 1000 years under 0.1–99.9% frequency) in the greater part of river and lake ecosystems under natural conditions, although extreme high- and extreme low-flow years have occurred (with 5–10% and 90–95% frequency). As soon as flow conditions close to the average discharge year occur (40–60% frequency), the ecosystem is restored to an equilibrium condition. The ecosystem's homeostasis is provided by the rhythmic variation of its hydrological characteristics in a vertical dimension (flow magnitude), in a horizontal dimension (season) and in frequency (from year to year). In the River Ob, years 1967 and 1968 were close to 95% frequency. The restoration of the meadow ecosystem occurred 1–2 years later, the fish productivity recovered between 4 and 10 years for separate species (pike, sturgeon, salmon).

Rivers systems play an important role in human life as water supply, power resources, transport and sanitary systems. One of the most important factors preserving river ecosystem stability is the channel-forming discharge. Antropovsky (1970) calculated that in the majority of flood plain rivers, the channel-forming discharge is near to the maximum values of 50% frequency. If this condition is not fulfilled the vertical erosion is replaced by the lateral one (Chalov, 1979). For the computation of the channel width compatible with dynamic stability, Karasyov (1975) introduced the following formula:

$$B \le 3.65(Hd)^{0.25}\left(\frac{H}{i}\right)^{0.5} \tag{13.22}$$

where

B = river width (m),
Q = discharge (m³/s),

H = mean depth (m),
i = slope (‰),
d = value of the particles belonging to the channel-forming sediment fraction.

The resistance to change and reliability of the ecosystem as a whole can be expressed by the probability of its reliable functioning during long periods under given boundary conditions. The stability of the river system depends on the stability of its separate components. The reliability and stability of ecosystems can be described by probability distribution curves, comprised within 0.1–99.9% frequency intervals. Such distribution can be employed to characterize the river flow, the oxygen regime, the flood plain meadow harvest and the fish catch. Table 13.8 describes the parameters of frequency curves characteristic of some ecosystem components within the middle Ob. The relative dispersion expressed by the coefficient of variation is of the same order both for flow characteristics and for living components, e.g. fish catches, harvest of flood plain meadows, mammal pelts. However, the coefficients of variation in fish catches and pelt storage are approximately two times larger; this can be explained by a dependence of living natural components on other physical factors such as temperature of water and air, light, pollution and from biotic interactions. Variation coefficients for flood plain vegetation practically coincide with flows produced by snowmelt. The lower limit of admissible changes can be estimated by a comparison of the degree of damage equal to natural components in a range of observed years. The majority of flood plains in spring fail to become inundated at a discharge frequency below 95%. Similarly no inundation occurs during floods of 99% frequency.

In winter, when many rivers are covered with ice, oxygen contents may decrease below 3 mg/l in years characterized by a 95% frequency discharge. In smaller rivers, 95% frequency years may result in a decrease in oxygen content to such a degree that the ice freezes together with the shoals and causes the destruction of salmon eggs. The upper limit, corresponding to the most favourable ecological conditions, is a year characterized by a 50% frequency discharge. Relating the 50% frequency of natural flow to the 25% frequency of ecological flow, and the 99% natural flow to the 95% ecological flow with a log-normal relationship, yields the frequency curve for an ecological flow. As soon as water

Table 13.8. Parameters of frequency curves.

N_{in} order	Component of river ecosystems	Mean value	Coefficient of variation	Coefficient of asymmetry
1	Maximum of snowmelt flood	18,192 m³/s	0.27	0.57
2	Minimum of winter flow	832 m³/s	0.23	1.20
3	Sturgeon catches	18.8 t/year	0.59	1.03
4	White salmon catches	24.5 t/year	0.89	1.9
5	Pike catches	118 t/year	0.52	0.74
6	Flood plain meadows hay harvest	2.5 t/year	0.29	1.40
7	Mammals	24,500 pelts/year	0.52	1.24

levels change, many physical and chemical properties change as well. Therefore, ecological flow serves as a complex index, including all the mentioned hydrological characteristics.

The quantification of ecological flow includes the following operations:

- Calculation of matrices of average monthly and annual discharges.
- Calculation of natural annual flows (using a modified Alekseev's method from Fashchevsky and Tamela, 1976).
- Estimation of the matrices of natural monthly flows (by Moklyak's method, 1976) or analysis of short series of within-year observations.
- Relative (within-year) distribution of ecological flow.

For rivers with mean ecological significance, the following formula permits the coefficient of decrease to introduce into the monthly values of discharge:

$$\tau = 1 - \frac{Q_{95\%}^{min}}{Q_{95\%}^{max}} \qquad (13.23)$$

where

τ = coefficient of flow decrease,
$Q_{95\%}^{min}$ = minimum monthly discharge of 95% frequency,
$Q_{95\%}^{max}$ = maximum monthly discharge of 95% frequency.

In Belarus, where most of the ameliorated rivers are straightened, discharges are not accompanied by water level increases. The analysis shows that when velocities grow by 1.5–2 times, water depth decreases. Therefore, under the conditions of ameliorated rivers, it is necessary to calculate ecological discharges and levels taking into account the modification of the natural hydrological network, since the stage–discharge relationship is altered. Ecological flow makes up of a greater part of the water budget of a given river basin: i.e. 65–92% of annual flow (Fashchevsky and Fashchevskaya, 2003).

Ecological Criteria for the Siting and Construction of Reservoirs

Water resources are very variable in surface, unlike terrestrial biomes, forests and mineral resources. Water level fluctuations cause expansion and retraction of aquatic ecosystems not withstanding administrative boundaries. In so doing, they create new landscapes and new habitats. The greatest changes in riverine ecosystems have happened due to reservoir construction and its consequent flow regulation, drainage and deforestation. Reservoir construction can have positive economic but negative ecological consequences. The positive side is very clear: energy production, water supply to industrial centres, irrigation and waterway improvement, recreation. The negative impact is perhaps not as evident but real experience shows the following picture. In upstream reaches:

- Development of wind abrasion.
- Restructuring of new banks and riparian area transformation.

- Flooding of large territories including forest, meadows, agricultural lands and settlements.
- Change of water quality (dissolved oxygen, nutrients, etc.).
- Change of thermal regime and in particular the formation of ice.
- Accumulation of pollutants in bottom deposits.
- Change of water levels and water velocity regimes.
- Hindrance of fish migration and separation from spawning grounds.

In downstream reaches:

- Flood plain desiccation as a result of changes in water regime (including cessation of flood plain inundation, artificial baseflow increase).
- Change in water quality.
- Increased erosion capacity owing to clarified water inflows from upstream reaches.
- Reduction of water levels due to an increase in the erosion capacity of the current.
- Change in thermal regime (can be warmer or colder depending on the depth of water uptake).
- Reduction of channel-forming and flood-forming discharges.
- Change in the icing regime.
- Change in local climatic conditions (increase in humidity, wind velocity).

Direct water withdrawal from rivers produces lower impacts than the construction of a reservoir, but a significant effect is still apparent on the flood plain below the point of abstraction, especially in small waterbodies. Deforestation can be caused by extensive forest clearance, or by forest flooding due to reservoir construction. In the first case, an intensification of erosion processes and washoff from the surface of the basin causes of a large volume of soil to be transferred to the river and the reservoir. This process speeds the silting up of the reservoir and changes the natural regime of suspended and bottom sediments in the river. In the second case, a worsening of the water quality occurs due to the rotting of wood. In both cases a reduction in the oxygen balance occurs. Many investigations have been made establishing ecological criteria for the siting of reservoirs and dams (Fashchevsky, 2002). These can be summarized as follows.

1. A land-use coefficient, calculated as the ratio of the flooded area and the potential installed capacity or output:

$$K_u = \frac{\text{flooded surface (km}^2)}{\text{potential installed capacity } (\times 10^6 \text{ GW})} \tag{13.24}$$

2. Coefficient of reservoir widening owing to wind erosion, measured from the initial bank width under mean stage to the bank width after 20 years, applied to reservoirs located in friable soils:

$$K_w = \frac{\text{forecasted water area (km}^2)}{\text{inital water area (km}^2)} \tag{13.25}$$

3. Coefficient of dissolved oxygen reduction within reservoirs during different seasons, in comparison to the natural during a 75% frequency of annual river discharge:

$$K_{fox} = \frac{\text{forecasted } O_2 \text{ concentration in the reservoir (mg/l)}}{\text{natural } O_2 \text{ concentration within the stream (mg/l)}} \qquad (13.26)$$

4. A eutrophication coefficient, calculated as the ratio of the forecasted total phosphorus (TP) concentration in the reservoir after 10 years of impoundment to the forecasted TP concentration in the reservoir during the first year of impoundment:

$$K_{eu} = \frac{\text{forecasted TP concentration after 10 years of impoundment (mg/l)}}{\text{forecasted TP concentration during first year of impoundment (mg/l)}}$$

$$(13.27)$$

5. Shallowing coefficient, calculated as the ratio of the water surface of the reservoir under the 2 m isobath to the total water surface at the 50% frequency level:

$$K_{sh} = \frac{\text{water surface under the 2 m isobath (km}^2)}{\text{reservoir water surface (km}^2)} \qquad (13.28)$$

6. Thermal stratification coefficient, calculated as the Froude number, computed for different reservoir water levels (95, 75, 50, 25% frequency):

$$Fr = \frac{V^2}{gh_0} \qquad (13.29)$$

where

V = mean velocity in the reservoir at the corresponding water level (m/s),
g = acceleration due to gravity (m/s^2),
h_0 = reservoir depth under the corresponding water level (m).

7. The water exchange coefficient is the ratio of reservoir inflow to reservoir volume, representing the renewal frequency of reservoir volume:

$$K_{wexch} = \frac{\text{50\% frequency annual flow } (\times 10^6 \text{ m}^3)}{\text{reservoir volume at mean water level } (\times 10^6 \text{ m}^3)} \qquad (13.30)$$

8. Maximum discharge smoothing coefficient, calculated as the ratio of the distance from the dam in the river section, where maximum discharges reach 50% frequency of the natural flow (in a year of 25% frequency natural inflow to the reservoir), to the distance from the dam to the river mouth:

$$K_{sm} = \frac{\text{distance from dam to section where } Q_{max} = 50\% \text{ (km)}}{\text{distance from dam to mouth (km)}} \qquad (13.31)$$

9. The flood plain development coefficient is the ratio of the weighted mean value of the inundated flood plain surface under the highest water level (1% frequency) to a weighted mean channel width:

$$K_{fd} = \frac{\text{width of water surface at 1\% frequency level (km)}}{\text{width of water surface within channel (km)}} \qquad (13.32)$$

10. Ecological flow, a quantitative indicator of the release flows which are ecologically necessary to sustain processes downstream. It takes into account discharges and water levels below the dam during all phases of the hydrological regime and in years characterized by different frequencies.

The main principles for the computation of ecological flow and water levels are given in equation (13.3).

14 Palaeohydrology: the Past as a Basis for Understanding the Present and Predicting the Future

L. STARKEL

Department of Geomorphology and Hydrology, Institute of Geography, Polish Academy of Sciences, Kraków, Poland

Introduction

All continents are drained by flowing water. Rivers are the most characteristic spatial systems in which the transmission of energy and matter from higher to lower elevations and towards oceans is realized. Existing fluvial systems and ecosystems connected to them not only reflect the current climatically con-trolled regime of energy and matter exchange, but also include elements inher-ited from the past – such as substrate composition, drainage pattern and river valley shape. Highly important influences on this were the transformations which occurred in the upper Quaternary, due to a shift of the climatic–vegetation zones and the effects of early human impact (Starkel, 1990b; Gregory *et al.*, 1995).

Fluvial Systems and Water Circulation

Transport and storage

Global water resources total $1350–1370 \times 10^6$ km^3, of which 97.4% is stored in the oceans, 2% in the ice caps and only 0.009‰ as atmospheric vapour (L'vovič, 1974; Walling, 1987). Less than 120×10^3 km^3 of precipitation falls annually over the continents, out of which $43–49 \times 10^3$ km^3 of water flows to oceans and lakes as surficial or underground runoff. Fluvial systems transport between 13.5 and 22×10^9 t of matter in suspension and as bedload, and more than 3.7×10^9 t of dissolved matter (L'vovič, 1974; Milliman and Meade, 1983).

Single river basins may cover up to several million km^2. Environmental conditions associated with their lithology, tectonics and relief, as well as climate, vegetation and land use, become more complex with increasing basin size.

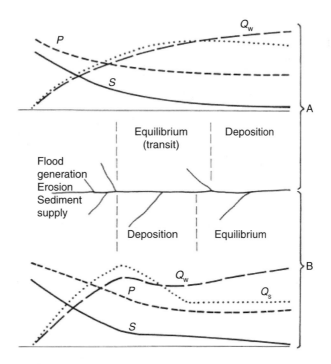

Fig. 14.1. Models of a river basin: (A) with gradual change from erosion to deposition downstream (after Schumm, 1977); (B) with deposition in the middle reach (at the mountain foreland) (after Starkel, 1990a,b). Key: *P*, precipitation; *S*, channel slope; Q_w, river discharge; Q_s, sediment load.

The precipitation regime as well as transport and storage conditions may thus differ within single large catchments.

Even simple catchments, flowing smoothly from their highland sources down to flat lowlands, cannot be considered totally uniform. Three segments of the river course may be distinguished based on changes in relief and hydrological regime (Schumm, 1977) (Fig. 14.1): (i) upper reaches, with their steeper slopes, higher gradient and higher precipitation, are characterized by sudden flood waves and intense erosion; (ii) middle reaches are rather at equilibrium between erosion and aggradation; and (iii) lower reaches are characterized by deposition. At the same time, higher grounds and depressions in the middle reaches frequently result in deposition. Further downstream, sediment deposition zones are often the consequence of sea level changes (Starkel, 1990b). Fluvial systems may appear even more complex, when tectonic and orographic controls as well as palaeogeographic transformations of the drainage pattern are taken into consideration (Starkel, 1979, 1989).

The role of tectonic and orographic factors

The relative size and relief of river reaches and their spatial pattern in the catchment depend on their relationship with factors such as geomorphology, the orientation of mountain ranges, neotectonic tendencies and distance from the ocean. Five major types of basins may be distinguished, superimposed on various morpho-tectonic units showing trends either to uplift or to subsidence (Starkel, 1990b):

1. River basins with headwaters in uplifted mountains and middle and lower courses in the lowlands. The influence of the mountains on the fluvial regime downstream depends on the spatial relationship between the two morphological structures. Examples are the Amazon and the Brahmaputra; this basin type is the most common.
2. Rivers which start in extensive plains characterized by low gradients and which do not receive floods and heavy bedload from their headwaters. Sea level changes tend to play a leading role in their evolution. Examples are the Volga and the Dniepr.
3. Rivers in the uplifted plateaus of Africa and America, which are occupied by extensive, flat catchments, draining to the sea across escarpments with rapids and waterfalls. Examples are the Zambezi, the Congo and the Iguassu.
4. Rivers in some interior plateaux, which drain seasonal runoff into large basins that are unconnected with the ocean. Examples are the Chari, the Amu Darya and the Coopers.
5. Rivers which cross two or more mountain ranges or uplands in their course and are separated by subsiding depressions characterized by aggradation. Examples include the Yang-tse and the Danube.

In Europe the first and second types are the most frequent ones next to rivers draining the coastal mountains (Ibbeken and Schleyer, 1991). River channels are straight or sinuous and incised in the upper (mountain) reaches, braided at the mountain foreland, meandering in the plains and anastomosing in the lowest reaches; finishing in deltas or tidal depressions (Fig. 14.2).

Climatic and vegetation controls of runoff and sediment load regimes

Typologies of fluvial regimes consider the main source of running water and sediment (rain, snowmelt, melting ice), the duration of flow (perennial, seasonal or episodic), as well as the magnitude and frequency of floods (Parde, 1955; Keller, 1962; Ľvovič, 1974; Hayden, 1988).

Figure 14.3 illustrates the interrelationship between precipitation and mean annual temperature with runoff regimes superimposed. It indicates a close relationship between fluvial systems, climate and vegetation. The north–south transect across Europe and Africa crosses zones of differing precipitation, vegetation and runoff regime (Fig. 14.4), reflected in changes of sediment load and channel pattern (cf. Ľvovič, 1974; Schumm, 1981; Jansson, 1982).

In the humid tropics, which are characterized by seasonal floods (especially in monsoon areas), river channels are mainly anastomosing or meandering. Rivers have a very high dissolved load and, after deforestation, tend to develop high suspended load and show a tendency towards braiding (Starkel, 1972; Gupta, 1988).

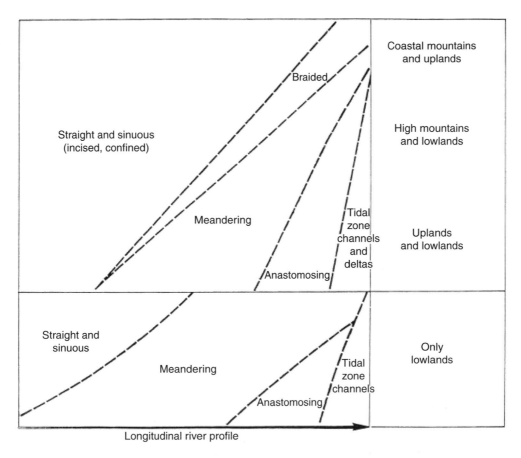

Fig. 14.2. Various channel forms in the longitudinal profiles of European rivers, depending on type of relief. (After Starkel, 1995b.)

In the arid zone, episodic flash floods control the high bedload transport and braided channel pattern (Schick, 1988; Baker *et al.*, 1995).

In semi-arid areas, seasonal or episodic rains cause flash floods, which carry high suspended load, varying quantities of bedload and low dissolved load values (Ľvovič, 1974; Jansson, 1982). Transitional channels between the braided and the meandering type are most typical.

In the temperate zone, transport fluxes are dominated by dissolved load during normal years and by suspended load (particularly in deforested areas) during years with extreme floods (Schumm, 1977; Froehlich, 1982). Rivers of forested catchments have mainly meandering or sinuous channels (Starkel, 1995b).

In the permafrost zone, rivers are characterized by ice-jam floods, medium bedload, low suspended and dissolved loads, and channels transitional between the meandering and the braided type (Woo, 1986; Maizels, 1995). Conversely, rivers supplied with glacial melt waters in the Arctic and in high mountain areas

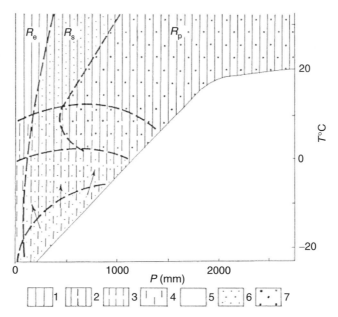

Fig. 14.3. Runoff regimes on a mean temperature (*T*)–precipitation (*P*) plot. Key: 1, rainy floods; 2, rainy and snowmelt floods; 3, snowmelt floods; 4, icemelt floods; 5, episodic runoff (R_e); 6, seasonal runoff (R_s); 7, perennial runoff (R_p). (After Starkel, 1990b.)

have very high bedload and suspended loads and are typically of a braided pattern (Church, 1988; Maizels, 1989).

In reality, river systems characteristic of different climatic zones show a much greater diversity. Moving downstream, the sediment yield decreases and the river channel changes (Schumm, 1977). Especially in arid and semi-arid areas, water loss and decreases in river discharge downstream are very frequent (Thornes, 1976). A great diversity of situations within one climatic–morphogenetic zone is also related to tectonic, orographic and lithologic factors as well as to effects of land use (Gregory, 1983; Gregory and Walling, 1987). In the middle and higher latitudes, formerly covered by the ice sheets, the flow and sediment load regimes are regulated to a large extent by transfluent lakes, which establish local base levels (Starkel, 1979, 1995a).

The role of floods in the transformation of river channels and flood plains

Intensive rain is one of the most frequent causes of floods (Parde, 1955; Starkel, 1976). Local heavy downpours, with an intensity exceeding 1–3 mm/min and a duration of several minutes or hours, cause intensive overland flow, piping and flash floods. Continuous rains, lasting from 1 to several days, and of a mean intensity of 5–20 mm/h, may cover areas of many thousands of km². The saturation of the ground triggers the inundation of large river basins. Regular rainy

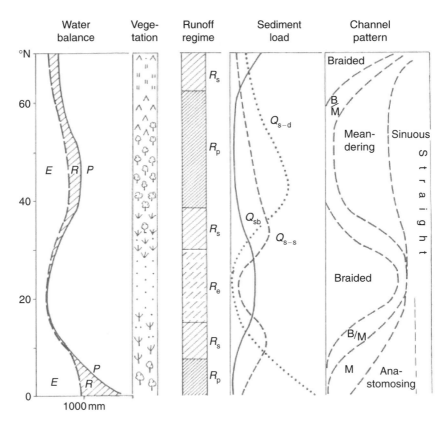

Fig. 14.4. Elements of fluvial systems in a European–African transect. Key: *P*, precipitation; *R*, runoff; *E*, evaporation; R_p, perennial runoff; R_s, seasonal runoff; R_e, episodic runoff; Q_{s-b}, bedload; Q_{s-s}, suspended load; Q_{s-d}, dissolved load. (After Starkel, 1990b.)

seasons in large catchments cause a general rise in the water level (such as in the Amazon river).

Snowmelt floods are characteristic of regions with regular snow cover and especially within areas with frozen grounds (Parde, 1955; Woo, 1986). In situations where rapid warming, simultaneous rainfall and ice-jams accompany floods, these may have a catastrophic magnitude (Church, 1988; Yamskich, 1993). In Iceland, the rapid melting of glaciers, connected with volcanic eruptions, creates floods called 'jokulhlaups' (Thorarinsson, 1957; Maizels, 1995).

Cataclysmic floods are connected with the failure of dams of various origins, such as dams created by glaciers, by landslides or man-made. Former floods, of the order of 200,000 m³/s in the Mississippi valley (Teller, 1990), were created by drainage of the ice-dammed Lake Agassiz during deglaciation. The Missoula flood in the north-western USA drained up to 17×10^6 m³ water/s from an ice-blocked lake (Baker, 1994).

Spatial patterns of floods may be very complex (Hayden, 1988; Waylen, 1995). At low latitudes, under barotrophic conditions, a seasonal monsoonal circulation is characteristic and floods are mainly related to the ENSO (El Niño Southern Oscillation) pattern. In the subtropical desert belt, storms and floods are very rare but can be spectacular. At mid-higher latitudes, under baroclinic conditions, the zonal circulation dominates and flood rains are influenced by the pattern of mountain ranges. Under more continental climates, the role of snowmelt seasons increases.

Three main types of rivers may be distinguished by considering the role of floods in the transformation of river channels and flood plains (Starkel, 1995b):

1. Rivers with frequent small to medium floods, and with extreme events repeated every 10–100 years, which disturb the equilibrium of the channel shape.
2. Rivers with frequent small to medium floods without extreme events.
3. Rivers with a very irregular regime, and with a varying frequency of rare extreme events, which dictate the transformation of the valley floor.

Rivers of the first group, with headwaters located in mountains, are typical of humid regions, especially in Mediterranean and monsoonal areas. The transformation of valley floors responds to the alternation of frequent small floods and rare extreme floods. Extreme floods with a heavy bedload cause the scouring of channels, their straightening or avulsion, and aggradation. In contrast, smaller floods cause the relaxation of the former profile, so that the river may return to its previous equilibrium and channel shape (Schumm, 1977; Froehlich and Starkel, 1987). Both types of flood lead jointly to the formation of more mature river channel and flood plain forms.

Rivers of the second group are restricted to lowlands. Flood plains may be inundated, but due to the low gradient and the sediment load, morphological changes are very small. Only ice-jam floods may create more substantial transformations (Woo, 1986).

Rivers of the third group are typical of arid and semi-arid zones. Rare but heavy downpours create flash floods, which may reach the stage of hyperconcentrated flows or even debris flows. Such floods carrying coarse debris are able to reshape river channels and, due to their long recurrence interval under desert conditions, the traces of such floods are visible after centuries or millennia (Costa, 1988; Schick, 1988; Baker *et al.*, 1995). Such cataclysmic floods create totally new landscape features.

A similar process leading to the formation of new channel shapes may be achieved by the clustering of floods during several consecutive years, as has been observed in the temperate zone in the Polish Carpathians (Soja, 1977; Starkel, 1996a) or in monsoonal areas of the foreland of the Bhutanese Himalaya (Starkel and Sarkar, 2002).

Large polyzonal rivers and their 'allochthonous' regime

Most European rivers flowing across lowland areas are controlled by the regime of their headwaters in the Alps, Carpathians or other mountains. Rivers flowing from high mountains reflect the regime of several vertical climatic belts including the nival zone. As a result, snowmelt or icemelt floods may occur in deep

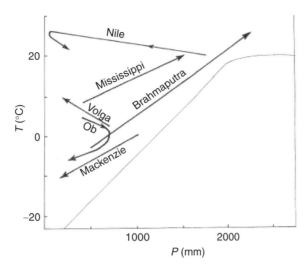

Fig. 14.5. Longitudinal profiles of polyzonal rivers superimposed on a temperature (*T*)–precipitation (*P*) plot. (After Starkel, 1990b.)

valleys (e.g. in the Alps or Tian-Shan mountains) (Tricart *et al.*, 1962; Starkel, 1976).

Large continental rivers cross the boundaries between morphogenetic zones, where the runoff regime of the lower course shows azonal features created in the headwaters; among these the rapid rises or decreases in river discharge and a high sediment load (Fig. 14.5).

The best examples of such polyzonal rivers flowing from humid or semi-humid regions to drier ones, and showing water loss as the river proceeds downstream, are the Nile, the Euphrates, the Amu Darya and the Colorado. An opposite trend, due to rainfall rising in the downstream portion of the catchment, can be found in the Mississippi and the Hoang-ho. At higher latitudes, the role of snowmelt floods is distinctive. Rivers flowing to the Arctic Ocean (e.g. the Ob, the Yenisei, the Lena and the Mackenzie) are subject to frequent ice-jam floods in the lower portion of their catchment, due to a delayed snowmelt and river-ice break-up season (Starkel, 1979; Yamskich, 1993). In contrast, rivers flowing towards lower latitudes show a downstream decline in the snowmelt component of flood runoff (e.g. the Mississippi and the Volga).

Methods of Palaeofluvial Reconstruction

Methods of reconstruction of the past hydrological budget include retrodiction as well as modelling (CLIMAP, 1976; COHMAP, 1988; Kutzbach, 1992). There are specific methods focusing on retrodiction of the precipitation regime, which are based on vegetation analysis (Klimanov, 1990), on the reconstruction of lake level changes and lake storage (Kutzbach, 1980; Street-Perrott and Harrison, 1985), and on the reconstruction of past runoff and fluvial regime (Rotnicki, 1991).

The archive of past fluvial history preserved in fluvial sediments and forms is rich, but in comparison to lake, bog or ice environments, fluvial records are discontinuous and present many hiatuses (Starkel, 1996b). The reason for this is the position of valley floors, which have been exposed to extreme floods and which, instead of being formed by a steady deposition in a vertical accretion mode, bear the influence of flood activities causing frequent re-bedding and scouring and the lateral shifting of the river channel and the active flood plain. The presence of cut-off meanders and avulsions indicates that, to complete the fluvial history of many rivers, it is necessary to study either the whole system of parallel cuts and fills (Schirmer, 1983; Starkel, 1983) or to focus on rocky canyons and on slack-water deposits (Baker, 1987).

The existence of dozens of palaeochannels of similar age and size in a single river catchment, such as the Vistula river (Kalicki, 1991; Starkel et al., 1991), documents phases of past intensive flooding activity. Large palaeochannels of Late-glacial age were formed by discharges several times greater than the present ones (Szumański, 1983; Sidorchuk et al., 2003). These phases of increased fluvial activity are also documented on one hand by a higher rate of overbank deposition and by a coarser sediment texture (Knox, 1983, 1995) and on the other hand by clusters of subfossil oaks which are considered good indicators of frequent flooding (Becker, 1982; Krapiec, 1992).

One source of information concerning bankfull and mean annual discharges is provided by the morphology of palaeochannels including width, depth, meander radius and length and channel gradient. Past discharges can be estimated from meandering palaeochannels by means of various equations (Dury, 1964; Schumm, 1977; Rotnicki, 1991).

Unfortunately, the estimation of the majority of these parameters can represent a challenge (Gregory and Maizels, 1991; Soja, 1994). In the case of palaeomeanders, we may start by assuming the existence of a single thread channel. It is often very difficult to determine bankfull discharge, channel depth or the historical base level of the coarse (armoured) horizon; there may be several such layers and the upper ones may have formed after abandonment of the main channel. Furthermore, well-developed palaeomeanders may represent fluvial regimes dating back several centuries and actually pre-dating organic deposition in cut-off portions of the channel (Starkel, 1996b).

The identification of braided palaeochannels in meandering river reaches indicates a totally different past fluvial regime; this case is frequently observed in many valleys of the present-day temperate zone, which during the Last Cold Stage were occupied by the periglacial zone or were fed by glacial melt waters (Maizels, 1983; Kozarski, 1991).

In the case of braided palaeochannels, errors in the reconstruction of past discharges may reach several hundred per cent (Maizels, 1983) because it is difficult to determine which branches of the braided river were active at a given time. Similarly, grain size comparisons may provide ambiguous information on past flood discharges (Church, 1978).

The analysis of coarse channel facies and fine slack-water deposits (Baker, 1987) is used to reconstruct the water level and discharge of extreme flood events especially in rocky canyons of the arid zone. Baker and his collaborators have reconstructed the frequency of palaeofloods by analysing slack-water deposits in various parts of the globe, such as the south-western USA, India, the Near East, China, Siberia and Australia (Baker *et al.*, 1988, 1995).

Additional information on palaeorunoff comes from lake records in closed depressions like palaeolake Chad (Kutzbach, 1980). In the arid zone, the rings of living trees may be calibrated with runoff data and may serve to reconstruct the palaeorunoff (Stockton *et al.*, 1990).

Changes in Water Storages and Fluxes from the Last Cold Stage to Present

Water storages and fluxes during the maximum expansion of ice sheets (20,000–18,000 years BP)

At the peak of the last glaciation, changes in the global water balance were small, but the areal distribution of the ice sheet at its maximum expansion varied substantially (CLIMAP, 1976; Gates, 1976; Kutzbach, 1983; COHMAP, 1988). Total global precipitation was up to 14% lower than today. The CLIMAP model hindcasts lower precipitation and drier ecosystems for all latitudinal zones of the northern hemisphere with the exception of the south-western USA (Frenzel *et al.*, 1992). Drier conditions in the equatorial zone were due to a decreased air pressure gradient and decreased evaporation from the cooler oceans. The deep water circulation in the North Atlantic (the so-called NADW; cf. Broecker *et al.*, 1989; Goslar, 1996) was not as active as at present and large ice sheets extended over North America, Europe and the margins of the Arctic. The volume of water stored in the ice was at least three times greater than it is at present (Fig. 14.6) implying that, during the earlier growth of the ice caps, precipitation must have been at least double that at present (Lockwood, 1979). The Eurasian continent was covered by permafrost up to about 70% and the fluvial regime was mainly controlled by it (Woo, 1986; Starkel, 1995a). The drainage pattern was under the influence of ice dams. Streamways (pradolinas) developed along ice sheet margins. Large dammed lakes in some cases reached above the watersheds to supply rivers flowing towards the south, like the Volga or the Mississippi. Within the closed basins of central Asia, runoff and the expansion of lakes were facilitated by permafrost conditions (Klimek and Rotnicki, 1977; Vipper *et al.*, 1987).

The course of shifts in climatic–vegetation zones at the Pleistocene–Holocene transition

Past changes in the seasonal distribution of the solar radiation at various latitudinal belts were instrumental in causing hydrological change, including the reactivation of the NADW, the melting of ice sheets and the rise of sea levels (Kutzbach, 1983; Starkel, 1993). Associated with these changes was the rise of annual precipitation in the inter-tropical zone between 12,000 and 6000 years BP (COHMAP, 1988) causing higher runoff and higher lake levels (Kadomura, 1995). A supplementary

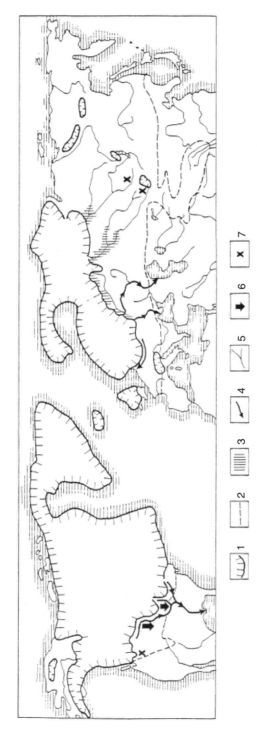

Fig. 14.6. Palaeohydrology of the northern hemisphere for the last 18,000 years. Key: 1, ice sheets; 2, southern limit of permafrost; 3, ice-dammed lakes; 4, outflows of meltwaters to the south; 5, other rivers; 6, extreme outflows of meltwaters; 7, local registered cataclysmic floods. (Based on Starkel, 1995a.)

factor in the continuous global rise of rainfall was the increase of heat and water exchange between both hemispheres as well as increases in volcanic activity (Bryson, 1989).

At higher latitudes of the northern hemisphere, the distinct warming accompanied by smaller precipitation rises was caused by the increasing solar radiation, which reached 8% more than baseline in July (and 8% less than baseline in January) at 9000 years BP, as reconstructed by Kutzbach (1983). This was accompanied by the melting of ice sheets, the retreat of the permafrost and a northward expansion of vegetation (Fig. 14.7). These changes had a profound impact on the transformation of fluvial systems over the whole globe. At least

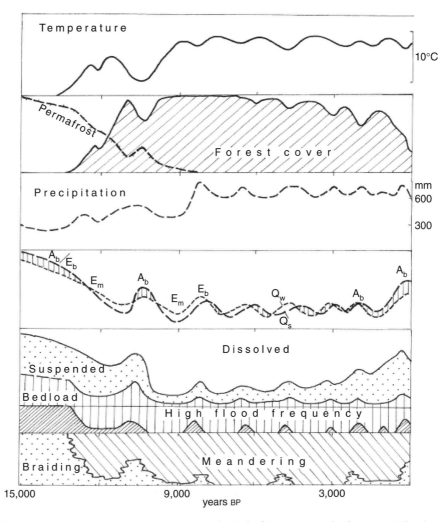

Fig. 14.7. Changes of various palaeohydrological parameters in the upper Vistula basin during the last 15,000 years. Ab, aggradation by braided river; Eb, erosion by braided river; E_m, erosion by meandering river; Q_w, water discharge; Q_s, sediment load. (Based on Starkel, 1990a with some changes.)

six different sequences of change in fluvial regime, connected with the shift of climatic–vegetation zones, have taken place during the last 18,000 years (Starkel, 1990a,b):

1. Arid/semi-arid to arid. In Africa and in the Near East, deserts are characterized by a large number of inactive river systems. These were transfluent, especially between 12,000 and 6000 years BP (Pachur and Kröpelin, 1987; Baker *et al.*, 1995).
2. Semi-arid to semi-humid. In Australia and the Mediterranean region, this trend is documented by a shift from a braided to a meandering river pattern (Schumm, 1977).
3. Semi-arid (or semi-humid) to arid. In the south-western USA, an opposite trend from meandering towards braided channels was recognized, accompanied by episodic floods (Baker, 1983).
4. Cold dry (with permafrost) to temperate (forest). This type is most extensive in the present-day temperate forest belt, where climate warming, precipitation rise and forest expansion caused a decline in sediment transport, a change from snowmelt to rainy floods, and a transformation from braided to meandering channels (Schumm, 1965; Falkowski, 1975; Starkel, 1983; Knox, 1995) (Figs 14.7 and 14.8). Changes which followed during the Holocene were of much lower

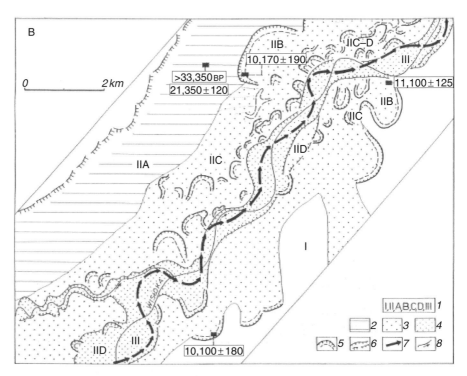

Fig. 14.8. Fragment of the Wisloka river valley at the mountain foreland. Key: 1, numbers of terraces (I – Pleniglacial); 2, pre Late-glacial level (IIA); 3, Holocene flood plains (IIB–IID); 4, actual flood plain formed in 18th–19th century; 5, palaeochannels; 6, terrace edges, scarps; 7, river course from AD 1780; 8. present-day river course. (Based on Starkel *et al.*, 1983; Starkel 1995b).

amplitude (*see* sections on 'Patterns of hydrological variations at lower and higher latitudes during the Holocene' and 'Second-order phases characterized by higher flood frequencies' below).

5. Cold (with permafrost) to boreal (with permafrost). Extensive large catchments over Siberia and Alaska showed only slight changes in the degradation of frozen grounds. Therefore shifting from braided to meandering channels was not as intense and frequent younger fills formed starting from 5000–4000 years BP are probably connected with the re-advancement of the permafrost (Baulin *et al.*, 1984).

6. Glacial to temperate. Previously glaciated regions bear the sign of former basins blocked by ice sheets, of valleys created by glacial melt water, and of new river systems formed in ice-free areas. In most cases, present-day larger rivers consist of a variety of reaches moulded by different factors acting at different times (Starkel, 1979; Fig. 14.9). Rivers such as the Vistula (Starkel *et al.*, 1991,

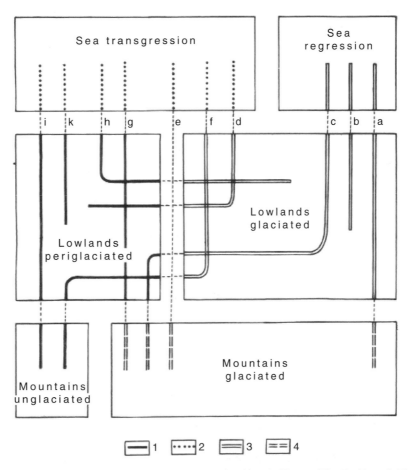

Fig. 14.9. Types of river valleys in the temperate zone. Key: a, Pite; b, Kyronjoki; c, Saskatchewan; d, Nieman; e, Austerdalen; f, Vistula; g, Rhine; h, Dniepr; I, Maas; k, Mississippi. 1, by periglacial processes in cold stage; 2, under Flandrian transgression; 3, after deglaciation; 4, by mountain glaciers. (After Starkel, 1979.)

Starkel, 1994) or the Saskatchewan proceeded downstream incising their channels, in association with the retreat of the ice. Many ice marginal streamways became abandoned and underwent paludification (Starkel, 1990a,b). Southward sloping river valleys, which carried melt water are now occupied, undersized streams (Dury, 1964). Even the great Mississippi and the Volga carved wide valley floors, which are far too extensive in the context of present-day discharges (Kvasov, 1976; Knox, 1983). Former sandur plains, formed after the glacial retreat, are now cut mainly by small creeks (Maizels, 1995).

In areas left free of ice, the new channels formed by superimposed river systems are composed of reaches derived from subglacial channels, lake plains, sandurs, transfluent lakes and gorges (some of which cut in the bedrock). Their longitudinal profiles are far from being at equilibrium (Koutaniemi, 1991). Many of these river morphologies, due to glacial rebound, show elongation downstream (Mansikkaniemi, 1991).

In various climatic regions, the metamorphosis of river valleys during the Pleistocene–Holocene transition was highly complex. A common factor across the globe was the eustatic rise (excluding areas of glacial rebound), expressed by the shortening of longitudinal profiles during the Flandrian transgression, which caused not only the submergence of a ~120 m vertical belt, but also the upstream shift of the aggradation zone.

Rapid transformations in water circulation during the Pleistocene–Holocene transition

The melting of ice sheets and the shift of latitudinal climatic–vegetation belts was interrupted by a severe disturbance of the hydrological regime, which was due to a number of concurring factors such as the Younger Dryas event and cataclysmic floods.

The Younger Dryas lasted about 1150 years (from 12,500 to 11,350 years BP) and was closely related to the routing of melt water from the Laurentian ice sheet. During this period the melt water flowed from southern Canada to the St. Lawrence Gulf, whereas before and after the ice lakes drained to the Mississippi river (Teller, 1990). This disturbance is clearly visible in the re-advancement of glaciers and of the permafrost, in the retreat of the forests and also in the rapid change of several river channel parameters (Starkel, 1990a).

Of greater importance were abrupt changes that lasted just over several decades, such as the cooling period at the beginning of the Younger Dryas and the warming at its end. Based on the ^{18}O curve and other records, the drop and the later rise in mean annual temperature was of the order 3–4°C (Lotter and Zbinden, 1989; Różanski et al., 1992; Goslar, 1996). The warming, in particular, caused a rapid decline in sediment loads and a change in trend from braiding and large meanders to small-sized ones; this is clearly noticeable in Polish river valleys (Kozarski and Rotnicki, 1977; Starkel, 1983; Szumański, 1983). In Sweden, the rate of ice retreat increased rapidly from 10–50 m/year to 250 m/year (Mörner, 1976).

In the north-western USA, other perturbations in fluvial systems, connected with late-glacial cataclysmic floods, were documented for the first time by Baker (1983), who calculated a peak discharge of 21×10^6 m³/s for the Missoula flood, associated with rapid flows from this ice-dammed lake. Later, similar late-glacial floods were discovered along Siberian rivers such as the Ob (Rudoy and Baker, 1993) and the Yenisei, where the water level fluctuated over 30 m during ice-jam floods (Yamskich, 1993).

Patterns of hydrological variations at lower and higher latitudes during the Holocene

During the Holocene there is evidence of distinct differences in precipitation and runoff in different climatic–vegetation regions. In the equatorial zone, as well as in the present arid zone, the rise in precipitation started before 12,000 years BP and continued with short breaks until 5000–4000 years BP (Fig. 14.10). In southern Asia, it was a time of very intense monsoonal activity (Kutzbach, 1983). High water levels and frequent floods were also recorded during the lower part of the Holocene (COHMAP, 1988; Baker *et al.*, 1995).

Towards the north, especially on the Asian transect, there was a distinct change, although from central Europe to Siberia and Mongolia the climate was warm but relatively dry during the early Holocene. Around 8500–8000 years BP, these regions experienced an increase in humidity and flood frequency with the reactivation of westerly winds and the spread of deciduous woodland (Starkel, 1983, 1995a). Progressive cooling occurred more or less simultaneously after

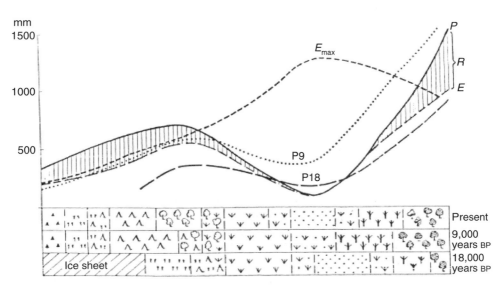

Fig. 14.10. The course of precipitation (*P*), runoff (*R*) and evaporation (*E*) in the European–African transect for the last 18,000 years BP (P18), 9000 years BP (P9) and at present (P). E_{max}, present-day potential evaporation. (Based on Starkel, 1995a.)

5000–4000 years BP at lower latitudes, leading to the restoration of permafrost and to a further change in the fluvial regime in Eastern Siberia and on the Tibetian Plateau.

Second-order phases characterized by higher flood frequencies

Detailed studies in the Vistula river basin, as well as in other European and North American river valleys, lead to the alternative hypotheses of either rhythmic phases with higher frequency of extreme floods (Starkel, 1983, 2003) or distinct discontinuities in the activity of fluvial systems (Knox, 1983). These secondary phases are manifested in the middle reaches of river systems by the presence of sequences of cut and fill with frequent avulsions, and by the presence of systems of abandoned palaeochannels in valleys with highland headwaters (Fig. 14.11). Synchronous phases from about 8500 to 8000 years BP, and a twofold phase between 5200 and 4200 years BP, are particularly recognizable (Starkel, 1983; Kalicki, 1991; Starkel *et al.*, 1996). All these phases are witnessed by distinct clusterings of subfossil oak trees (Becker, 1982; Krapiec, 1992). Lowland rivers generally do not show such a clear reaction to heavy rainfall; however in the Polish lowland, the dating of abandoned meanders bears an influence due to such phenomena (Turkowska, 1988; Kozarski, 1991).

Each high flood frequency phase lasting 200–500 years is characterized by a higher rate of deposition, a change of facies, a coarsening of grain sizes and a significant change in the morphology of river channels (Starkel, 1983, 2003; Starkel *et al.*, 1991). Channels tended to widen and straighten, then to develop single meander cut-offs and finally the braided channels became either more deeply incised or the process was interrupted by channel avulsion (Fig. 14.12). The last phase of these frequent floods coincided with the Little Ice Age. During more stable periods, braided or widely incised meandering channels were transformed again by lateral migration into a freely meandering system.

In Europe, the Holocene fluctuations in fluvial regime coincided with other climatic and hydrological changes (Fig. 14.13). Phases of high flood frequency correlate well with features such as the advancement of Alpine glaciers, lake level fluctuations, rates of peat bog growth, landslide and debris flow frequency and the precipitation of calcareous tufa (Starkel, 1985; Magny, 1993). It is thought that these fluctuations are related to the position of the jet stream, as it is visible in discontinuities connected with the shift of various air masses on the North American continent (Knox, 1983, 1995). A possible coincidence with phases of high volcanic activity may not be excluded (Bryson, 1989; Nesje and Johannessen, 1992; Starkel, 1995a).

In arid and semi-arid regions, rare episodic floods or their clustering related to ENSO events played a leading role in the reshaping of river channels and flood plains (Baker *et al.*, 1995). Relaxation times under such conditions may be very long. Therefore, many river channels still preserve features inherited from extreme events, which took place several centuries or millennia ago.

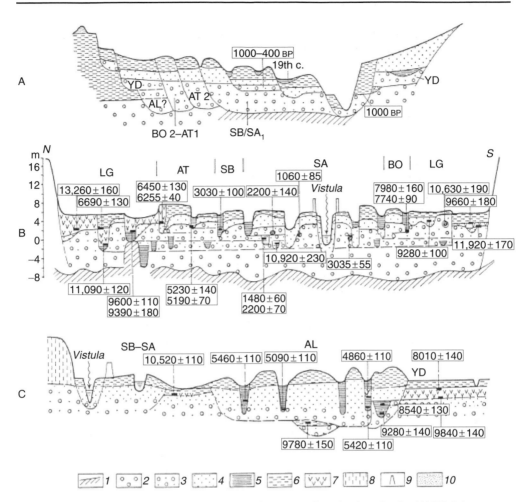

Fig. 14.11. Examples of Holocene cuts and fills in the Sandomierz Basin: (A) Wisloka at the mountain foreland (after Starkel, 1995a); (B) Vistula downstream of Cracow (after Kalicki, 1991); (C) Vistula at the Grobla forest (after Starkel *et al.*, 1991). Key: 1, bedrock; 2, gravels; 3, sands and gravels; 4, sands; 5, palaeochannel fills; 6, overbank loams; 7, peat; 8, loess; 9, embankment; 10, eolian sand; IP, Interpleniglacial; AL, Alleröd; YD, Younger Dryas; AT, Atlantic phase; SB, sub-boreal; SA, sub-Atlantic.

Human Modification of Fluvial Systems

The impact of deforestation on runoff and sediment load

Human impact started in the Holocene with forest clearance, land cultivation and grazing. Such changes commenced at the ecotone between steppe and forest in the early Holocene and later expanded across forested regions as well as across steppes and semi-deserts (Goudie, 1981; Starkel, 1987). During the first Landnam phases, these changes occupied only small areas with more fertile soils

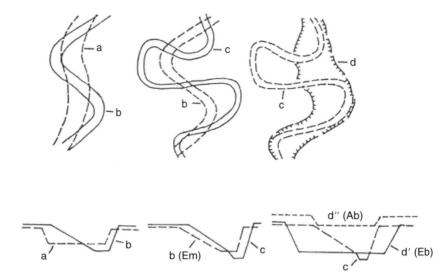

Fig. 14.12. Changes of river channel during rhythmic fluctuations of the hydrological regime in the Holocene, in plan view and in cross-section. Key: a, straightened
channel of braided river; b, sinuous transitional channel; c, deeper meandering
channel (Em); d, next phase of straightening and braiding with tendency towards
erosion (d′ – Eb) or to aggradation (d″ – Ab). (Based partly on Starkel, 1983.)

and shifting cultivation facilitated forest regeneration. Later, population growth
and new agricultural techniques caused the transformation of fluvial regime in
larger river catchments. Increased runoff caused an increase in flood frequency,
whereas increased slope wash, piping, mass movements and deflation caused a
rise in sediment yield with the overloading of rivers and a tendency to aggradation
(Fig. 14.14).

A very rapid response to deforestation and agriculture has been well documented for various sites dating from the Neolithic period (Wasylikowa *et al.*,
1985; Starkel, 1987). Thick alluvial deposits accumulated during Roman times
in the Mediterranean region (Vita-Finzi, 1969), while 1–1.5 m thick flood loam
deposited in the north-eastern USA between the 19th and 20th century (Wolman,
1967; Knox, 1977).

During the 20th century, expansion of land cultivation and overgrazing to all
zones and regions (except polar and high montane belts) caused the acceleration
of denudation processes and increases in sediment load by 100- to 1000-fold
(Jansson, 1982; Walling, 1987). In monsoonal areas, slope wash and landslip
caused changes in the trend of landscape evolution and aggradation replaced
erosion (Froehlich and Starkel, 1987). In semi-arid areas, perennial rivers
changed to an intermittent regime and periodic rivers changed to episodic ones.
The disturbance of slope and channel equilibrium is most frequent during extreme
events (Selby, 1987) and since there is no time for relaxation, many river reaches
have undergone transformation from meandering to braided or from aggradation
to down-cutting.

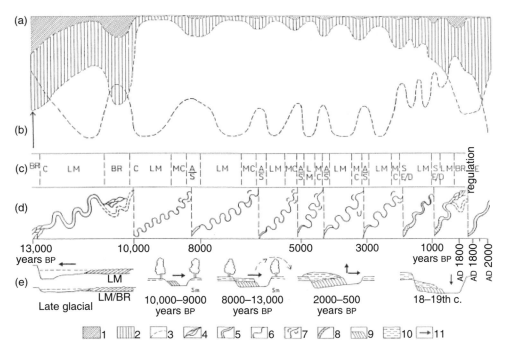

Fig. 14.13. Model of rhythmic changes and thresholds in the evolution of river and flood plains during the last 13,000 years (elaborated by L. Starkel). Key: (a) relative fluctuations of transport and delivery of bedload and suspended load; (b) fluctuations in flood frequency (Kalicki, 1991; Starkel 1994); (c) main directions of changes; (d) rhythmic changes of channel parameters, various cycles are separated by threshold changes in the fluvial system; (e) schematic channel cross-sections and directions of their transformation during various phases of the late Vistulian and Holocene. Key: BR, braided channels; C, concentration of channels; LM, lateral migration; MC, meander cut-off; A, avulsion; S, straightening; E, down-cutting; D, aggradation; 1, bedload; 2, suspended load; 3, curve of flood frequency; 4, braided channel; 5, large palaeomeanders; 6, small palaeomeanders; 7, cut-off meanders; 8, incision of the straightened channel; 9, channel bars; 10, overbank deposits; 11, directions of channel changes.

The Little Ice Age – the last phase with frequent floods and human impact

Historical remains, instrumental data as well as geomorphological–sedimentological records show that the period of the so-called Little Ice Age (AD 1450–1850) was characterized by a higher frequency of extreme hydrological events. This period experienced very unstable weather (Grove, 1988; Pfister, 1992) with heavy floods (Pavese *et al.*, 1992; Zhang and Wu, 1990). In the upper Vistula basin, frequent floods caused the transformation of river channels from meandering to braided (Szumański, 1983; Klimek and Starkel, 1974; Fig. 14.8). During this period, in the higher mountains, frequent debris flows, avalanches and local floods have also been recorded (Grove, 1988; Kotarba, 1992). Decadal or multi-decadal clusterings, occurring before AD 1600, around AD 1700 and

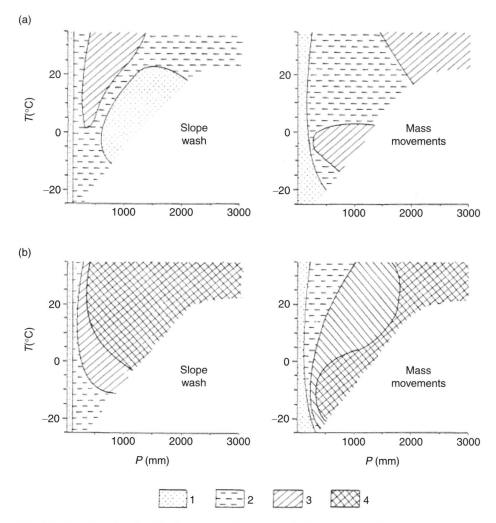

Fig. 14.14. Morphoclimatic diagrams of intensity of slope wash and mass movements under natural condition (a) and accelerated by human activity (b). Rate of processes: 1, low; 2, temperate; 3, high; 4, very high. (Based on concept of Wilson, 1968 and Starkel, 1989.)

between AD 1800 and 1820, have been recorded, which seem to coincide not only with advances of alpine glaciers but also with increases in the frequency of stronger El Niño events (Quinn and Neal, 1992). The coincidence of these clusterings with increased volcanic activity is possibly significant (Bradley and Jones, 1992).

In Europe, this period was also characterized by increased deforestation and, commencing with 18th century population growth, by the cultivation of slopes and the introduction of potatoes and other plants, which accelerated erosion. The resultant increases in suspended load and bedload were probably an additional factor causing the widening of river channels and a tendency to braiding (Starkel, 1990a).

Human impact on water storages and transport fluxes during the 20th century

Present-day changes to fluvial systems include a great variety of direct and indirect activities, which in many cases have totally transformed water budgets and runoff regimes, and caused drastic deterioration in water quality (Gregory, 1987, 1995; Walling, 1987; Starkel, 1987; 1993).

In the tropical lowlands (Brazil) and in mountain areas (India, China), continuous deforestation processes have caused an increase in sediment load in previously stable areas. In semi-arid and in arid regions, the water deficit has resulted in the need for irrigation of cultivated areas. On a global scale, 2500–4000 km^3 of water are used annually for this purpose (Shiklomanov, 1990). The total annual water consumption amounts to more than 12% of global annual runoff, exceeding 5000 km^3. More than half (about 2900 km^3) of this amount does not return to the freshwater cycle and may be classified as 'irreversible' water loss (Shiklomanov, 1990). The greatest demands and losses exist in regions of highest water deficit as well as in overpopulated and industrialized areas, where the use of water resources has risen to 25–75% of runoff (Fig. 14.15). Such high water demands are related to several types of interventions in fluvial systems. One of them is the construction of reservoirs, which store more than 5000 km^3 of water and which regulate river discharge. Various flood defences have been constructed along river channels, because increased irregularity of runoff in deforested and urbanized basins has caused rises in flood frequency and damage to flood plains of all climatic zones (Starkel, 1993; Gregory, 1995). The most recent floods in the Mississippi, Rhine and Oder river valleys may serve as examples. A more dramatic situation is associated with the impact of very irregular rainfall patterns on runoff and water deficits in the arid zone. To produce 1 t of biomass in this zone, cultivation consumes five times more water than in the humid areas. Irrigation has led to the lowering of the groundwater table and to desiccation of the Aral Sea, Lake Balkash and other waterbodies (Shiklomanov, 1990). In overpopulated areas, during seasons of low river discharge, water pollution has passed the threshold of ecological catastrophe (L'vovič, 1974). Pessimistic prognoses made in the early 1970s considered that the proportion of unused flood runoff would have reached 55% and that more than 80% of the outflow reaching the ocean would have become polluted (Fig. 14.16).

Engineering works in humid areas play a counterproductive role in preventing sustainable agricultural development. Constructions of concrete river channels and dense networks of drainage canals increase flood waves and cause the rapid lowering of the groundwater table, leading under some circumstances to the desiccation of marshes and peat bogs and to the lowering of lake water levels.

Human activity may totally change the functioning of fluvial systems. Each reservoir creates a local base level for the catchment upstream, but in systems such as the Nile, the downstream reduction in sediment load promotes the incision of river channels and restricts overbank and deltaic deposition.

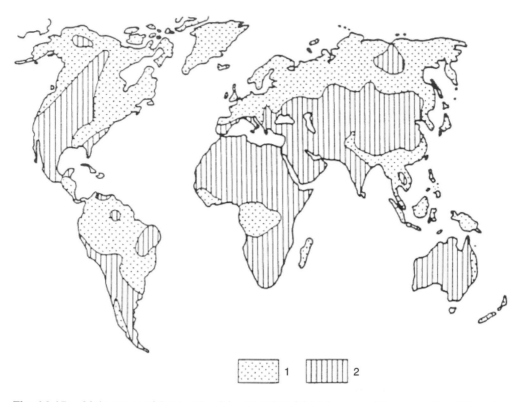

Fig. 14.15. Main areas of the surplus (1) and deficit (2) of river runoff based on the difference between precipitation and potential evaporation in predicted climate change models. (Simplified from Shiklomanov, 1990; after Starkel, 1993.)

On a global scale, precipitation regimes are still controlled by the zonal air mass circulation, but the runoff and evaporation regimes have been changed to such an extent that the resilience of fluvial systems and especially of river channels and flood plains is perturbed. The combined effects of these changes influence the position of ecotonal zones and water resources within large river catchments.

Knowledge of the Past Promotes a Better Understanding of Present and Future Conditions in Fluvial Systems

The information that has been drawn from the past is indicative of the great diversity of existing fluvial systems, depending on the age of the different inherited elements and the different stages of evolution and transformation undergone, including human impact. The role of past extreme events, as well as potential future changes under the impact of expected global warming, may be more clearly understood from a long timescale perspective (Starkel, 1995, 1996b). Such an issue should be addressed when considering the chance of full recovery of water resources and river valleys as natural runoff axes and ecological corridors.

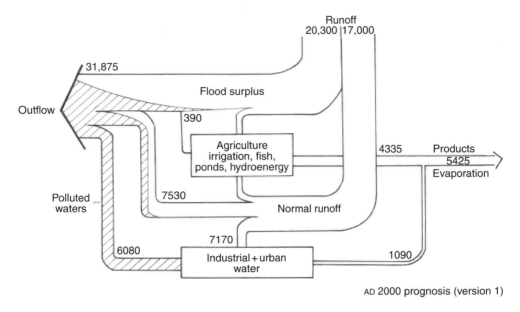

Fig. 14.16. Pessimistic prognosis of a future human water budget; values in km^3.
(After L'vovič, 1974; redrawn from Starkel, 1995a.)

The parallel existence of different river systems

The study of forms and sediment histories of river valleys and the reconstruction of their history indicate that different lithologies and different stages of evolution may exist within a single climatic zone (Starkel, 1983). Deep narrow canyons in resistant limestones may exist next to wide mature valleys cut in marls or shales; similar situations are realized with respect to the type or scale of human intervention. For example, the Rhine and Oder river valleys are characterized by regulated channels whereas the channels of the Nieman and Bug rivers meander across an undisturbed flood plain. In the context of hydrological as well as eco-hydrological issues, each fluvial system has individual features which may be significant in water management planning.

Incorporation of past geomorphological elements

Given an apparently uniform river valley, various geomorphological elements such as terraces, palaeochannels, steep or gentle slopes, and sediments covering the flood plain and the channel itself, have been inherited from various time periods and are incorporated into current dynamic river systems. The only features that are adapted to current runoff regimes are the natural river channels. By analysing such rivers, their channel and their flood plain in the context of a longitudinal profile, we may observe the incorporation of reaches of various origin and age. This is especially well pronounced in areas of deglaciation and is also

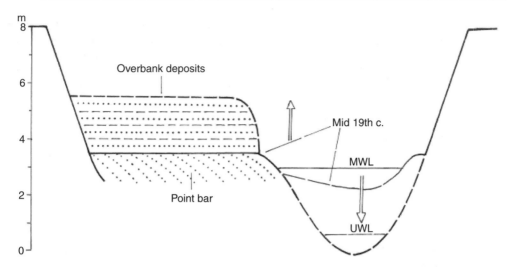

Fig. 14.17. Model of the creation of the 5 m high flood plain in the Wisloka valley (Carpathian foreland) after channel regulation, which started in 1860, showing parallel scouring of the river channel and up-building of former bars by overbank deposits (MWL, mean water level).

valid in the case of fragmentary regulation works such as those of the Vistula river valley. In this river basin, the upper and the lower course were regulated and, as a result, river channels are incised; whereas the middle course retains a semi-natural braided pattern, as it was in the 19th century. In the upper course, the effect is the transformation of a formerly low flood plain – with channel bars only 1 m above the mean water level – into a 5 m high terrace, which floods relatively rarely. This resulted from a down-cutting of up to 2 m that took place during the last century, as well as from a 1–3 m thick overbank deposition (Klimek and Starkel, 1974; Fig. 14.17). The Vistula river indicates that the impact of all engineering works regulating the fluvial regime should be assessed starting from headwater areas, where runoff is formed before the flood wave proceeds downstream.

Knowledge of present-day long-term trends in fluvial systems

In every type of water management, knowledge of long-term trends is of great importance. River valleys and channels are in various stages of evolution and a knowledge concerning present-day trends in relation to down-cutting, aggradation or lateral migration of the channel is needed. To establish whether a braided river may be in a phase of aggradation or erosion depending on the sediment load balance (Starkel, 1983), we should focus on the present-day frequency of extreme events and their relaxation times, which may be different from reach to reach. Of particular relevance are phases characterized by high flood frequency and clusterings of extreme events, which lead to long-term evolutionary changes (Starkel, 1995a, 1999). In the Polish Carpathians, during the

20th century, a change from aggradation and lateral erosion to down-cutting occurred in several valleys during consecutive flood years (1958, 1959, 1960 or 1970, 1972–1974). In the deforested Darjeeling Himalayas, progressive upstream aggradation is been caused by rare extreme floods (two or three in one century) causing deposition, alternating with annual floods which produce slow down-cutting (Froehlich and Starkel, 1987). Regulated river channels and water reservoirs delay the natural processes of valley floor transformation by erosion and aggradation, which in tectonically active areas is ultimately controlled by the uplift of high mountains and the subsidence of tectonic depressions.

A global warming perspective on fluvial systems

Growing water demands and water pollution suggest an increasing freshwater deficit (Shiklomanov, 1990) while deserts and semi-deserts continue expanding (Rapp, 1987). Under a continuous increase in carbon dioxide and other greenhouse gases, global atmospheric circulation models (GCM) predict changes in the distribution of rainfall and evaporation. A sharpening contrast in water resource availability between humid and dry zones is expected (Brouwer and Falkenmark, 1989; Houghton *et al.*, 1990; Kaczmarek *et al.*, 1996); however, perspectives presented by various scenarios in single regions of the globe differ greatly. For example, under a doubling of carbon dioxide, Hulme *et al.* (1990) predicted a rise in winter precipitation and a fall in summer rainfall across southern Europe. Shiklomanov (1990) forecasted a rise of runoff in northern Russia and a distinct drop in the arid south. Under a more continental climate, a decline of winter water storage in snow was proposed by Falkenmark (1990). In the case of the Volga river, summer runoff has been predicted to decline by 40% and winter runoff to rise by 60% (Shiklomanov, 1990). A consequence of climate warming could be that snowmelt floods may become very rare and be replaced by rain-induced floods. In the steppe zone and even in temperate forest zones, many perennial rivers could thus switch to a periodic regime.

Expected consequences of the slow annual rise in sea levels are a landward shift of coastal swamps and an increased marine abrasion (Boorman, 1990). Growing water demands will also cause shrinkage of water resources and an acceleration of water circulation. These mechanisms may feed back into a continuous desiccation increase and lead to an acceleration of the shifting of climatic–vegetation belts (Melillo *et al.*, 1990). In other situations, current stability equilibria may pass the threshold whereby large flood events may exceed resisting forces and create new valley floor morphologies and new riverine ecosystems.

A better knowledge of the long-term causes and effects of droughts and floods may be required as the predicted increase in frequency of these extreme events may play a key role in the economic development of many regions (Baker, 1991; Starkel, 1993; Gregory, 1995) and in the transformation of runoff, sediment load and biomass production in river valleys.

Acknowledgements

The author would like express cordial thanks to Dr Angela Gurnell from the Department of Geography, Kings College London, UK and Dr Nic Pacini from the Department of Ecology, University of Calabria, Italy for comments and improving the English manuscript; as well as to Mrs Maria Klimek for the drawings and Mrs Barbara Gnela for retyping the manuscript.

15 Ecohydrology: Understanding the Present as a Perspective on the Future – Global Change

I. WAGNER

European Regional Centre for Ecology u/a UNESCO, Lodz, Poland

Introduction

Sustainable development, based on environmental security, is not possible without maintenance of the ability of ecosystems to adapt to a changing environment (Janssen, 1998). This means that human interventions must be at a level that maintains the equilibrium between the natural systems conditioning the Earth's capacity – atmosphere, hydrosphere, lithosphere and cryosphere – with terrestrial and aquatic ecosystems, as expressed in Agenda 21 of the 1992 Rio Convention. The basic processes linking these systems – energy flow, water and matter cycling – have already been handicapped by uncontrolled and rapid human development, and redirected from their natural pathways into artificial, human-induced tracks. This has progressively diminished their integrity and resulted in losses of the regulatory and supporting services of related ecosystems (Zalewski and Wagner, 2005; Wagner *et al.*, 2008). This impacts not only on environment quality, but also on the performance and development of societies (Millennium Ecosystem Assessment, 2005).

Disruption of the carbon cycle is probably the most evident proof, both to the public and decision makers, of human intervention on biogeochemical cycles at the global scale. Analyses of ice cores show that the global atmospheric concentration of carbon dioxide, one of the greenhouse gases, was over 379 ppm by 2005 (IPCC, 2007a), which is far in excess of the natural range (180–300 ppm) of the last 650,000 years. The Scientific Expert Group on Climate Change and Sustainable Development of the United Nations Foundation (UN Foundation, 2007) estimates that 75–85% of this increase has come from the intensified burning of fossil fuels, mainly in developed countries. The other reason, accounting for 15–25% of the carbon dioxide release, is deforestation and other land-cover changes, taking place mainly in developing countries in the tropics.

Global changes in climate

Increased accumulation of carbon dioxide in the atmosphere is one of the major factors altering the Earth's energy budget and increasing the average global surface temperature (IPCC, 2007a). The latter is now about 0.8°C higher than 250 years ago, with most of the increase having occurred in the 20th century. The latest reports conclude that, even if carbon dioxide emissions were stopped instantaneously, the current atmospheric concentrations of greenhouse gases and particles, given the slow equilibration of changes in atmospheric composition with the oceans, will result in a further rise of average global surface temperature by 0.4–0.5°C (IPCC, 2007a,b; UN Foundation, 2007). If the emissions continue to grow by mid-range projections, the temperature may rise by 0.2–0.4°C per decade throughout the 21st century and would continue to rise thereafter. The cumulative warming by the year 2100 would be approximately 3–5°C over pre-industrial conditions.

Observations collected in recent years from all world regions show that regional changes in temperature have already affected hydrological, ecological and human systems. According to the IPCC report (2007b), such anomalies as heavy precipitation, floods, drought, heat waves, tropical storms and wildfires that have appeared or increased in frequency in many regions result from global warming. Cold regions are suffering reductions in the extent of summer sea ice (Arctic), large increases in summer ice sheet melting (Greenland) and destabilization through enlargement and increased numbers of glacial lakes and rock avalanches in mountains (West Antarctic). Many northern and upland hydrological systems are experiencing increased runoff and earlier spring peak discharge, while warming is affecting their thermal structure and water quality. Northern Europe, northern and central Asia, and both Americas have all faced increased precipitation over the last 100 years. On the other hand, drying of some southern regions in Asia, the Mediterranean, southern Africa and the Sahel have exacerbated already existing problems relating to low water accessibility (IPCC, 2007b). Further warming and increases in evapotranspiration may cause land desiccation and result in decreased runoff and water availability despite a precipitation increase. The number and intensity of extreme hydrological events during the 21st century – high flows and deep droughts – may increase considerably.

Rising water temperatures and related changes in ice cover, salinity, oxygen levels and water circulation have already contributed to global shifts in ranges and abundance of algae, zooplankton and fish in high-latitude oceans and high-latitude and high-altitude lakes, and to earlier migrations of fish in rivers. New analyses show that 15–37% of a sample of 1103 land plants and animals might eventually become extinct as a result of climate changes expected by 2050 (Thomas *et al.*, 2004). The loss of biodiversity by global warming will be mostly caused by shifts of physical characteristics of ecosystems and shrinking of suitable habitats. Other species will not be able to reach the suitable habitats because of increasing disconnections and disintegration of climate and landscape.

Confronting the reasons that brought about this situation will necessitate several strategies mitigating both the existing and forecasted consequences. Most of them propose reducing carbon dioxide emissions and enlarging carbon dioxide sinks by, for example, crop management, carbon sequestration and reforestation (IPCC, 2007c). According to Ripl *et al.* (unpublished report) however, preventing a climatic catastrophe becomes sometimes more a form of trade with dry greenhouse gas emission certificates (mainly carbon dioxide and methane) on a national and global scale. This approach, although essential, will certainly not solve the problem of climate change on its own. The reasons for decreasing global security are multifaceted and result from long-term degradation of landscape, and thus water cycles and the Earth's radiation and energetic balance. These problems have to be first approached at the level of individual regions by applying system solutions to fully address global climate changes and their mitigation.

Spatial complexity of non-climate pressures on aquatic ecosystems

The last years have brought significant improvements in understanding and modelling of the global radiation balance and physical processes of energy transport and exchange in the atmosphere. However, the complex relationships between human and natural drivers often make predictions for climate change at regional and local scales unsatisfactory. At any moment in time, the changing global conditions are superimposed upon hydrological and ecological drivers, are at different temporal and spatial (global, regional, local) scales, and are creating a complex organizational pattern within a catchment. These include regional and local hydrological and climatic processes (such as cloud formation, wind speed and direction, groundwater recharge, percolation, landscape retentiveness), land-use albedo, and the effect of vegetation on elements of the water cycle and its role in energy transformations.

The above complexity of impacts weakens the accuracy of projections of climate change effects on aquatic ecosystems. Being natural receivers of water, weathering and erosion products and pollutants generated within a catchment, freshwaters react particularly strongly to disintegration of the hydrological and biogeochemical pathways and ecological degradation in adjacent terrestrial ecosystems. The autonomy of these processes enclosed within a catchment makes each catchment case-specific in its response to global warming and sets limits to any generalization. Progressive, independent catchment degradation, owing to the impacts of other, powerful, non-climatic agents of 'global change', may act either in synergistic or antagonistic ways and thus enhance or mitigate the expected effects of the changing climate. The most important socio-economic drivers superimposed upon the ecological and hydrological ones include changing demography, technology development, dynamics of economic markets and resources demands, political and social institutions, culture, knowledge and information exchange (Redman *et al.*, 2004). They stand at the beginning of a chain of causal links transformed though 'pressures' (emissions, waste, resource use and land use) creating pressure on nature and the environment, including climate change, to 'states' (physical, chemical and biological) and 'impacts' on ecosystems, human health and functions. These eventually may lead to political 'responses' such as prioritization, target setting and indicators, policies, regulations, or possibly socio-economic constraints and losses (Ohl *et al.*, 2007).

Finally, 'local-scale' pressures have the most limited range of influence in the spatial context, although usually of considerable importance due to high potency and persistence. Point-source pollution may have acute effects on biota; physical degradation and simplification of habitats, such as river canalization, impact directly on ecosystem structure. Both examples lead to consequent degradation of biotic functions, which enlarge ecosystems' vulnerability to other stressors, including climate change.

Water and temperature as major 'driving forces' for ecohydrological processes

Availability of water – its abundance, seasonal variability and predictability of hydrological processes – and temperature are the two major determinants ('driving forces') of biota dynamics in both terrestrial and aquatic ecosystems (Zalewski, 2002). Water and temperature determine, among others, the primary production, composition, structure and biological diversity of ecosystems, energy flow, the range of global biomass, the pattern of ecosystem succession and the type of climax biome on the globe (Varlygin and Bazilevich, 1992; Zalewski *et al.*, 2003; Fig. 15.1). On the other hand, the vegetation cover provides feedbacks on temperature and the water cycle, being the natural regulatory force for regional climate (Ripl and Wolter, 2002; Zalewski *et al.*, 2003).

The interlinks between energy flow, matter pathways and productivity of biological systems, on one hand, and temperature and catchment water dynamics, on the other, are effective self-regulating mechanisms under relatively undisturbed climatic and environmental conditions, with high naturalness of landscape and steady landscape and ecosystem processes. This creates the backbone for the use of the 'dual regulation' *sensu* Zalewski (2006) between hydrological and

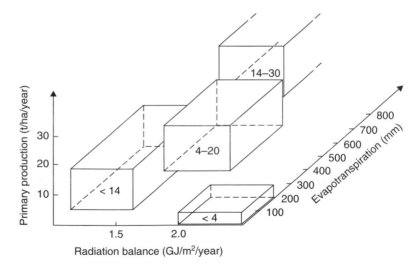

Fig. 15.1. Energy and water as driving forces for primary productivity in global scale. (After Zalewski *et al.*, 1997.)

biota dynamics as a management strategy. It can be used on a local scale, for regulation of some ecosystem functions to increase their capacity against human impact, within the boundaries of the ecosystem steady state. According to the Ecohydrology Concept (Zalewski *et al.*, 1997; Zalewski, 2000, 2006), such an approach may compensate for environmental degradation in some catchment sections and enhance delivery of ecosystem services at a catchment level. The approach combines several measures operating at different structural levels of a catchment (Zalewski, 2000), and is basically concerned with dislocation of nutrients from the available into unavailable pool to designated trophic levels of ecosystems (Zalewski, 1999; Krauze and Wagner, 2008). This makes ecohydrology an integral component of basin-scale management, especially under high environmental degradation.

More extreme and unpredictable patterns of abiotic conditions as a consequence of global warming pose a question about the direction of changes in the dynamics of the ecohydrological interrelationships. Change of the hierarchy of driving factors may constrain or augment the efficiency of ecohydrological measures. Ecohydrology will operate under new or evolving steady-state conditions, and possibly limited resistance or resilience of ecosystems. Quantification of the water mesocycle and analyses of energy–water–biota interactions under the new environmental framework will be fundamental for both scientific survey and management.

Changing Water Cycle and the 'Terrestrial/Atmospheric' Phase of Ecohydrology

Vegetation interactions with climate and water

The cooperation among hydrologists, plant physiologists and soil scientists provided a background for the understanding and use of landscape properties in the regulation of the water cycle (e.g. Eagelson 1982, 2002; Baird and Wilby, 1999; Rodriguez-Iturbe, 2000). At a regional scale, vegetation modifies meteorological conditions due to interception of solar energy, partitioning of intercepted energy into the different components of the heat balance and the differentiation of particular heat balance flux intensities within the mosaic structure of the landscape (Ryszkowski, 1992). Plant cover stabilizes microclimate by reducing temperature extremes and wind speeds, supporting snow accumulation and stabilizing the heat budget. Local landscape diversity may enhance mesoscale circulation, similar to that of breezes between land and waterbodies. Evaporation and condensation of water through woody vegetation is the most effective 'cooling effect' of energy dissipation from Earth and it occurs mostly in vegetated, water-retentive areas. The evapotranspiration, thus the 'cooling' potential of the continents, can be bigger than that of the oceans (Ripl *et al.*, unpublished report). The accumulated effect of regional impacts can thus contribute considerably to global improvement of climate. Regional-scale changes of terrestrial vegetation cover in key regions may cause significant changes not only at regional but also at global scale, such as changes in the tropics influencing the high-latitude northern hemisphere winter climate via atmospheric teleconnections. According to Franzen (1994), vegetation cover could even have played a significant role in the Quaternary climate variations at global scale. Analysis of air trapped in Arctic ice cores revealed that carbon dioxide and methane concentrations varied significantly during the Quaternary, from mean values of about 200 ppmv and 300 ppmv, respectively, in glacials rising to over 300 ppmv and 700 ppmv during interglacials. These fluctuations were closely correlated with temperature changes. The author suggests that the decrease of greenhouse gas concentration could have resulted from the intensive growth of peat deposits in the temperate zone. Contrary to a relatively balanced carbon budget of other world ecosystems, the carbon budget of peat ecosystems is mostly positive, due to low microbiological activity. Approximately 0.16 Gt of atmospheric carbon per year is accumulated in all the peat bogs in the world.

Responding to temperature changes, vegetation feeds back on the water cycle locally through evapotranspiration, interception, regulation of runoff, infiltration and storage in soil and biomass. Moisture availability and cloud production are significantly higher over forests, densely vegetated and transpiring, than over pastures and agricultural lands. Patchiness plays a decisive role, decreasing evapotranspiration rates on fields located between shelter belts compared with those in open areas (Ryszkowski et al., 1997). Evapotranspiration decreases considerably in catchments with highly transformed land cover (deforested) and, together with intensified runoff, leads to increased water yield in a short timescale (Eamus et al., 2006). However, degradation of vegetation cover ultimately leads to serious reductions in humidity, soil moisture and retentiveness and groundwater recharge. Change of the albedo effect and the ability of the landscape to produce clouds and enhanced winds reduce precipitation. Ultimately, deforestation, especially in the low–medium precipitation and high temperature regions, leads to destabilization of the water cycle, increased incidents of droughts and floods, and in a long-term perspective to desertification and decrease of water yield.

Vegetation represents an easily managed landscape element, which can modify climatic factors at the catchment scale to a certain extent. Maintaining or re-establishing diversified landscapes can thus mitigate the effects of climate changes on the water cycle, such as desiccation in some regions or excessive runoff and hydropeaking in others. The extent of water storage in the landscape determines the rate of organic matter decay and soil weathering, regulating nutrient availability for plant growth and the productivity of terrestrial biological systems, in this way strengthening the regional climate-regulating mechanisms of plants. Re-planting of vegetation may additionally serve as a natural carbon dioxide sink. Forests retain today the largest amount of biologically fixed carbon, 500,000 million t which is equivalent to two-thirds of the atmosphere carbon. There is thus potential for lowering carbon dioxide concentrations in the atmosphere by phytoremediation (carbon sequestration), transforming about 15,000 million t of carbon into wood each year, which could be considerably increased by proper landscape management. According to the 'green feedback concept', understanding the role of plants in the control of water and biogeochemical cycles can improve not only the status of water resources, but also ecosystem goods and services, such as biodiversity, agricultural production, bio-energy generation and have positive social feedback, by potential increase of employment opportunities (Zalewski *et al.*, 2003).

Export from catchments and freshwater supply

Modification of precipitation patterns due to climate changes will directly influence runoff and thus the hydrological characteristics of rivers (Wagner and Zalewski, 2000; Zalewski *et al.*, 2000). The duration and intensity of high and low flows, as well as the number of extreme hydrological events and their amplitude will increase. The rate and timing of nutrient transport in both the terrestrial and aerosol phase, as well as the distribution, timing and contribution of particular sources contributing to the overall supply of freshwaters, are directly controlled by the water cycle (e.g. Shanahan and Somlyody, 1995; Wagner and Zalewski, 2000; Walling *et al.*, 2000; May *et al.*, 2001). Therefore the timing and intensity of nutrient supply may change in the future. In catchments where precipitation is predicted to exceed evapotranspiration, more intensive surface and associated subsurface runoff will take place, increasing water and nutrient supply. This will be intensified in catchments with degraded land cover due to deforestation, landscape drainage and wetland loss to cultivation. Open nutrient cycles, caused by landscape leaching, reduced retentiveness in biomass and mineralization of organic soils, will intensify their loss to freshwater. Extending of the growing season due to global warming may additionally expand the time of agricultural activities during a year, which may cause more nutrient leaching from agricultural areas (Hillbricht-Ilkowska, 1993).

The processes of nutrient transport within the river system are to a great extent determined by hydrological parameters, and may differ depending on the river section, area of drained catchment, land use and sources of pollution. The highest loads are however usually observed during the rising hydrograph stage and they tend to decrease before the maximum discharge and during the hydrograph fall (e.g. Wagner and Zalewski, 2000; Wagner-Lotkowska, 2002). The regularity of this pattern can be disturbed in the case of flash floods of short duration and high amplitude, which introduce an element of uncertainty in predicting the system response to the modified hydrological pattern of river systems. In general, a foreseen increase in frequency of extreme hydrological events may intensify the nutrient export from some catchments and their downstream transport by rivers, posing a greater risk of eutrophication for reservoirs and coastal areas.

The pathways and intensity of matter flow will change along the management intensity gradient from natural systems, where perturbations in water and biogeochemical states are driven primarily by climate variability, to systems where disturbance induced by human activity has been significant. Some of the cycles operating individually at local or regional scales, such as nitrogen and phosphorus cycles, have cumulative effects constituting a global issue. Nitrogen and phosphorus fertilizers are applied to large areas of the Earth and the sum of the regional excesses and runoffs of rivers produces a global effect. Additionally nitrogen, in the gaseous part of its cycle, can be spread globally by atmospheric circulation (US NAP, 2000).

Wetlands are natural barriers to arrest non point-source pollution and provide a natural protection between aquatic ecosystems and disrupted terrestrial cycles of water and matter. They are also among the most productive ecosystems in the world (Mitsch and Gosselink, 2000), and are able to stabilize hydrological disturbances, support biodiversity across river and landscape, and efficiently store and transform nutrients (Naiman *et al.*, 1989). For example, Petersen *et al.* (1992) report that the efficiency of ecotone zones of 19–50 m width may reach nitrogen removal rates of up to 78–98% in surface waters and 68–100% in groundwater. Other authors estimate reduction efficiencies at a level of 50–90% for nitrogen and 25–98% for phosphorus (in groundwater) depending on the initial concentrations, the buffer zone width, soil type, ecotone slope and interactions between plants and other organisms.

Being to a great extent dependent on a permanent or temporary supply of water, the efficiency of buffering functions of wetlands may be constrained in changed environmental conditions. Forecast increased variability of water levels will impact factors that are crucial for wetland functioning: seasonal pulsing (Wagner and Zalewski, 2000; Mitsch *et al.*, 2005), depth and hydroperiod (Mitsch and Jørgensen, 2004), groundwater recharge (e.g. Lucassen *et al.*, 2005) and connectedness with landscape and waterbodies (e.g. Li *et al.*, 2005). Long-term seasonal desiccation may possibly diminish their protective functions. Loss of the biotic structure and soil retentiveness may lead, especially during heavy rains after long periods of drought, to nutrient 'bypassing' of degraded vegetation zones and diminish their buffering functions.

Effects of Increasing Temperatures in Lakes and Reservoirs

Nutrient cycling and internal load

The metabolic rate of biota, which is strongly dependent on the thermal conditions of the environment, will accelerate even at a relatively small temperature increase. Together with the possible increased external loading from a catchment, it will accelerate energy flow and matter cycling within aquatic ecosystems. Temperature increase will intensify internal loading. Nutrients may be released at a greater rate and over a longer time, as a consequence of intensification of a variety of temperature-regulated processes such as microbial autochtonic/allochtonic organic matter decomposition, enzyme activity, release from bottom sediments and animal excretion. In some lakes and reservoirs, higher temperatures will increase oxygen demand, which may inhibit aerobic organic matter decomposition and favour anoxic breakdown, which is less efficient. On the other hand, the internal load in such ecosystems may be periodically enhanced due to chemical phosphorus release from iron complexes of the bottom sediments, which intensify in low oxygen and redox conditions. In rivers and shallow reservoirs, where the oxygen content is usually high due to mixing or wind action, the breakdown is directly dependent on temperature and may accelerate internal loading (Zalewski and Wagner, 1998; Fig. 15.2). At the same time, the experimental relationship between pH and decomposition rate shows the process is inhibited in acid or alkaline conditions. Thus the temperature effect will be most prominent in ecosystems with pH close to neutral or those situated in limestone geologies or those highly eutrophicated, where the pH can increase seasonally due to high primary production. In ecosystems affected by acid rains or situated on granite, pH will be the limiting factor for matter cycling.

Phosphorus recycling is also connected with enzyme activities. Cyanobacteria and algae produce different types of phosphohydrolases (phosphatases) that release reactive phosphorus from phosphoesters in periods of inorganic phosphorus shortage. This mechanism may periodically increase their competitiveness over other groups of phytoplankton (Romanowska-Duda *et al.*, 1998; Tarczynska *et al.*, 1999; Trojanowska *et al.*, 2001). The activity of the enzyme rises with temperature; thus it is probable that even if phosphorus concentrations were reduced below the natural level of 30 μg P/l, enhanced phosphorus cycling due to phosphatase activity may maintain eutrophication symptoms at a higher level.

Higher temperatures will also maintain high phosphorus recirculation due to animal excretion. Zooplankton (*Daphnia pulex*) recycle between 1.5 and 2.5% of the total-body phosphorus per hour. Gulati *et al.* (1995) calculated that the amount of phosphorus regenerated by zooplankton in different temperatures in Lake Vechten accounted for between 22 and 239% of phytoplankton phosphorus demand.

Energy flow

Energy flow and matter cycling accelerate even at a relatively small temperature increase; warming by 1°C enhances ecosystem productivity by 10–20% at all

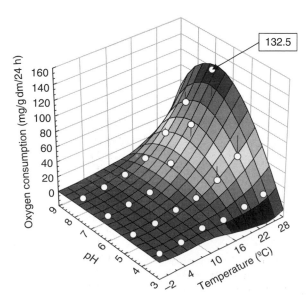

Fig. 15.2. Relationship between temperature, pH and decomposition processes of organic matter in a river supplying a drinking water reservoir in Poland. (After Zalewski and Wagner, 1998.)

trophic levels (Regier and Meisner, 1990). The main paths of energy and nutrient cycling in aquatic ecosystems are through poikilotherms, evolutionarily adapted to the climatic characteristics of regions.

The relationship between temperature and energy requirements of poikilotherms is described by a right-skewed function. To maintain living processes at conditions of higher energy needs, fish must increase their food intake. At certain temperature ranges, rates of respiration may accelerate faster than those of assimilation, resulting in an energy deficiency and body size decrease (Elliott, 1996; Moore *et al.*, 1996). Changing consumption rates for a warm/temperate climate fish are shown in Fig. 15.3. At a temperature of about 10°C, the daily amount of food is roughly equal to 1.2% of its body weight, and only 0.4% of the consumption is converted into the biomass. The highest growth is observed at 30°C: daily consumption reaches up to 9% of body weight, which transforms to roughly 3% increase in weight. Further temperature increments dramatically decrease daily food intake, which at 36°C falls to about 4% of body weight. This amount of food does not cover the needs of the high-rate metabolic processes and the fish loses about 1% of body weight per day (Kitchell *et al.*, 1977; Zalewski, 1988).

Changes of the pathways and intensity of the energy flow are accompanied by quantitative and qualitative changes of inter- and intraspecific relationships

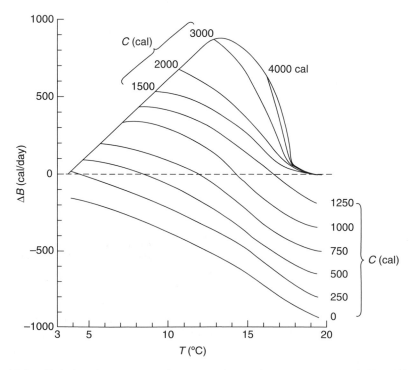

Fig. 15.3. Relationship between the mean change in energy content (ΔB, cal/day) and temperature (T, °C) at fixed levels of energy intake (C, cal/day) for trout of initial weight close to 50 g. (After Elliott, 1976.)

within the cascade of phytoplankton/primary producers, invertebrate and fish communities (Gucinski *et al.*, 1990; Hillbricht-Ilkowska, 1993). This may possibly lead to shifts in dominant species, destabilization of the ecosystem equilibrium and its shift to another steady state. In the case of zooplanktivorous fish, the intensification of food consumption may reduce the density of zooplankton. The resulting decrease in fish food base may further inhibit their growth. Overlapping of changing abiotic conditions, such as increasing temperature and decreasing dissolved oxygen content, may be an additional stressor (Fig. 15.4), contributing to lowering of biodiversity and ecosystem functions.

The situation is similar in the case of zooplankton. The mean body size of a *Daphnia pavula* is positively correlated with temperature at lower ranges (below 15°C), while its further rise increases energy demand, reduces energy available for growth and leads to body size decrease. This affects the fitness of an individual by leading to lower egg production and smaller clutches (Orcutt *et al.*, 1983; Moore and Folt, 1993; Moore *et al.*, 1996). However, acceleration of development rates at high temperatures may compensate for these processes and result in increased zooplankton population growth. Increasing energy demand has a

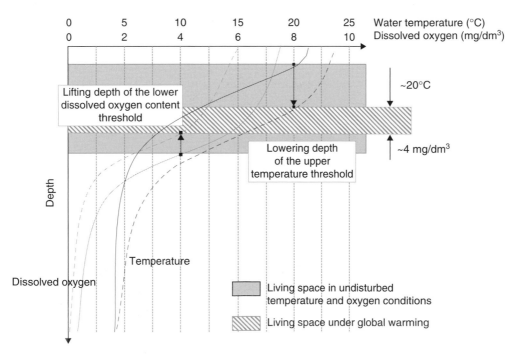

Fig. 15.4. Changes of overturn and stratification pattern in dimictic lakes (a) resulting from climate warming. Increase of mean water temperature results in earlier time of spring overturn, delayed autumn overturn and longer period of summer stratification (b). If minimum water temperature in winter is +4°C, there will be only one overturn (shift into monomictic regime; c). Further warming may lead to elimination of overturns and summer stratification will be maintained all year (shift to amictic regime; d). M, overturn; SL, summer stratification; SZ, winter stratification.

less prominent effect on smaller zooplankton, as their food-collecting efficiency increases faster than their energy requirements. The threshold food concentration (a food level that should be ingested to exactly balance the processes of respiration and assimilation to maintain constant biomass of zooplankton) is higher and increases more rapidly with temperature rise for large zooplankton. The small species also avoid predation from visual fish predators better. Therefore they may be favoured over the large species during global warming. Temperature increase by only 2–3°C may alter zooplankton community structure (Moore and Folt, 1993; Moore *et al.*, 1996).

Concerning interactions down the trophic pyramid, increase of temperature enhances the zooplankton filtering rate and may reduce phytoplankton density (Zalewski *et al.*, 1990). On the other hand, decrease in zooplankton body size and shifts in community structure may release phytoplankton from the grazer stress and have the opposite result. Decreased body length of *D. pulex* of 0.2 mm (about 7% of mean body size of the population) lowers the filtering rate by two-fold, as the result of physical constraints. This matters especially in the case of cyanobacterial blooms which, occurring in colonies (e.g. *Microcystis* spp.) or flakes of hundreds of filaments aligned in parallel (e.g. *Aphanizomenon* spp.), are often too large for zooplankton to feed on (Holm *et al.*, 1983). Flakes larger than <0.2 mm wide and <1.5 mm long are not used as a food.

Cyanobacteria may benefit from increasing internal nutrient recirculation by small zooplankton species and individuals, which recycle more nutrients per body unit than larger ones. They may induce a shift from phosphorus to nitrogen limitation of phytoplankton growth in low-nutrient lakes and support cyanobacteria because of their nitrogen-fixing ability, which gives them a competitive advantage over other phytoplankton species.

The acceleration of biological processes will have different effects on aquatic ecosystem functioning, depending on its hydrological type, characteristics and complexity. In the case of rivers, the elevated nutrient supply may be assimilated by primary producers and retained in the biotic structure in natural or restored ecosystems with intermediate complexity of ecotone zones, high biomass of macrophytes and diversified river bed structure. The ability of these rivers to reduce nutrient loads because of the effect of the physical, chemical and biological processes, i.e. self-purification, is high and will increase with temperature. In contrast, regulated rivers that are devoid of well-developed biotic structure and ecotone zones possess a substantially lower likelihood of nutrient trapping and self-purification. In this case, increasing water temperature will result in progressive lowering of water quality.

A river's ability to self-purify and retain nutrients is important not only because of the enhancement of its productivity and maintenance of biodiversity, but also because of the quality of water in reservoirs and lakes situated downstream. Increase in temperature will be especially dangerous for shallow, non-stratified lakes and reservoirs, where temperature over 18°C and higher nutrient availability will lead to an elongation of the heating period and the intensity of appearance of algal and cyanobacterial blooms.

The above processes may to a certain extent regulate hydrological factors such as water retention time, which is positively correlated with cyanobacteria biomass (Tarczynska *et al.*, 2001), or the hydrological pattern of inflow to reservoir (Wagner and Zalewski, 2000; Zalewski *et al.*, 2000). Changes of wind speed and direction contributing to the pattern of lake and reservoir mixing and also thermal stratification may modify the biomass by destabilization of water stability (Reynolds, 1984; Meyer *et al.*, 1994; Varis and Somlyódy, 1994). Higher temperature raises cyanobacteria bloom toxicity. Experiments performed by Sivonen (1990) and Lethimaki *et al.* (1994) showed that production of cyanobacterial toxins reaches the highest rate at temperatures of 15–25°C. A toxin produced by microcistis – Microcistina-LR – is ten times more toxic than strychnine and its toxicity may be enhanced by the presence of other toxins. Intensity and toxicity of cyanobacterial blooms, already present in all continents, may intensify posing restrictions on water resources use for humans.

Ecohydrology as a Tool for Mitigating the Effects of Global Change

Analyses of extended observations, supported by modelling studies, have indicated (with a greater than 66% probability) that the global changes observed so far in recent years in hydrological and ecological systems have been driven mostly by human impacts related to the development of industrial civilization over the past 200 years (IPCC, 2007a,b). Changes of the predicted magnitude in terms of average temperature increase and other global effects will adversely impact agriculture, forestry, fisheries, the availability of fresh water, the geography of disease, the habitability of human settlements, and more. Even over the next decade, the growing impacts of climate change will make it difficult to meet the United Nations' Millennium Development Goals (UN Foundation, 2007). Following the idea of the 'anthropocene', mankind's activities may remain a major geological and morphological force and could dominate the control of the globe's functioning for many millennia or millions of years to come (Crutzen and Stoermer, 2000; Meybeck, 2003).

The progressing resynchronization of natural regulatory processes and the lessened ability of biological systems to adapt to rapid changes of large magnitude impact global environmental security. Reversing this situation has various benefits, including ones of environmental, aesthetic, cultural and economic nature. According to United Nations' estimates (UNEP, 2004), 40% of the global economy is based on biological products and processes. Poor societies living in areas of low agricultural productivity depend strongly on biodiversity. The safety, well-being and quality of life of developed societies depend to a great extent on the 'regulating' services of ecosystems. The ability of ecosystems to regulate climate and microclimate, flooding mitigation, water purification and others, depends on their potential to fully perform their natural functions. They also support other services such as nutrient cycling, primary productivity, soil formation, and the water cycle. These have been seriously handicapped over the last decades due to environmental degradation (Millennium Ecosystem Assessment, 2005).

Comprehending the extensive threats resulting from changing climate, the report published by the United Nations Foundation (UN Foundation, 2007) calls for a twofold approach to confront global climate change, by avoiding the unmanageable and managing the unavoidable. Mitigation measures preventing the degree of climate change from becoming unmanageable (e.g. reductions in emissions of greenhouse gases and black soot) have to be accompanied by elaborating and implementing adaptation measures, reducing the harm from climate change that proves unavoidable. These have to include improved management of natural resources and preparedness/response strategies to cope with future climatic conditions that will be fundamentally different from those experienced for the last 100 years.

However the strategy needs to go beyond elimination of the escalating threats and should, at the same time, enlarge the opportunities for the environment by re-activating its self-protective mechanisms (Zalewski, 2002). Water is a key element in this process, by mainstreaming and driving several interdependent cycles in a catchment: biota dynamics, biodiversity, climate regulation, and social and economic systems. This approach has been used by UNESCO (United Nations Educational, Scientific and Cultural Organization) in designing the approach towards water resources investigation and management in Phase VII (2008–2013) of the International Hydrological Programme (UNESCO, 2006). Water and temperature drive the functioning of the fundamental climate-regulating factor: vegetation. The dynamics of the water and water vapour cycles on the continents, controlled through vegetation, is an important factor in the process of physical dissipation of the Sun's energy. Its effect may be superior to greenhouse gases in shaping regional climate, and build up into a larger scale (Ripl, 2003).

Considering the diversity of the scale and character of different pressures, Zalewski (2004) suggests that the management actions towards sustainable water resources management in the face of global change should take in the following dimensions and scales:

- The global scale – reduction of energy use, in the sense of carbon dioxide emission from fossil fuels, as the preventive measure against global climate change, which increases the variability and unpredictability of water resources.
- The regional and local scales – reduction in the growth of materials use per capita, as a method reducing pollution and waste generation, which affect water quality.
- Related to all ranges of scales – enhancement of the absorbing capacity of ecosystems.

The rapid environmental changes and the limited ability of ecosystems to anticipate or counteract environmental extremes require replacing the conservative, sartorial water management with an adaptive approach, based on an understanding and use of evolutionarily established relationships between hydrological/abiotic characteristics of ecosystems and biota functioning. Such an approach appears to be more flexible to deal with the increasing uncertainty of appearance, scale, reversibility and damage potential of water-related risks, using the resilience of the natural systems against the pressures (Krauze and Wagner, 2008).

Increasing public involvement in decision making, and reconciling and merging between environmental and socio-economic policies, technical solutions and social needs may accelerate the process of implementation of the new strategies. The experience from the UNESCO and UNEP (United Nations Environment Programme) Demonstration Project in the Pilica river valley (Wagner-Lotkowska *et al.*, 2004) showed that, especially in the case of increasing tensions between society and water environmental resources, the efficient stimulation for public involvement is the assumption of 'ecosystem services first' (Zalewski, 2004). A visible incentive for society at the first stage of the implementation process is essential for their acceptance. Adaptive assessment management, taking into account possible drivers and adapting to them, then becomes the best way not only for understanding the complexity of the ecohydrology feedbacks and revising the new management strategy, but also for ensuring that societal needs are met.

Uncertainty of the predicted changes in the global environment and their regional divergence raise concerns about building a convincing, unified strategy for mitigating the impacts of global climate change and ensuring environmental security on a global scale. The powerful impact of local and regional pressures brings them strongly into the picture as essential or even supreme elements of the strategy. Inclusion of ecohydrological measures involving landscape restoration and the dual regulation of hydrological and biological processes in aquatic eco-systems into regional management may have a twofold effect: (i) mitigating the effects of global climate change on ecosystems and society; and (ii) decelerating or mitigating the global climate changes themselves. Quantification of the water mesocycle and related biogeochemical processes within the boundaries delimited by a catchment will be fundamental for the first objective. This understanding can be applied not only for maintenance, but also for enlarging the ability of some ecosystems to deal with increased level of stress. This approach, already tested in several semi-natural systems (e.g. Wagner-Lotkowska *et al.*, 2004; Agostinho *et al.*, 2005; Wolanski *et al.*, 2005), can be crucial in the face of the additional and highly unpredictable stressor: global warming. The second objective will closely relate to the local re-establishment of the degraded processes. Understanding the energy–water–biota interactions and the implementation of their mechanisms based on this understanding (e.g. by maintaining and restitution of landscape patchiness and vegetation cover) will be fundamental for stabilizing regional climate elements, such as temperature, wind and humidity, and will strongly contribute to carbon sequestration, adding up for a global effect.

References

Acreman, M.C. (2001) *Hydro-ecology: Linking Hydrology and Aquatic Ecology.* IAHS
    Publication No. 266. Centre for Ecology and Hydrology, Wallingford, UK.
Acreman, M.C. and Hollis, G.E. (1996) *Water Management and Wetlands in sub-Saharan
    Africa.* IUCN, Gland, Switzerland, 250 pp.
Acreman, M.C. and Howard, G. (1996) The use of artificial floods for floodplain restora-
    tion and management in sub-Saharan Africa. *IUCN Wetlands Programme Newsletter*
    12, 20–25.
Acreman, M.C., Farquharson, F.A.K., McCartney, M.P., Sullivan, C., Campbell, K.,
    Hodgson, N., Morton, J., Smith, D., Birley, M., Knott, D., Lazenby, J., Wingfield, R.
    and Barbier, E.B. (2000) *Managed Flood Releases from Reservoirs: Issues and Guid-
    ance. Report to DFID and the World Commisson on Dams.* Centre for Ecology and
    Hydrology, Wallingford, UK.
Adams, A. (1999) A grassroots view of Senegal river development agencies: OMVS,
    SAED. Submission to the World Commission on Dams, Regional Consultation in
    Cairo, 8–9 December.
Adams, S.M., McLean, R.B. and Huffman, M.M. (1982) Structuring of a predator popula-
    tion through temperature-mediated effects on prey availability. *Canadian Journal of
    Fisheries and Aquatic Science* 39, 1175–1184.
Adams, W.M. (1992) *Wasting the Rain: Rivers, People and Planning in Africa.* Earthscan,
    London, 255 pp.
Admiraal, W., Breugem, P., Jacobs, D.M. and De Reuyter van Steveninck, D. (1990)
    Fixation of dissolved silicate and sedimentation of biogenic silicate in the lower river
    Rhine during diatom blooms. *Biogeochemistry* 9, 175–185.
Agostinho, A.A., Gomes, L.C, Verıssimo, S. and Okada, E.K. (2005) Flood regime, dam
    regulation and fish in the Upper Paraná River: effects on assemblage attributes,
    reproduction and recruitment. *Reviews in Fish Biology and Fisheries (2004)* 14,
    11–19.
Aho, J.M. and Terrell, J.W. (1986) *Habitat Suitability Index Models and Instream Flow
    Suitability Curves: Redbreast Sunfish.* Biological Report 82(10.119). US Fish and
    Wildlife Service, Fort Collins, Colorado.

Albaret, J.J. (1994) *Peuplements de poissons, ressources halieutiques, pisciculture dans le delta du fleuve Senegal: impact des modifications de l'environnement.* ORSTOM/ CRODT, Dakar, 47 pp.

Albaret, J.J. and Diouf, P.S. (1994) Diversité des poissons des lagunes et des estuaires ouest-africains. In: Teugels, G.G., Guégan, J.F. and Albaret, J.J. (eds) Proceedings of a Symposium on Biological diversity of african fresh- and brackish water fishes. *Annales du Musée Royal de l'Afrique Centrale, Zoologie* 275, 165–177.

Alcamo, J.M. (1983) Development and testing of an estuarine phytoplankton model. In: Lauenroth, W.K., Skogerobe, G.V. and Flug, M. (eds) *Analysis of Ecological Systems: State-of-the-Art in Ecological Modeling.* Elsevier Scientific Publishing Co., Amsterdam, pp. 575–592.

Alfredsen, K. (1997) A modeling system for estimation of impacts on fish habitat. In: Holly, F.M. and Alsaffar, A. (eds) *Proceedings of the 27th Congress of the International Association for Hydraulic Research*, San Francisco, California, 10–15 August, pp. 883–888.

Alimov, A.F. (1989) *Introduction to Productive Hydrobiology.* Hydrometeoizdat, Leningrad, Russia (in Russian).

Alkemade, J.R.M., Wiertz, J. and Latour, J.B. (1996) *Kalibratie van Ellenbergs milieu-indicatiegetallen aan werkelijk gemeten bodemfactoren (Calibration of Ellenberg's Environmental Indicator Values to Measured Soil Factors).* RIVM, Bilthoven, The Netherlands.

Allan, J.D. (1995) *Stream Ecology. Structure and Function of Running Waters.* Chapman and Hall, London.

Allan, J.D., Ericksson, D.L. and Fay, J. (1997) The influence of catchment land use on stream integrity across multiple spatial scales. *Freshwater Biology* 37, 149–161.

Alvey, N.G., Banfield, C.F., Baxter, R.I., Gower, J.C., Krzanowski, W.J., Lane, P.W. and Kinson, G.N. (1977) GENSTAT: A General Statistical Program. Rothamsted Experimental Station, Harpenden, UK.

Ambühl, H. (1959) Die Bedeutung der Strömung als ökologischer Faktor. *Schweizerische Zeitschrift für Hydrobiologie* 21, 133–264.

Amon, R.M. and Benner, R. (1996) Photochemical and microbial consumption of dissolved organic carbon and dissolved oxygen in the Amazon river system. *Geochimica et Cosmochimica Acta* **60**, 1783–1792.

Andrews, P., Groves, C.P. and Horne, J.F.M. (1975) Ecology of the Lower Tana River floodplain (Kenya). *Journal of the East African Natural History Society and National Museum* 151, 1–31.

Anon. (1992) Agenda 21. United Nations Conference on Environment and Development (UNCED), Rio de Janeiro, Brazil.

Anon. (2004) *The East African*, May.

Antropovsky, V.I. (1970) Connection of channel process with suitable factors. *Proceedings of the State Hydrological Institute* 183, 70–80 (in Russian).

Armitage, P.D. (1989) The application of a classification and prediction technique based on macroinvertebrates to assess the effects of river regulation. In: Gore, J.A. and Petts, G.E. (eds) *Alternatives in Regulated River Management.* CRC Press, Boca Raton, Florida, pp. 267–293.

Armitage, P.D., Pardo, I. and Brown, A. (1995) Temporal constancy of faunal assemblages in mesohabitats – application to management. *Archiv für Hydrobiologie* 133, 367–387.

Arthington, A.H., King, J.M., O'Keeffe, J.H., Bunn, S.E., Day, J.A., Pusey, B.J., Bluhdorn, D.R. and Tharme, R. (1991) Development of an holistic approach for assessing environmental flow requirements of riverine ecosystems. In: Pigram, J.J. and Hopper,

B.P. (eds) *Water Allocation for the Environment: Proceedings of an International Seminar and Workshop*. Centre for Water Policy Research, University of New England, Armidale, New South Wales, Australia, pp. 69–76.

Baird, A.J. (1999) Modelling. In: Baird, A.J. and Wilby, R.L. (eds) *Eco-hydrology. Plants and Water in Terrestrial and Aquatic Environments*. Routledge, London, pp. 300–345.

Baird, A.J. and Wilby, R.L. (eds) (1999) *Eco-hydrology. Plants and Water in Terrestrial and Aquatic Environments*. Routledge, London, 402 pp.

Bajkiewicz-Grabowska, E. (1985) Factors affecting nutrient budget in lakes of r. Jorka watershed (Masurian Lakeland, Poland) I. Geographical description, hydrographic components and man's impact. *Ekologia polska* 33, 173–200.

Baker, E.K. and Harris, P.T. (1991) Copper, lead, and zinc distribution in the sediments of the Fly river delta and Torres strait. *Marine Pollution Bulletin* 22, 614–618.

Baker, V.R. (1983) Paleoflood hydrological analysis from slack-water deposits: Proceedings of the Holocene Symposium. *Quaternary Studies in Poland (Poznań)* 4, 19–26.

Baker, V.R. (1987) Palaeoflood hydrology and extraordinary flood events. *Journal of Hydrology* 96, 79–99.

Baker, V.R. (1991) A bright future for old flows. In: Starkel, L., Gregory, K.J. and Thornes, J.B. (eds) *Temperate Palaeohydrology*. John Wiley and Sons Ltd, Chichester, UK, pp. 497–520.

Baker, V.R. (1994) Global to modern changes in global river fluxes. In: Panel on Global Surficial Geofluxes, Board on Earth Sciences and Resources, National Research Council (eds) *Material Fluxes on the Surface of the Earth*. National Academy Press, Washington, DC, pp. 86–98.

Baker, V.R., Kochel, R.C. and Patton, P.C. (eds) (1988) *Flood Geomorphology*. John Wiley and Sons Ltd, Chichester, UK, 503 pp.

Baker, V.R., Bowler, J.M., Enzel, Y. and Lancaster, N. (1995) Late Quaternary palaeohydrology of arid and semi-arid regions. In: Gregory, K.J., Starkel, L. and Baker, V.R. (eds) *Global Continental Palaeohydrology*. John Wiley and Sons Ltd, Chichester, UK, pp. 203–231.

Balls, H., Moss, B and Irvine, K. (1989) The loss of submerged plants with eutrophication. I. Experimental design, water chemistry, aquatic plant and phytoplankton biomass in experiments carried out in ponds in the Norfolk Broad. *Freshwater Biology* 22, 71–97.

Balogun, I. (1996) Wetlands in Nigeria: characteristics and available options for their utilization and conservation. In: Leclerc, M., Capra, H., Valentin, S., Boudreault, A. and Coté, Y. (eds) *Ecohydraulics 2000. Proceedings of the 2nd International Symposium on Habitat Hydraulics*. INRS-Eau, Québec City, Canada, pp. 811–828.

Baras, E., Philippart, J.C. and Nindaba, J. (1996) Importance of gravel bars as spawning grounds and nurseries for European running water cyprinids. In: Leclerc, M., Capra, H., Valentin, S., Boudreault, A. and Coté, Y. (eds) *Ecohydraulics 2000. Proceedings of the 2nd International Symposium on Habitat Hydraulics*. INRS-Eau, Québec City, Canada, pp. 367–378.

Barendregt, A. (1993) Hydro-ecology of the Dutch polder landscape. PhD thesis, University of Utrecht, Utrecht, The Netherlands.

Barendregt, A., Wassen, M.J., de Smidt, J.T. and Lippe, E. (1986) Ingreep-effect voorspelling voor waterbeheer (Intervention-effect prediction for water management). *Landschap* 1, 41–55.

Barko, J.W. and James, W.F. (1998) Effects of surmerged aquatic macrophytes on nutrient dynamics, sedimentation, and resuspension. In: Jeppesen, E., Sondergaard, Ma., Sondergaard, Mo. and Christoffersen, K. (eds) *The Structuring Role of Submerged Macrophytes in Lakes. Ecological Studies*, Vol. 131. Springer, New York, New York, pp. 197–214.

Barmuta, L.A. (1989) Habitat patchiness and macrobenthic community structure in an upland stream in temperate Victoria, Australia. *Freshwater Biology* 21, 223–236.

Bartholow, J.M. and Waddle, T.J. (1986) *Introduction to Stream Network Habitat Analysis*. Instream Flow Information Paper No. 22, Biological Report 86(8). US Fish and Wildlife Service, Washington, DC.

Baulin, V.V., Belopukhova, Ye.B. and Danilova, N.S. (1984) Holocene permafrost in the USSR. In: Velichko, A.A. (ed.) *Late Quaternary Environments of the Soviet Union*. University of Minnesota Press, Minneapolis, Minnesota, pp. 87–94.

Bayha, K. (1978) *Instream Flow Methodologies for Regional and National Assessments*. Instream Flow Information Paper No. 7, FWS/OBS-78/61. US Fish and Wildlife Service, Washington, DC.

Bayley, P.B. (1995) Understanding large rivers: floodplain ecosystems. *BioScience* 45, 153–158.

Becker, B. (1982) Dendrochronologie und Paläoekologie subfossilen Baumstämme aus Flussablagerungen; ein Beitrag zur nacheiszeitlichen Auenetwicklung im südlichen Mitteleuropa. *Mitteilungen der Kommission für Quartärforschung der Osterreichischen Akademie der Wissenschaften* 5, 1–120.

Bedford, K.W., Sykes, R.M. and Libicki, C. (1983) *A Dynamic One-Dimensional, Riverine Water Quality Model*. Report DACW39-82-M-3548. US Army Engineer Waterways Experiment Station, Vicksburg, Mississippi.

Beecher, H.A. (1990) Standards for instream flows. *Rivers* 1, 97–109.

Behrendt, H. and Opitz, D. (1999) Retention of nutrients in river systems: dependence on specific runoff and hydraulic load. *Developments in Hydrobiology* 410, 111–122.

Behrendt, H., Kornmilch, M., Korol, R., Stronska, M. and Pagenkopf, W.-G. (1999) Point and diffuse nutrient emissions and transports in the Odra basin and its main tributaries. *Acta Hydrochimica Hydrobiologica* 27, 274–281.

Beintema, A.J. (1995) Management of the Djoudj National Park after completion of the Diama dam, Senegal. In: Roggeri, H. (ed.) *Tropical Freshwater Wetlands*. Kluwer Academic Publishers, Dordrecht, The Netherlands, pp. 173–178.

Bencala, K.E. (1993) A perspective on stream–catchment connections. *Journal of the North American Benthological Society* 22, 44–47.

Benndorf, J. and Pütz, K. (1987a) Control of eutrophication on lakes and reservoirs by means of pre-dams – 1. Mode of operation and calculation of the nutrient elimination capacity. *Water Research* 21, 829–838.

Benndorf, J. and Pütz, K. (1987b) Control of eutrophication on lakes and reservoirs by means of pre-dams – 2. Validation of the phosphate removal model and size optimisation. *Water Research* 21, 839–842.

Berg, V.A. (1974) About hydrodynamic conditions of the freezing-over and ice-transit on streams. *Proceedings of the Coordination Meetings in Hydraulic Engineering, Leningrad* 96, 35–78 (in Russian).

Bernez, I. and Haury, J. (1986) Downstream effects of hydroelectric impoundment on river macrophyte communities. In: Leclerc, M., Capra, H., Valentin, S., Boudreault, A. and Coté, Y. (eds) *Ecohydraulics 2000. Proceedings of the 2nd International Symposium on Habitat Hydraulics*. INRS-Eau, Québec City, Canada, pp. 13–24.

Betson, R.P. and McMaster, M. (1975) Non point source mineral water quality model. *Journal of the Water Pollution Control Federation* 47, 10.

Beugelink, G.P., Claessen, F.A.M. and Mülschlegel, J.H.C. (1992) *Effecten op natuur van grondwaterwinning t.b.v. Beleidsplan Drink- en Industriewatervoorziening en MER (Policy Analysis of the Effects on Nature of Groundwater Withdrawal by Drinking Water Companies and Industry)*. RIVM, Bilthoven and RIZA, Lelystad, The Netherlands.

Beuscher, J.H. (1967) *Water Rights*. College Printing and Publishing, Madison, Wisconsin.

Beven, K. and Germann, P. (1982) Macropores and water flow in soils. *Water Resources Research* 18, 1311–1325.

Bierman, V.J., Verhoff, F.H., Poulson, T.L. and Tenney, M.W. (1974) Multi-nutrient dynamic models of algal growth and species competition in eutrophic lakes. In: Middlebrooks, E.J., Falkenborg, D.H. and Maloney, T.E. (eds) *Modeling the Eutrophication Process*. Ann Arbor Science Publishers, Ann Arbor, Michigan, pp. 89–109.

Bierman, V.J., Dolan, D.M., Stoermer, E.F., Gannon, J.E. and Smith, V.E. (1980) *The Development and Calibration of a Spatially Simplified Multi-class Phytoplankton Model for Saginaw Bay, Lake Huron*. Great Lakes Environmental Planning Study, Contribution No. 33. Great Lakes Basin Commission, Ann Arbor, Michigan, p. 126.

Binns, N.A. (1982) *Habitat Quality Index Procedures Manual*. Wyoming Game and Fish Department, Cheyenne, Wyoming.

Binns, N.A. and Eiserman, F.M. (1979) Quantification of fluvial trout habitat in Wyoming. *Transactions of the American Fisheries Society* 108, 215–228.

Blöschl, G. and Sivapalan, M. (1995) Scale issues in hydrological modelling: a review. *Hydrological Processes* 9, 251–290.

Bodeanu, N. (1988) Structure et dynamique de l'algoflore unicellulaire dans les eaux du littoral roumaine de la Mer Noire. *Cercetari Marine* 20/21, 19–250.

Boisclair, D. and Sirois, P. (1993) Testing assumption of fish bioenergetics models by direct estimation of growth, consumption, and activity rates. *Transactions of the American Fisheries Society* 122, 784–796.

Bonnell, M. (2002) Ecohydrology – a completely new idea? *Hydrological Sciences* 47, 809–810.

Boorman, I.A. (1990) Impact of sea level changes on coastal areas. In: Boer, M.M. and de Groot, R.S. (eds) *Landscape–Ecological Impact of Climatic Change*. IOS Press, Amsterdam, pp. 379–391.

Bornette, G., Amoros, C. and Lamouroux, N. (1998) Aquatic plant diversity in rivereine wetlands: the role of connectivity. *Freshwater Biology* 39, 267–283.

Bournaud, M. and Cogerino, L. (1986) Les microhabitats aquatiques des rives d'un grand cours d'eau: approche faunistique. *Annales de Limnologie* 22, 285–294.

Bouman, A.C. (1992) *Waterplanten in het Naardermeer (Water Plants in Lake Naardermeer)*. Vereniging tot Behoud van Natuurmonumenten, 's-Graveland, The Netherlands.

Bovee, K.D. (1982) *A Guide to Stream Habitat Analysis Using the Instream Flow Incremental Methodology*. Instream Flow Information Paper No. 12, FWS/OBS-82/26. US Fish and Wildlife Service, Washington, DC.

Bovee, K.D. (1986) *Development and Evaluation of Habitat Suitability Criteria for Use in the Instream Flow Incremental Methodology*. Instream Flow Information Paper No. 21, Biological Report 86(7). US Fish and Wildlife Service, Washington, DC.

Bovee, K.D. (1996) Perspectives on two-dimensional river habitat models: the PHABSIM experience. In: Leclerc, M., Capra, H., Valentin, S., Boudreault, A. and Coté, Y. (eds) *Ecohydraulics 2000. Proceedings of the 2nd International Symposium on Habitat Hydraulics*. INRS-Eau, Québec City, Canada, pp. B149–BB162.

Bovee, K.D., Gore, J.A. and Silverman, A.J. (1978) *Field Testing and Adaptation of a Methodology to Measure 'In-stream' Values in the Tongue River, Northern Great Plains (NGP) Region*. EPA-908/4/78-004A. US Environmental Protection Agency, Washington, DC.

Bowes, G. and Reiskind, J. (1987) Inorganic carbon concentrating systems from an environmental perspective. In: Biggens, J. (ed.) *Progress in Photosynthesis Research*, Vol. 4. Martinus Nijhoff Publishers, Dordrecht, The Netherlands, pp. 345–352.

Bowes, G., Holaday, A.S. and Haller, W.T. (1979) Seasonal variation in the biomass, tuber density and photosynthetic metabolism of *Hydrilla* in three Florida lakes. *Journal of Aquatic Plant Management* 17, 61–65.

Bowlby, J.N. and Imhof, J.G. (1989) Alternative approaches in predicting trout populations from habitat in streams. In: Gore, J.A. and Petts, G.E. (eds) *Alternatives in Regulated River Management*. CRC Press, Boca Raton, Florida, pp. 317–330.

Bradley, R.S. and Jones, P.D. (1992) Records of explosive volcanic eruptions over the last 500 years. In: Bradley, R.S. and Jones, P.D. (eds) *Climate Since A.D. 1500*. Routledge, London, pp. 606–622.

Bravard, J.-P., Amoros, C. and Pautou, G. (1986) Impact of civil engineering works on the successions of communities in a fluvial system. *Oikos* 47, 92–111.

Bravard, J., Kondolf, G.M. and Piégey, H. (1999) Environmental and societal effects of channel incision and remedial strategies. In: Darby, S.E and Simon, A. (eds) *Incised River Channels: Processes, Forms, Engineering and Management*. John Wiley and Sons Ltd, Chichester, UK, pp. 303–342.

Breil, P. and Capra, H. (1996) Likelihood expression of hydrological structuring events for brown trout population dynamics: methodological approach. In: Leclerc, M., Capra, H., Valentin, S., Boudreault, A. and Coté, Y. (eds) *Ecohydraulics 2000. Proceedings of the 2nd International Symposium on Habitat Hydraulics*. INRS-Eau, Québec City, Canada, pp. 421–431.

Brett, J.R. and Groves, T.D.D. (1979) Physiological energetics. In: Hoar, W.S., Randaln D.J. and Brett, J.R. (eds) *Fish Physiology*, Vol. VIII. Academic Press, New York, New York, pp. 279–352.

Brewer, C.A. and Parker, M. (1990) Adaptations of macrophytes to life in moving water: upslope limits and mechanical properties of stems. *Hydrobiologia* 194, 133–142.

Brierley, G.J. and Fryirs, K. (2000) River styles, a geomorphic approach to catchment characterization: implications for river rehabilitation in Bega catchment, New South Wales, Australia. *Environmental Management* 25, 661–679.

Brittain, J.E. and Eikeland, T.J. (1988) Invertebrate drift – a review. *Hydrobiologia* 166, 77–93.

Broecker, W.S., Kennett, J.P., Flower, B.P., Teller, J.T., Trumbore, S., Bonani, G. and Wolfli, W. (1989) Routing of meltwater from the Laurentide Ice Sheet during the Younger Dryas cold episode. *Nature* 341, 318–321.

Broodbakker, N.W. (1990) *Naar een natuurconvenant voor het Naardermeer (Towards a Covenant for Nature of Lake Naardermeer)*. ECOTEST, Amsterdam.

Brooker, M.P. (1982) *Conservation of Wildlife in River Corridors, Part I. Methods of Survey and Classification, Part II*. Welsh Water Authority, Brecon, UK.

Brookes, A. (1995) *Changing river channels*. John Wiley and Sons, Chichester, UK.

Brouwer, F. and Falkenmark, M. (1989) Climate-induced water availability changes in Europe. *Environmental Monitoring and Assessment* 13, 75–98.

Brown, J., Ribeny, F., Wolanski, E. and Codner, G. (1981) A mathematical model of the hydrologic regime of the upper Nile basin. *Journal of Hydrology* 51, 97–107.

Brown, J.M.A., Dromgoole, F.I., Towsey, M.W. and Browse, J. (1974) Photosynthesis and photorespiration in aquatic macrophytes. In: Bieleski, R.L., Ferguson, A.R. and Cresswell, M.M. (eds) Mechanisms of Regulation of Plant Growth. *Royal Society of New Zealand Bulletin* 12, 343–349.

Bruijnzeel, L.A. (1990) *Hydrology of Moist Tropical Forests and Effects of Conversion: A State of Knowledge Review*. UNESCO International Hydrology Programme. UNESCO, Paris.

Brussock, P.P. and Brown, A.V. (1991) Riffle-pool geomorphology disrupts longitudinal patterns of stream benthos. *Hydrobiologia* 220, 109–117.

Brusven, M.A. and Prather, K.V. (1974) Influence of stream sediments on distribution of macrobenthos. *Journal of the Entomological Society of British Columbia* 71, 24–32.

Brusven, M.A. and Rose, S.T. (1981) Influence of substrate composition and suspended sediment on insect predation by the torrent sculpin, *Cottus rhotheus*. *Canadian Journal of Fisheries and Aquatic Science* 38, 1444–1448.

Bryson, R.A. (1989) Late Quaternary volcanic modulation of Milankovitch climate forcing. *Theoretical and Applied Climatology* 39, 115–125.

Brzezinski, M.A. (1985) The Si:C:N ratio of marine diatoms: interspecific variability and effects of some environmental variables. *Journal of Phycology* 21, 347–357.

Buffagni, A., Crosa, G.A., Harper, D.M. and Kemp, J.L. (2000) Using macroinvertebrate species assemblages to identify river channel habitat units: an application of the functional habitats concept to a large, unpolluted Italian river (River Ticino, northern Italy). *Hydrobiologia* 435, 213–225.

Bunyard, P. (1996) Gaia in Action. *Science of the Living Earth*, Floris Books, Edinburgh.

Burch, J.B. (1982) *Freshwater Snails (Mollusca: Gastropoda) of North America*. EPA-600/3-82-026. US Environmental Protection Agency, Cincinnati, Ohio.

Burkholder, J-A.M. (1992) Phytoplankton and episodic suspended sediment loading: phosphate partitioning and mechanisms for survival. *Limnology and Oceanography* 37, 974–988.

Burt, T.P., Heathwaite, A.L. and Trudgill, S.T. (1993) *Nitrate: Processes, Patterns, and Management*. John Wiley and Sons Ltd, New York, New York, 350 pp.

Calow, P. (1985) Adaptive aspects of energy allocation. In: Tytler, W.P. and Calow, P. (eds) *Fish Energetics: New Perspectives*. Croom Helm, London, pp. 13–31.

Canale, R.P. and Effler, S.W. (1989) Stochastic phosphorus model for Onondaga Lake. *Water Research* 23, 1009–1016.

Canton, S.P. and Ward, J.V. (1981) Emergence of Trichoptera from Trout Creek, Colorado, USA. In: Moretti, G.P. (ed.) *Proceedings of the Third International Symposium on Trichoptera*. Dr. W. Junk, The Hague, The Netherlands, pp. 39–45.

Caraco, N.F., Cole, J.J. and Likens, G.E. (1989) Evidence for sulphate-controlled phosphorous release from sediments of aquatic systems. *Nature* 341, 316–318.

Cardwell, H., Jager, H.I. and Sale, M.J. (1996) Designing instream flows to satisfy and human water needs. *Journal of Water Resources Planning and Management* 122, 356–363.

Carpenter, S.R. and Lodge, D.M. (1986) Effects of submersed macrophytes on ecosystem processes. *Aquatic Botany* 26, 341–370.

Carroll, L. (1871) *Through the Looking-glass and What Alice Found There*. Macmillan, London.

Casparie, W.A. (1972) *Bog Development in Southeastern Drenthe (The Netherlands)*. Rijks University, Groningen, The Netherlands, 271 pp.

Casper, S.J. and Krausch, H.D. (1980) Pteridophyta und Anthophyta Bd. 23. In: Ettl, H., Gerloff, J. and Heynig, H. (eds) *Die Süßwasserflora von Mitteleuropa*. Gustav Fischer Verlag, Stuttgart, Germany.

Casper, S.J. and Krausch, H.D. (1981) Pteridophyta und Anthophyta Bd. 24. In: Ettl, H., Gerloff, J. and Heynig, H. (eds) *Die Süßwasserflora von Mitteleuropa*. Gustav Fischer Verlag, Stuttgart, Germany.

Castelle, A.J., Johnson, A.W. and Conolly, C. (1994) Wetland and stream buffer size requirements – a review. *Journal of Environmental Quality* 23, 878–882.

Cellot, B., Doleolivier, M.J., Bornette, G. and Pautou, G. (1994) Temporal and spatial environmental variability in the upper Rhone river and its floodplain. *Freshwater Biology* 31, 311–325.

Chalov, R.S. (1979) *Geographical Investigations of Channel Process*. Moscow University Publisher, Moscow (in Russian).

Chambers, R.M. and Odum, W.E. (1990) Porewater oxidation, dissolved phosphate and the iron curtain. *Biogeochemistry* 10, 37–52.

Chang, Y.-T., Tong, W.-F. and Janauer, G.A. (1998) The sporogenesis of Isoetes taiwanensis DeVol. In: *Proceedings of the 10th EWRS Symposium on Aquatic Weeds*, Lisbon. European Weed Research Society, Doorwerth, The Netherlands.

Chapin, F.S. III, Hoel, M., Carpenter, S.R., Lubchenco, J., Walker, B., Callaghan, T.V., Folke, C., Levin, S.A., Mäler, K.-G., Nilsson, C., Barrett, S., Berkes, F., Crépin, A.-S., Danell, K., Rosswall, T., Starrett, D., Xepapadeas, A. and Zimov, S.A. (2006) Building resilience and adaptation to manage Arctic change. *Ambio* 35, 198–202.

Chapra, S.C. (1975) Comment on 'An empirical model for estimating the retention of phosphorus in lakes'. *Water Resources Research* 11, 1033–1034.

Chapra, S.C. and Canale, R.P. (1991) Long term phenomenological model of phosphorus and oxygen for stratified lakes. *Water Research* 25, 707–715.

Cherkinsky, S.N. (1971) *Sanitary Conditions of Sewage Outfalls*. Stroiizdat, Moscow (in Russian).

Cheslak, E.F. and Jacobson, A.S. (1990) Integrating the instream flow incremental methodology with a population response model. *Rivers* 1, 264–288.

Child, G., Parris, R. and Le Riche, E. (1971) Use of mineralised water by Kalahari wildlife and its effects on habitats. *East African Wildlife Journal* 9, 125–142.

Chisholm, I.M. and Hubert, W.A. (1986) Influence of stream gradient on standing stock of brook trout in the Snowy Range, Wyoming. *Northwest Science* 60, 137–139.

Chow, V.T (1981) *Open-channel Hydraulics*, 2nd edn. McGraw Hill, New York, New York, 680 pp.

Chu, G.C., Townley-Smith, L., Murrell, S. and Lennox, T. (1998) Impacts of stocking rate intensity on pasture dugout water quality. *The Grazing Gazette* 7(1), 3.

Church, M. (1978) Palaeohydrological reconstructions from a Holocene valley fill. *Fluvial Sedimentology* 5, 743–772.

Church, M. (1988) Floods in cold climates. In: Baker, V.R., Kochel, R.C. and Patton, P.C. (eds) *Flood Geomorphology*. John Wiley and Sons Ltd, Chichester, UK, pp. 205–229.

Chutter, F.M. (1969) The effects of silt and sand on the invertebrate fauna of streams and rivers. *Hydrobiologia* 34, 57–76.

Chynoweth, D.P. (1987) Biomass conversion options. In: *Aquatic Plants for Water Treatment and Resource Recovery*. Magnolia Press, Orlando, Florida, pp. 621–642.

Ciborowski, J.J.H. and Clifford, H.F. (1983) Life histories, microdistribution and drift of two mayfly (Ephemeroptera) species in the Pembina River, Alberta, Canada. *Holarctic Ecology* 6, 3–10.

Ciborowski, J.J.H. and Craig, D.A. (1989) Factors influencing dispersion of larval black flies (Diptera: Simuliidae): effects of current velocity and food concentration. *Canadian Journal of Fisheries and Aquatic Science* 46, 1329–1341.

Ciborowski, J.J.H., Pointing, P.J. and Corkum, L.D. (1977) The effect of current velocity and sediment on the drift of the mayfly *Ephemerella subvaria* McDunnough. *Freshwater Biology* 7, 567–572.

Clark, R.O. (1982) What is IAWM? *Aquatics* 4(8), 18.

Clements, F. (1916) *Plant Succession: An Analysis of the Development of Vegetation*. Carnegie Institution, Washington, DC.

CLIMAP (1976) The surface of ice-age Earth. *Science* 191, 1131–1144.

CNRH (1998) *Estrategia para la Gestión Integral de los Recursos Hídricos del Ecuador*. Consejo Nacional de Recursos Hídricos, Quito.

Cohen, J.E. (1995) Population growth and earth's human carrying capacity. *Science* 269, 341–346.

COHMAP (1988) Climatic changes of the last 18 000 years: observations and model simulations. *Science* 241, 1043–1052.

Coleman, N.L. (1981) Velocity profiles with suspended sediment. *Journal of Hydrological Research* 19, 211–229.

Colenbrander, H.J., Blumenthal, K.P., Cramer, W., Volker, A. and Wesseling, J. (1989) *Water in The Netherlands*. Lakerveld, The Hague, The Netherlands.

Coles, D. (1956) The law of the wake in the turbulent boundary layer. *Journal of Fluid Mechanics* 1, 191–222.

Conley, D.L., Schelske, C. and Stoermer, E.F. (1993) Modification of the biogeochemical cycle of silica with eutrophication. *Marine Ecology Progress Series* 101, 179–192.

Connell, J.H. (1978) Diversity in tropical rain forest and coral reefs. *Science* 199, 1302–1310.

Cooper, A.B. and Cooke, J.B. (1984) Nitrate loss and transformation in two vegetated headwater streams. *New Zealand Journal of Marine and Freshwater Research* 18, 441–450.

Cooper, J.R., Gilliam, J.W., Daniels, R.B. and Robarge, W.P. (1987) Riparian areas as filters for agricultural sediment. *Soil Science Society of America Journal* 51, 416–420.

Coordinatiecommissie Onkruidonderzoek (1984) *Report on Terminology in Weed Science in the Dutch Language Area*, 2nd edn. Coordinatiecommissie Onkruidonderzoek, NRLO, The Netherlands.

Corrarino, C.A. and Brusven, M.A. (1983) The effects of reduced stream discharge on insect drift and stranding of near shore insects. *Freshwater Invertebrate Biology* 2, 88–98.

Correll, D.L. (1997) Buffer zones and water quality protection: general principles. In: Haycock, N.E., Burt, T.P., Goulding, K.W.T. and Pinay, G. (eds) *Buffer Zones: Their Processes and Potential in Water Protection*. Quest Environmental, Harpenden, UK, pp. 7–16.

Costa, J.E. (1988) Rheologic, geomorphic and sedimentologic differentiation of water floods, hyperconcentrated flows and debris flows. In: Baker, V.R., Kochel, R.C. and Patton, P.C. (eds) *Flood Geomorphology*. John Wiley and Sons Ltd, Chichester, UK, pp. 113–122.

Coyne, M.S., Gilfillen, R.A., Rhodes, R.W. and Blevins, R.L. (1995) Soil and fecal coliform trapping by grass filter strips during simulated rain. *Journal of Soil and Water Conservation* 50, 405–408.

Crisman, T.L., Chapman, L.J., Chapman, C.A. and Kaufman, L.S. (2003) *Conservation, Ecology, and Management of African Fresh Waters*. University Press of Florida, Gainesville, Florida, 514 pp.

Crosa, G.A. and Buffagni, A. (1996) L'habitat idraulico quale elemento per la gestione degli ambienti fluviali. *Atti della Societe Italiana di Ecologia* 17, 581–583.

Crutzen, P.J. and Stoermer, E.F. (2000) The 'Anthropocene'. *IGBP Newsletter* 41, 17–18.

Cuker, B.E., Gama, P.T. and Burkholder, J.M. (1990) Type of suspended clay influences lake productivity and phytoplankton community response to phosphorus loading. *Limnology and Oceanography* 35, 830–839.

Culp, J.M., Wrona, F.J. and Davies, R.W. (1986) Response of stream benthos and drift to fine sediment deposition versus transport. *Canadian Journal of Zoology* 64, 1345–1351.

Cummins, K.W. (1974) Structure and function of stream ecosystems. *BioScience* 24, 631–641.

Cummins, K.W. (1975) The importance of different energy sources in freshwater ecosystems. In: Reichle, D.E., Franklin, J.F. and Goodall, D.W. (eds) Productivity of world ecosystems. *Proceedings of the National Academy of Sciences*, Washington, DC. 166 pp.

Cummins, K.W. and Klug, M.J. (1979) Feeding ecology of stream invertebrates. *Annual Review of Ecology and Systematics* 10, 147–172.

Cummins, K.W., Minshall, W.G., Sedell, J.R., Cushing, C.E. and Petersen, R.C. (1984) Stream ecosystem theory. *Verhandlungen des Internationalen Vereins für Limnologie* 22, 1818–1827.

Cushman, R.M. (1985) Review of ecological effects of rapidly varying flows downstream from hydroelectric facilities. *North American Journal of Fisheries Management* 5, 330–339.

Dagg, J.L. (2002) Unconventional bed mates: Gaia and the selfish gene. *Oikos* 96, 182–186.

Davies, B.R. and Wishart, M.J. (2000) River conservation in the countries of the Southern African development community (SADC). In: Boon, P.J., Davies, B.R. and Petts, G.E. (eds) *Global Perspectives on River Conservation: Science, Policy and Practice*. John Wiley and Sons Ltd, Chichester, UK, pp. 179–204.

Davies, B.R., Thoms, M.C., Walker, K.F., O'Keeffe, J.H. and Gore, J.A. (1994) Dryland rivers: their ecology, conservation, and management. In: Calow, P. and Petts, G.E. (eds) *The Rivers Handbook*, Vol. 2. Blackwell Scientific Publications, London, pp. 484–511.

Davis, J.A. and Barmuta, L.A. (1989) An ecologically useful classification of mean and near-bed flows in streams and rivers. *Freshwater Biology* 21, 271–282.

Dawkins, R. (1976) *The Selfish Gene*. Oxford University Press, Oxford.

Dawkins, R. (1982) *The extended Phenotype*. W.H. Freeman, Oxford.

Dawson, F.H. (1978) The seasonal effects of aquatic plant growth on the flow of water in a stream. In: *Proceedings of the 5th EWRS Symposium on Aquatic Weeds*. European Weed Research Society, Doorwerth, The Netherlands, pp. 71–78.

De Lange, W. J. (1996) Groundwater modelling of large domains with analytic elements. PhD thesis, Delft University of Technology, Delft, The Netherlands.

Demars, B.O.L. and Harper, D.M. (2005) Water column and sediment phosphorus in a calcareous lowland river and their differential response to point source control measures. *Water Air and Soil Pollution* 167, 273–293.

Den Hartog, C. (1970) *The Sea-grasses of the World*. Verhandelingen der Koninklijke Nederlands Akademie Wetenschappen, Afdelung Natuurkunde. Tweede Reeks 59/1. North-Holland Publishing Co., Amsterdam.

Den Hartog, C. and Segal, S. (1964) A new classification of the water-plant communities. *Acta botanica Neerlandica* 13, 367–393.

Descy, J.-P. (1992) Qualité des eaux de surface en Wallonie. *Tribune de l'eau* 45, 53–57.

Diehl, S. (1988) Foraging efficiency of three freshwater fish: effects of structural complexity and light. *Oikos* 53, 207–214.

Diehl, S. and Kornijow, R. (1998) Influence of submerged macrophytes on trophic interactions among fish and macroinvertebrates. In: Jeppesen, E., Sondergaard, Ma., Sondergaard, Mo. and Christoffersen, K. (eds) The *Structuring Role of Submerged Macrophytes in Lakes. Ecological Studies*, Vol. 131. Springer, New York, New York, pp. 24–46.

Dillaha, T.A., Reneau, R.B., Mostaghimi, S. and Lee, D. (1989) Vegetated filter strips for agricultural nonpoint source pollution control. *Transactions of the American Society of Agricultural Engineers* 32, 513–519.

Di Toro, D.M. and Conolly, J.P. (1980) *Mathematical Models of Water Quality in Large Lakes; Part 2. Lake Erie*. EPA-600/3-80-065. US Environmental Protection Agency, Washington, DC, 231 pp.

Di Toro, D.M., Thomann, R.V., O'Connor, D.J. and Mancini, J.L. (1977) Estuarine phytoplankton biomass models; verification analysis and preliminary applications. In: Goldberg, E.D. (ed.) *The Sea: Ideas and Observations on Progress in the Study of the Seas.* John Wiley and Sons Inc., New York, New York, pp. 969–1019.

Dokulil, M. and Janauer, G.A. (1989) Die Neue Donau als Badegewässer – Nutzungsprobleme nach Errichtung der Staustufe Freudenau. *Perspektiven Spezial* 10–14.

Donabaum, K., Pall, K., Teubner, K. and Dokulil, M. (2004) Alternative stable states, resilience and hysteresis during recovery from eutrophication – a case study. *SILNEWS* 43, 1–4.

Donigan, A.S. (1988) Selection, appraisal and validation of experimental models. Paper presented at the *International Symposium on Water Quality Modelling of Agricultural Non-Point Sources*, Utah State University, Logan, Utah, June.

Dorioz, J.M., Pilleboue, E. and Fehri, A. (1989) Dynamique du phosphore dans les bassins versants: importance des phénomènes de rétention dans le sédiments. *Water Research* 23, 147–158.

Dudley, T.L. and Anderson, N.H. (1987) The biology and life cycles of *Lipsothrix* spp. (Diptera: Tipulidae) inhabiting wood in western Oregon streams. *Freshwater Biology* 17, 437–451.

Duff, J.H. and Triska, F.J. (1990) Denitrification in sediments from the hyporheic zone adjacent to a small forested stream. *Canadian Journal of Fisheries and Aquatic Sciences* 47, 1140–1147.

Dugdale, R.C., Wilkerson, F.P. and Minas, H.J. (1995) The role of a silicate pump in driving new production. *Deep-Sea Research* 42, 697–719.

Dunne, T. (1978) Field studies of hillslope flow processes. In: Kirkby, M.J. (ed.) *Hillslope Hydrology.* John Wiley and Sons Ltd, Chichester, UK, pp. 227–293.

Dunne, T., Moore, T.R. and Taylor, C.H. (1975) Recognition and prediction of runoff-producing zones in humid regions. *Hydrological Sciences Bulletin* 20, 305–327.

Dunne, T., Mertes, L.A.K., Meade, R.H., Richey, J.E. and Forsberg, B.R. (1998) Exchanges of sediment between the flood plain and channel of the Amazon river in Brazil. *Geological Society of America Bulletin* 110, 450–467.

Dury, G.H. (1964) *Principles of Underfit Streams.* Professional Paper No. 452A. US Geological Survey, Washington, DC, 67 pp.

Dussart, G.B.J. (1987) Effects of water flow on the detachment of some aquatic pulmonate gastropods. *American Malacological Bulletin* 5, 65–72.

Dwasi, J.A. (2002) *Trans-boundary Environmental Issues in East Africa: An Assessment of the Environmental and Socio-economic Impacts of Kenya's Policies on Tanzania.* World Research Institute, Washington, DC.

Dynesius, M. and Nilsson, C. (1994) Fragmentation and flow regulation of river systems in the northern third of the world. *Science* 266, 753–762.

Eagelson, P.S. (1982) Ecological optimality in water limited natural soil–vegetation systems. 1. Theory and hypothesis. *Water Resources Research* 18, 325–340.

Eagelson, P.S. (2002) *Ecohydrology: Darwinian Expression of Vegetation Form and Function.* Cambridge University Press, Cambridge, UK.

Eamus, D., Hatton, T., Cook, P. and Colvin, C. (2006) *Vegetation Function, Water and Resource Management.* CSIRO Publishing, Melbourne, New South Wales, Australia.

Eeles, C.W.O., Robinson, M. and Ward, R.C. (1990) Experimental basins and environmental models. In: Hooghart, J.C., Posthumus, C.W.S. and Warmerdam, P.M.M. (eds) *Hydrological Research Basins and The Environment.* TNO Committee on Hydrological Research, The Hague, The Netherlands, pp. 3–12.

Einsele, G. and Lempp, C. (1992) Das Verwitterungsprofil und seine Bedeutung für den Boden-schutz in Gebieten mit Festgesteins-Untergrund (The weathering profile and its role in soil protection in hard-rock catchments). In: Rosenkranz, D., *et al.* (eds) *Bodenschutz*, Beitrag 1150. Erich Schmidt Verlag, Berlin, pp. 1–27.

Ellenberg, H. (1952) *Wiesen und Weiden und ihre standörtliche Bewertung (Grasslands and Meadows and Their Site Significance)*. Eugen-Ulmer, Stuttgart, Germany.

Ellenberg, H. (1979) Zeigerwerte der Gefässpflanzen (ohne *Rubus*) (Indicator values of vascular plants (except for *Rubus*)). *Scripta Geobotanica* 9, 1–122.

Ellenberg, H. (1991) Zeigerwerte der Gefässpflanzen (ohne *Rubus*) (Indicator values of vascular plants (except for *Rubus*)). *Scripta Geobotanica* 18, 9–166.

Elliott, J.M. (1976) The energetics of feeding, metabolism and growth of brown trout (*Salmo trutta*) in relation to body weight, water temperature and ration size. *Journal of Animal Ecology* 45, 923–948.

Engelberg, J. and Boyarsky, L.L. (1979) The noncybernetic nature of ecosystems. *American Naturalist* 114, 317–324.

Engelund, F. (1970) Instability of erodible beds. *Journal of Fluid Mechanics* 42, 225–244.

Engle, D.L. and Melack, J.M. (1993) Consequences of riverine flooding for seston and the periphyton of floating meadows in an Amazon floodplain lake. *Limnology and Oceanography* 38, 1500–1520.

Environment Agency (1997) *River Habitat Survey, Field Survey Guidance Manual*. Environment Agency, Warrington, UK.

Environment Agency (2004) *Hydroecology: integration for modern regulation*. Environment Agency for England and Wales, Bristol, UK.

Ertsen, A.C.D. (1998) Ecohydrological response modelling. Predicting plant species response to changes in site conditions. PhD thesis, University of Utrecht, Utrecht, The Netherlands.

Evans, J.W. and Noble, B.L. (1979) The longitudinal distribution of fishes in an east Texas stream. *American Midland Naturalist* 101, 333–343.

Everts, C.H. (1973) Particle over-passing on flat granular boundaries. *Journal of Waterways, Harbors, and Coastal Engineering Division, American Society of Civil Engineers* 99, 425–438.

Facey, D.E. and Grossman, G.D. (1990) The metabolic cost of maintaining position for four North American stream fishes: effects of season and velocity. *Physiological Zoology* 63, 757–776.

Facey, D.E. and Grossman, G.D. (1992) The relationship between water velocity, energetic costs, and microhabitat in four North American stream fishes. *Hydrobiologia* 239, 1–6.

Fair, P. and Meeke, L. (1983) Seasonal variations in the pattern of photosynthesis and possible adaptive response to varying light flux regimes in *Ceratophyllum demersum* L. *Aquatic Botany* 15, 81–90.

Falkenmark, M. (1990) Hydrological shifts as part of landscape–ecological impact of climate change. In: Boer, M.M. and de Groot, R.S. (eds) *Landscape–Ecological Impact of Climatic Change*. IOS Press, Amsterdam, pp. 194–217.

Falkowski, E. (1975) Variability of channel processes of lowland rivers in Poland and changes of the valley floors during the Holocene. *Biuletyn Geologiczny UW Warszawa* 19, 45–78.

FAO (1999) Integrated resource management for sustainable inland fish production. 23rd Committee on Fisheries, Rome, 15–19 February 1999. FAO, Rome; available at http://www.fao.org/UNFAO/BODIES/COFI/COFI23/w9880e.htm

Fardeau, J.C. and Frossard, E. (1991) Processus de transformation du phosphore dans les sols de l'Afrique de l'ouest semi-aride: application du phosphore assimilable. In: Tiessen, H. and Frossard, H. (eds) *Proceedings of the SCOPE/UNEP Workshop on Phosphorus Cycling in Terrestrial and Aquatic Ecosystems*, Nairobi, 18–22 March, Vol. 4, pp. 108–128.

Fashchevsky, B.V. (1966) About temperature of the river waters in the Altai Mountains. *Questions of Siberian Geography* 6, 135–141 (in Russian).

Fashchevsky, B.V. (1986) Hydrological aspects of floodplain landscape conservation. In: Bachourin, G.V. (ed.) *Hydrological Investigations of Landscapes*. Naouka, Novosibirsk, Russia, pp. 57–64 (in Russian).

Fashchevsky, B.V. (1988) *Criteria of Ecological Flow. Problems and Technical Decisions of Nature-protected Measures under Water Management of Construction*. Souzvodproect, Moscow, pp. 28–32 (in Russian).

Fashchevsky, B.V. (1989) *The Justification of Admissible Flow Regulation*. Belniinti, Minsk, Belarus (in Russian).

Fashchevsky, B.V. (1990) Problems of ecosystems rate-setting for water regimes of streams and lakes under anthropogenic impact. In: *Seminar on Ecosystems Approach to Water Management*. UN Economic Commission for Europe, Oslo, pp. 1–9 (in Russian).

Fashchevsky, B.V. (1994) Mapping of ecological and free river flows in the CIS (former USSR). In: Seuna, P., Gustard A., Arnell, N.W. and Cole, G.A. (eds) *FRIEND: Flow Regimes from International Experimental and Network Data*. IAHS Publication No. 187, Centre for Ecology and Hydrology, Wallingford, UK, pp. 247–253.

Fashchevsky, B.V. (1996) *The Fundamentals of Ecological Hydrology*. Ecoinvest, Minsk, Belarus (in Russian).

Fashchevsky, B.V. (2002) Environmental consequences of reservoir creation and criteria for justification their construction. In: *Proceedings of the 9th International Conference on the Conservation and Management of Lakes*, Otsu City, Shiga Prefecture, Japan, 11–16 November, Session 5, pp. 312–315.

Fashchevsky, B.V. and Fashchevskaya, T. (2003) Water management budget as a basis for assessing water priorities in a catchment. In: Bloschl, G., Franks, S., Kumagai, M., Misiake, K. and Rosbjerg, D. (eds) *Water Resources Systems – Hydrological Risk, Management and Development*. IAHS Publication No. 281. Centre for Ecology and Hydrology, Wallingford, UK, pp. 322–326.

Fashchevsky, B.V. and Tamela, J.A. (1976) Algorithm and computation program for the calculation of flow characteristics at different frequency. *Complex Use of Water Resources* 3, 37–42 (in Russian).

Feddes, R.A., Koopmans, R.W.R. and van Dam, J.C. (1997) *Agrohydrology*. Department of Water Resources, Wageningen Agricultural University, Wageningen, The Netherlands.

Feldmeth, C.R. (1970) The respiratory energetics of two species of stream caddis fly larvae in relation to water flow. *Comparative Biochemistry and Physiology* 32, 193–202.

Filipeck, S., Keith, W.E. and Giese, J. (1987) The status of the instream flow issue in Arkansas. *Proceedings of the Arkansas Academy of Science* 41, 43–48.

Fleischer, S. and Stibe, L. (1989) Agriculture kills marine fish in the 1980s. Who is responsible for fish kills in the year 2000? *Ambio* 18, 347–350.

Forbes, S.A. (1887) The lake as a microcosm. *Bulletin of the Peoria Science Association, Illinois Natural History Survey Bulletin* 15, 537–550.

Foret, J.A. (1974) An integrated program for alligator weed control in rice irrigation canals. *Proceedings of the Southern Weed Science Society*, Startville, Missisippii, 282 pp.

Forman, R.T.T. (1995) Some general principles of landscape and regional ecology. *Landscape Ecology* 10, 133–142.

Forman, R.T.T. and Gordon, M. (1986) *Landscape Ecology*. John Wiley and Sons Ltd, New York, New York, 618 pp.

Frahm, J.-P. and Frey, W. (1992) *Moosflora*, 3rd edn. Ulmer, Stuttgart, Germany.

Franzen, L.G. (1994) Are wetlands the key to the ice-age cycle enigma? *Ambio* 23, 300–308.

Freeman, M.C., Bowen, Z.H. and Crance, J.H. (1997) Transferability of habitat suitability criteria for fishes in warmwater streams. *North American Journal of Fisheries Management* 17, 20–31.

Frenzel, B., Pecsi, M. and Velichko, A.A. (1992) *Paleogeographical Atlas of the Northern Hemisphere*. Hungarian Academy of Sciences, Budapest.

Friedman, J.M. and Lee, V.J. (2002) Extreme floods, channel change, and riparian forests along ephemeral streams. *Ecological Monographs* 72, 409–425.

Froehlich, W. (1982) The mechanism of fluvial transport and water supply into the stream channel in a mountainous flysch catchment (in Poland). *Prace Geograficzne (Instytut Geografii i Przestrzennego Zagospodarowania, PAN)* 143, 144.

Froehlich, W. and Starkel, L. (1987) Normal and extreme monsoon rains – their role in the shapping of Darjeeling Himalaya. *Studia Geomorphologica Carpatho-Balcanica* 21, 129–160.

Froelich, P.V. (1988) Kinetic control of dissolved phosphate in natural rivers and estuaries: a primer on the phosphate buffer mechanism. *Limnology and Oceanography* 33, 649–668.

Fukuoka, S. and Watanabe, A. (1997) Horizontal structure of flood flow with dense vegetation clusters along main channel banks. In: Holly, F.M. and Alsaffar, A. (eds) *Proceedings of the 27th Congress of the International Association for Hydraulic Research*, San Francisco, California, 10–15 August, pp. 1408–1413.

Gachter, R. and Meyer, J.S. (1993) The role of microorganisms in mobilization and fixation of phosphorus in sediments. *Hydrobiologia* 253, 103–121.

Garcia, D. and Martinez, P.V. (2005) Aquatic habitats, analysis and restoration. *Fifth International Symposium on Ecohydraulics*. IAHR-AIRH, Madrid.

Gassama, A. (1999) Résume de la communication sur le plan directeur de développement intégré pour la rive gauche de la Vallée du Fleuve Sénégal. Submission to the World Commission on Dams, Regional Consultation in Cairo, 8–9 December.

Gates, D.M. (1962) *Energy Exchange in the Biosphere*. Harper and Row, New York, New York.

Gates, W.L. (1976) The numerical simulation of ice-age climate with a global general circulation model. *Journal Atmospheric Sciences* 33, 1844–1873.

Genereux, D.P., Hemond, H.F. and Mulholland, P.J. (1993) Spatial and temporal variability in streamflow generation on the West Fork of Walker Branch Watershed. *Journal of Hydrology* 142, 137–166.

Gereta, E. and Wolanski, E. (1998) Wildlife–water quality interactions in the Serengeti National Park, Tanzania. *African Journal of Ecology* 36, 1–14.

Gereta, E., Wolanski, E., Borner, M. and Serneels, S. (2002) Use of an ecohydrological model to predict the impact on the Serengeti ecosystem of deforestation, irrigation and the proposed Amala weir water diversion project in Kenya. *Ecohydrology and Hydrobiology* 2, 127–134.

Gersabeck, E.F. Jr and Merritt, R.W. (1979) The effect of physical factors on the colonization and relocation behavior of immature black flies (Diptera: Simuliidae). *Environmental Entomology* 8, 34–39.

Gessner, F. (1955) *Hydrobotanik. Die physiologischen Grundlagen der Pflanzenver-breitung im Wasser*. VEB Deutscher Verlag der Wissenschaften. Berlin.

Gibson, R.J. (1983) Water velocity as a factor in the change from aggressive to schooling behavior and subsequent migration of Atlantic salmon smolt (*Salmo salar*). *Le Naturaliste Canadiene* 110, 143–148.

Gilliam, J.W. (1994) Riparian wetlands and water quality. *Journal of Environmental Quality* 23, 896–900.

Gilliam, J.W., Skaggs, R.W. and Doty, C.W. (1986) Control agriculture: an alternative riparian vegetation. In: Correll, D.W. (ed.) *Watershed Reserve Perspective.* Smithsonian Institution Press, Washington, DC, pp. 225–243.

Gilliam, J.W., Parsons, J.E. and Mikkelsen, R.L. (1997) Nitrogen dynamics and buffer zones. In: Haycock, N.E., Burt, T.P., Goulding, K.W.T. and Pinay, G. (eds) *Buffer Zones: Their Processes and Potential in Water Protection.* Quest Environmental, Harpenden, UK, pp. 54–65.

Giorgini A. and Zingales, F. (eds) (1986) *Agricultural Non-Point Source Pollution: Model Selection and Application.* Proceedings of the Venice Workshop, Italy, June 1984. Elsevier, Amsterdam.

Gleick, P.H. (1993) *Water in Crisis: A Guide to the World's Fresh Water Resources.* Oxford University Press, New York, New York.

Glozier, N.E., Culp, J.M. and Scrimgeour, G.J. (1997) Transferability of habitat suitability curves for a benthic minnow, *Rhinichthys cataractae. Journal of Freshwater Ecology* 12, 379–393.

Goldschmidt, U., Chovanec, A., Schmid, J., Keckeis, H., Rakowitz, G., Waidbacher, H., Straif, M., Janauer, G. and Schmidt, B. (2006) In: Goldschmidt, U. (ed.) *Fischökologische Untersuchung zur Funktionsfähigkeit antrhopogen gestalteter Flusshabitate an der Donau und ausgewählter Zubringer in den Regionen Wien und Györ.* Report MA-45. Municipality of Vienna, Agency for Water Engineering, Vienna, 30 pp.

Golley, F.B. (1993) *The Ecosystem Concept in Ecology: More than the Sum of the Parts.* Yale University Press, New Haven, Connecticut.

Golterman, H.L. (1980) Phosphate models – a gap to bridge. *Hydrobiologia* 72, 61–71.

Gopal, B. (1993) Aquatic weed problems and management in Asia. In: Pieterse, A.H. and Murphy, K.J. (eds) *Aquatic Weeds. The Ecology and Management of Nuisance Aquatic Vegetation.* Oxford University Press, Oxford, UK, pp. 318–340.

Gordon, N.D., McMahon, T.A. and Finlayson, B.L. (1992) *Stream Hydrology: An Introduction for Ecologists.* John Wiley and Sons Ltd, Chichester, UK, 526 pp.

Gore, J.A. (1977) Reservoir manipulations and benthic macroinvertebrates in a prairie river. *Hydrobiologia* 55, 113–123.

Gore, J.A. (1978) A technique for predicting the in-stream flow requirements of benthic macroinvertebrates. *Freshwater Biology* 8, 141–151.

Gore, J.A. (1983) Considerations of size related flow preferences among macroinvertebrates used in instream flow studies. In: Shuval, H.I. (ed.) *Developments in Ecology and Environmental Quality*, Vol. II. Balaban International Publishers, Jerusalem, pp. 389–398.

Gore, J.A. (1987) Development and application of macroinvertebrate instream flow models for regulated flow management. In: Craig, J.F. and Kemper, J.B. (eds) *Regulated Streams: Advances in Ecology.* Plenum Press, New York, New York, pp. 99–115.

Gore, J.A. (1989a) Models for predicting benthic macroinvertebrate habitat suitability under regulated flows. In: Gore, J.A. and Petts, G.E. (eds) *Alternatives in Regulated River Management.* CRC Press, Inc., Boca Raton, Florida, pp. 253–265.

Gore, J.A. (1989b) Case histories of instream flow analyses for permitting and environmental impact assessments in the United States. *South African Journal of Aquatic Sciences* 15, 194–208.

Gore, J.A. (1997) Water quality in the USA: evolving perspectives and public perception. In: Boon, P.J. and Howell, D.L. (eds) *Freshwater Quality: Defining the Indefinable?* The Stationery Office, Scottish Natural Heritage, Edinburgh, pp. 69–85.

Gore, J.A. (1999) Macroinvertebrates in instream flow management: issues of density, diversity, and taxonomic scale. In: *Proceedings of the International Conference on Modeling for the Twenty-First Century, Predicting Plant and Animal Occurrences: Issues of Scale and Accuracy*, Snowbird, Utah, June.

Gore, J.A. and Bryant, R.M. Jr (1990) Temporal shifts in physical habitat of the crayfish, *Orconectes neglectus* (Faxon). *Hydrobiologia* 199, 131–142.

Gore, J.A. and Judy, R.D. Jr (1981) Predictive models of benthic macroinvertebrate density for use in instream flow studies and regulated flow management. *Canadian Journal of Fisheries and Aquatic Science* 38, 1363–1370.

Gore, J.A. and Nestler, J.M. (1988) Instream flow studies in perspective. *Regulated Rivers* 2, 93–101.

Gore, J.A. and Pennington, W.M. (1988) Changes in larval chironomid habitat with distance from peaking hydropower operations. Presented at *Annual Meeting of the North American Benthological Society*, Tuscaloosa, Alabama, 20 May.

Gore, J.A., Statzner, B. and Resh, V.H. (1986) Physical habitat characteristics and microdistribution of final instars of *Hydropsyche angustipennis* (Curtis). Presented at *Fifth International Symposium on Trichoptera*, Lyon, France.

Gore, J.A., King, J.M. and Hamman, K.C.D. (1991) Application of the instream flow incremental methodology (IFIM) to southern African rivers. I. Protecting endemic fish of the Olifants River. *Water SA* 17, 225–234.

Gore, J.A., Layzer, J.B. and Russell, I.A. (1992) Non-traditional applications of instream flow techniques for conserving habitat of biota in the Sabie River of southern Africa. In: Boon, P.J., Petts, G.E. and Calow, P. (eds) *River Conservation and Management*. Wiley, New York, New York, pp. 161–177.

Gore, J.A., Crawford, D.J. and Addison, D.S. (1998) An analysis of artificial riffles and enhancement of benthic community diversity by Physical Habitat Simulation (PHAB-SIM) and direct observation. *Regulated Rivers* 14, 69–77.

Gore, J.A., Layzer, J.B. and Mead, J. (2001) Macroinvertebrate instream flow studies after 20 years: a role in stream and river restoration. *Regulated Rivers* 17, 527–542.

Goslar, T. (1996) Naturalne zmiany atmosferycznej koncentracji radiowegla w okresie szybkich zmian klimatu na przelomie vistulianu i holocenu. *Geochronometria (Zeszyty Naukowe Polit. Slaskiej, Gliwice)* 15, 196.

Goudie, A. (1981) *The Human Impact: Man's Role in Environmental Change.* Blackwell Scientific Publications, Oxford, UK.

Goulding, M., Smith, N.J.H. and Mahar, D.J. (1996) *Floods of Fortune – Ecology and Economy along the Amazon.* Columbia University Press, New York, New York.

Gregory, K.J. (ed.) (1983) *Background to Palaeohydrology. A Perspective.* John Wiley and Sons Ltd, Chichester, UK, 486 pp.

Gregory, K.J. (1987) River channels. In: Gregory, K.J. and Walling, D.E. (eds) *Human Activity and Environmental Processes*. John Wiley and Sons Ltd, Chichester, UK, pp. 207–235.

Gregory, K.J. (1995) Human activity and palaeohydrology. In: Gregory, K.J., Starkel, L. and Baker, V.R. (eds) *Global Continental Palaeohydrology*. John Wiley and Sons Ltd, Chichester, UK, pp. 151–172.

Gregory, K.J. and Maizels, J. (1991) Morphology and sediments: typological characteristics of fluvial forms and deposits. In: Starkel, L., Gregory, K.J. and Thornes, J.B. (eds) *Temperate Palaeohydrology*. John Wiley and Sons Ltd, Chichester, UK, pp. 31–59.

Gregory, K.J. and Walling, D.E. (eds) (1987) *Human Activity and Environmental Processes.* John Wiley and Sons Ltd, Chichester, UK, 466 pp.

Gregory, K.J., Starkel, L. and Baker, V.R. (eds) (1995) *Global Continental Palaeohydrology.* John Wiley and Sons Ltd, Chichester, UK.

Gremmen, N.J.M., Reijnen, M.J.S.M., Wiertz, J. and van Wirdum, G. (1990) A model to predict and assess the effects of groundwater withdrawal on the vegetation in the Pleistocene areas of the Netherlands. *Journal of Environmental Management* 31, 143–155.

Gril, J.J., Real, B., Patty, L., Fagot, M. and Perret, I. (1997) Grass buffer zones to limit contamination of surface waters by pesticid: research and action in France. In: Haycock, N.E., Burt, T.P., Goulding, K.W.T. and Pinay, G. (eds) *Buffer Zones: Their Processes and Potential in Water Protection*. Quest Environmental, Harpenden, UK, pp. 70–73.

Groffman, P.M. (1994) Denitrification in freshwater wetlands. *Current Topics in Wetland Biogeochemistry 1*. Wetland Biogeochemistry Institute, Louisiana State University, Baton Rouge, Louisiana, pp. 63–78.

Grove, A.T. (1972) The dissolved and solid load carried by some west African rivers: Senegal, Niger, Benue and Shari. *Journal of Hydrology* 16, 277–300.

Grove, J.M. (1988) *The Little Ice Age*. London, Routledge.

Grzybkowska, M., Szczepko, K. and Temaech, A. (1993) Macroinvertebrate drift in a large lowland river (Central Poland). *Acta Hydrobiologica* 35, 357–366.

Gucinski, H., Lackey, R.T. and Spence, B.C. (1990) Global climate change: policy implications for fisheries. *Fisheries* 15, 33–38.

Guerrin, F. (1991) Qualitative reasoning about an ecological process: interpretation in hydroecology. *Ecological Modeling* 59, 165–201.

Gulati, R.D., Martinez, C.P. and Siewertsen, K. (1995) Zooplankton as a compound mineralising and synthesizing system: phosphorus excretion. *Hydrobiologia* 315, 25–37.

Gupta, A. (1988) Large floods as geomorphic events in humid tropics. In: Baker, V.R., Kochel, R.C. and Patton, P.C. (eds) *Flood Geomorphology*. John Wiley and Sons Ltd, Chichester, UK, pp. 301–315.

Gurnell, A.M., Gregory, K.J. and Petts, G.E. (1999) The role of course woody debris in forest aquatic habitats: implications for management. *Aquatic Conservation: Marine and Freshwater Ecosystems* 5, 143–166.

Gustard, A. (1992) Analysis of river regimes. In: Calow, P. and Petts, G.E. (eds) *The Rivers Handbook*, Vol. 1. Blackwell Scientific Publications, Oxford, UK, pp. 29–47.

Gutknecht, D. (1996a) Abflussentstehung an Hängen – Beobachtungen und Konzeptionen (Runoff generation on hillslopes – observations and concepts). *Österreichische Wasser- und Abfallwirtschaft* 48, 134–144.

Gutknecht, D. (1996b) Process oriented modelling – linking engineering and scientific issues. In: Blöschl, G., Sivapalan, M., Gupta, V.K. and Beven, K.J. (eds) *Proceedings of the 4th International Workshop on Scale Problems in Hydrology*. Institute of Hydraulics, Hydrology and Water Resources, University of Technology Vienna, p. 11.

GWP (2000) *Integrated Water Resources Management*. TAC Background Paper No. 4. Global Water Partnership, Denmark.

Gzybkowska, M., Szczepko, K. and Temech, A. (1993) Macroinvertebrate drift in a large lowland river (Central Poland). *Acta Hydrobiologica* 35, 357–366.

Haag, K.H. (1984) Behavioural and physiological response of water hyacinth weevils (*Neochetina eichhorniae* and *Neochetina bruchi*) to herbicide application. In: *Proceedings Annual Meeting Entomological Society of America*, San Antonio, Texas, 9–13 December.

Hiwasaki, L. and Arico, S. (2007). Integrating the social sciences into ecohydrology: facilitating an interdisciplinary approach to solve issues surrounding water, environment and people. *Ecohydrology and Hydrobiology* 7, 3–9.

Hakala, I. (1998) Some observations concerning changes in the major nutrient loads to Lake Paajarvi during last 25 years. *Lammi Notes* 25, 4–7.

Halim, Y. (1991) The impact of human alterations of the hydrological cycle on ocean margins. In: Mantoura, R.F.C., Martin, J.M. and Wollast, R. (eds) *Ocean Margin Processes in Global Change*. John Wiley and Sons Ltd, Chichester, UK, pp. 301–327.

Haller, W.T. (1976) *Hydrilla, A New and Rapidly Spreading Aquatic Weed Problem*. Circular S-245. Florida Agricultural Experiment Station, IFAS, University of Florida, Gainesville, Florida, 13 pp.

Hamerlynck, O. and Duvail, S. (2003) *The Rehabilitation of the Delta of the Senegal River in Mauritania: Fielding the Ecosystem Approach*. IUCN, Gland, Switzerland, 88 pp.

Hamerlynck, O., Duvail, S. and Lemineould Baba, M. (1999) Reducing the environmental impacts of the Manantali and Diama dams on the ecosystems of the Senegal river and estuary: alternatives to the present and planned water management schemes. Submission to the World Commission on Dams, Regional Consultation in Cairo, 8–9 December.

Hannah, D., Wood, P.J. and Sadler, J.P. (2004) Ecohydrology or Hydroecology: a new paradigm? *Hydrological Processes* 18, 3429–3445.

Hanson, C.H. and Li, H.W. (1978) *A Research Program to Examine Fish Behavior in Response to Hydraulic Flow Fields – Development of Biological Design Criteria for Proposed Water Diversions*. Completion Report, Project C-7679, OWRT. US Department of the Interior, Washington, DC.

Hanson, G.C., Groffman, P.M. and Gold, A.J. (1994) Denitrification in riparian wetlands receiving high and low groundwater nitrate inputs. *Journal of Environmental Quality* 23, 917–922.

Hanson, P.C., Johnson, T.B., Schindler, D.E. and Kitchell, J.F. (1997) *Fish Bioenergetics 3.0 for Windows*. Sea Grant, University of Wisconsin, Madison, Wisconsin.

Harley, K.L.S., Fornoi, I.W., Kassulke, R.C. and Sands, D.P.A. (1984) Biological control of water lettuce. *Journal of Aquatic Plant Management* 22, 101–102.

Harmon, M.E., Franklin, J.F., Swanson, F.J., Sollins, P., Lattin, J.D., Anderson, N.H., Gregory, S.V., Cline, S.P., Aumen, N.G., Sedell, J.R., Lienkaemper, G.W., Cromack, K. Jr and Cummins, K.W. (1986) The ecology of coarse woody debris in temperate ecosystems. *Advances in Ecological Research* 15, 133–302.

Harper, D.M. (1992) *Eutrophication of Freshwaters*. Chapman and Hall, London.

Harper, D.M. and Brown, A.G. (eds) (1999) *The Sustainable Management of Tropical Catchments*. John Wiley and Sons Ltd, Chichester, UK, 381 pp.

Harper, D.M. and Smith, C.D. (1995) *Habitats in British Rivers: Biological Reality and Practical Value in River Management*. R&D Note No. 346. National Rivers Authority, Anglian Region, UK.

Harper, D.M., Smith, C.D. and Barham, P.J. (1992) Habitats as the building blocks for river conservation assessment. In: Boon, P.J., Calow, P. and Petts, G.E. (eds) *River Conservation and Management*. John Wiley and Sons Ltd, Chichester, UK, pp. 311–319.

Harper, D.M., Smith, C., Barham, P. and Howell, R. (1995) The ecological basis for the management of the natural river environment. In: Harper, D.M. and Ferguson, A.J.D. (eds) *The Ecological Basis for River Management*. John Wiley and Sons Ltd, Chichester, UK, pp. 219–238.

Harper, D.M., Smith, C.D., Crosa, G.A. and Kemp, J.L. (1998) The use of 'functional habitats' in the conservation, management and rehabilitation of rivers. In: Bretschko, G. and Helesic, J. (eds) *Advances in River Bottom Ecology*. Backhuys Publishers, Leiden, The Netherlands.

Haslam, S.M. (1978) *River Plants*. Cambridge University Press, London, 396 pp.

Haslam, S.M. (1987) *River Plants of Western Europe*. Cambridge University Press, Cambridge, UK, 512 pp.

Hawkes, H.A. (1975) River zonation and classification. In: Whitton, B.A. (ed.) *River Ecology*. Blackwell Scientific Publications, Oxford, UK, pp. 312–374.

Haycock, N.E. and Pinay, G. (1993) Nitrate retention in grass and poplar vegetated buffer strips during the winter. *Journal of Environmental Quality* 22, 273–278.

Haycock, N.E., Burt, T.P., Goulding, K.W.T. and Pinay, G. (eds) (1997) *Buffer Zones: Their Processes and Potential in Water Protection.* Quest Environmental, Harpenden, UK, 250 pp.

Hayden, B.P. (1988) Flood climates. In: Baker, V.R., Kochel, R.C. and Patton, P.C. (eds) *Flood Geomorphology.* John Wiley and Sons Ltd, Chichester, UK, pp. 13–26.

Haygarth, P. (1997) Agriculture as a source of phosphorus transfer to water: sources and pathways. *SCOPE Newsletter in Europe* June, 15.

Hecky, R.E., Mopper, K., Kilham, P. and Degens, E.T. (1973) The amino acid and sugar composition of diatom cell-walls. *Marine Biology* 119, 323–331.

Hejny, S. (1960) *Ökologische Charakteristik der Wasser-und Sumpfpflanzen in den slowakischen Tiefebenen (Donau- und Theißgebiet).* Verlag der Slowakischen Akademie der Wissenschaften, Bratislava, Czechoslovakia.

HELCOM (1996) Third periodic assessment of the state of the environment of the Baltic Sea: 1989–1993. *Baltic Sea Environ. Proc. 64B.* Helsinki Commission, Helsinki.

Helfield, J.M. and Naiman, R.J. (2001) Effects of salmon-derived nitrogen on riparian forest growth and implications for stream productivity. *Ecology* 82, 2403–2409.

Helfield, J.M. and Naiman, R.J. (2006) Keystone interactions: salmon and bear in riparian forests of Alaska. *Ecosystems* 9, 167–180.

Hellsten, S., Dieme, C., Mbengue, M. and Janauer, G.A. (1999) Typha control efficiency of a weed-cutting boat in the Lac de Guiers in Senegal: a preliminary study on mowing. *Hydrobiologia* 415, 249–255.

Henderson-Sellers, B. and Markland, H.R. (1987) *Decaying Lakes – The Origin and Control of Cultural Eutrophication.* John Wiley and Sons Ltd, New York, New York, 253 pp.

Herlihy, A.T., Stoddard, J.L. and Johnson, C.B. (1998) The relationship between stream chemistry and watershed land cover data in the Mid-Atlantic Region USA. *Water, Air and Soil Pollution* 105, 377–386.

Heródek, S., Istvánovics, V., Jolánkai, G., Csathó, P., Németh, T. and Várallyay, Gy. (1995) The P cycle in the Balaton catchment – a Hungarian case study. In: Tiessen, H. (ed.) *Phosphorus in the Global Environment.* SCOPE 54. John Wiley and Sons Ltd, Chichester, UK, pp. 275–300.

Higler, L.W.G. (1988) *A Worldwide Surface Water Classification System.* International Hydrology Programme. UNESCO, Paris, 212 pp.

Higler, L.W.G. (1993) The riparian community of north-west European lowland streams. *Freshwater Biology* 29, 229–241.

Higler, L.W.G. and Mol, A.W.M. (1984) Ecological types of running water based on stream hydraulics in The Netherlands. *Hydrobiology Bulletin* 18, 51–57.

Higler, B. and Statzner, B. (1988) A simplified classification of freshwater bodies in the world. *Verhandlungen des Internationalen Vereins für Limnologie* 23, 1495–1499.

Higler, L.W.G. and Verdonschot, P.F.M. (1992) Caddis larvae as slaves of stream hydraulics. In: *Proceedings of the 6th International Symposium on Trichoptera.* Adam Mickiewicz University Press, Poznań, Poland, pp. 57–57.

Hill, A.R. (1997) The potential role of in-stream and hyporheic environments as buffer zones. In: Haycock, N.E., Burt, T.P., Goulding, K.W.T. and Pinay, G. (eds) *Buffer Zones: Their Processes and Potential in Water Protection.* Quest Environmental, Harpenden, UK, pp. 115–127.

Hill, L.G. and Matthews, W.J. (1980) Temperature selection by the darters *Etheostoma spectabile* and *Etheostoma radiosum* (Pisces: Percidae). *American Midland Naturalist* 104, 412–415.

Hill, M.O. (1979) *TWINSPAN – A FORTRAN Program for Arranging Multivariate Data in an Ordered Two-Way Table by Classification of the Individuals and Attributes*. Cornell University Press, Cornell, New York.

Hillbricht-Ilkowska, A. (1988) Transport and transformation of phosphorus compounds in watershed of Baltic Lakes. In: Tiessen, H. (ed.) *Phosphorus Cycles in Terrestrial and Aquatic Ecosystems*. SCOPE/UNEP Proceedings. University of Saskatchewan, Saskatoon, Saskatchewan, Canada, pp. 193–206.

Hillbricht-Ilkowska, A. (1993) Lake ecosystems and climate changes. *Kosmos* 42, 107–121 (in Polish).

Hillbricht-Ilkowska, A. (1995) Managing ecotones for nutrient and water. *Ecology International* 22, 73–93.

Hillbricht-Ilkowska, A. (1999) Lake and landscape: ecological relations and protection measures. In: Zdanowski, B., Kamiński, M. and Martyniak, A. (eds) *Functioning and Protection of Freshwater Ecosystems in the Protected Regions*. Institute of Inland Fishery, Olsztyn, Poland, pp. 19–40 (in Polish).

Hillbricht-Ilkowska, A. (2002a) Links between landscape, catchment basin, wetland and lake: the Jorka river–lake system (Masurian Lakeland, Poland) as the study object. *Polish Journal of Ecology* 50, 411–426.

Hillbricht-Ilkowska, A. (2002b) Eutrophication rate of lakes in the Jorka river system (Masurian Lakeland, Poland): long-term changes and trophic correlations. *Polish Journal of Ecology* 50, 459–474.

Hillbricht-Ilkowska, A. (2002c) River–lake system in a mosaic landscape: main results and some implications for theory and practice from studies on the river Jorka system (Masurian Lakeland, Poland). *Polish Journal of Ecology* 50, 543–550.

Hillbricht-Ilkowska, A. and Bajkiewicz-Grabowska, E. (1991) Continous and discontinous processes of transport and transformation of matter along the river–lake system. *Verhandlungen des Internationalen Vereins für Limnologie* 24, 1767–1771.

Hillbricht-Ilkowska, A. and Kostrzewska-Szlakowska, I. (1996) Estimation of the phosphorus load and degree of endangerment of lakes along River Krutynia (Masurian Lakeland) and the relationship between the load and in-lake concentration of phosphorus. *Zeszyty Naukowe Komitetu Polskiej Akademii Nauk „Człowiek i Srodowisko"* 13, 97–124 (in Polish with English summary).

Hillbricht-Ilkowska, A. and Wisniewski, R. (eds) (1993) Lake–watershed dependences in diversified landscape (Suwalski Landscape Park, North-Eastern, Poland). *Ekologia polska* 41, 7–283.

Hillbricht-Ilkowska, A., Ryszkowski, L. and Sharpley, A.N. (1995) Phosphorus transfer and landscape structure: riparian sites and diversified land use patterns. In: Tiessen, H. (ed.) *Phosphorus in the Global Environment*. SCOPE 54. John Wiley and Sons Ltd, Chichester, UK, pp. 201–228.

Hillbricht-Ilkowska, A., Rybak, J. and Rzepecki, M. (2000) Ecohydrological research of lake–watershed relations in diversified landscape (Masurian Lakeland, Poland): study results and perspectives for management. *Ecological Engineering* 16, 91–98.

Hildrew, A.G. and Giller, P.S. (1994) Patchiness, species interactions and disturbance in stream communities. In: Giller, P.S., Hildrew, A.G. and Raffaelli, D.G. (eds) *Aquatic Ecology: Scale, Pattern and Process*. Blackwell Scientific Publications, Oxford, UK, pp. 21–62.

Hoare, R.A. (1980) The sensitivity to phosphorus and nitrogen loads of Lake Potorua, New Zeland. *Progress in Water Technology* 12, 897–904.

Hollis, G.E. (1994) Halting the reversing wetland loss and degradation: a geographical perspective on hydrology and land use. In: *Wetland Management, Proceedings of the*

International Conference, Institution of Civil Engineers, London, 2–3 June. Thomas Telford, London, pp. 181–196.

Hollis, G.E. (1996) Hydrological inputs to management policy for the Senegal River and its floodplain. In: Acreman, M.C. and Hollis, G.E. (eds) *Water Management and Wetlands in sub-Saharan Africa*. IUCN, Gland, Switzerland, pp. 155–185.

Hollis, G.E., Adams, W.M. and Aminu-Kano, M. (1993) *The Hadejia-Nguru Wetlands: Environment, Economy and Sustainable Development of a Sahelian Floodplain Wetland*. IUCN, Gland, Switzerland, 244 pp.

Holly, F.M. and Alsaffar, A. (eds) (1997) In: *Proceedings of the 27th Congress of the International Association for Hydraulic Research*, San Francisco, California, 10–15 August, pp. 1396–1401.

Holm, N., Ganf, G.G. and Shapiro, J. (1983) Feeding and assimilation rates of *Daphnia pulex* fed on *Aphanizomenon flos-aquae*. *Limnology and Oceanography* 28, 677–687.

Holmes, N.T.H. and Newbold, C. (1984) *River Plant Communities – Reflectors of Water and Substrate Chemistry*. Focus on Nature Conservation No. 9. Nature Conservancy Council, Shrewsbury, UK, pp. 4–73.

Holmes, N.T.H. and Whitton, B.A. (1975) Makrophytes of the River Tweed. *Transactions of the Botanical Society of Edinburgh* 42, 369–381.

Holmes, N.T.H. and Whitton, B.A. (1977) The macrophytic vegetation of the River Tees in 1975: observed and predicted changes. *Freshwater Biology*, 7, 43–60.

Hosmer, D.W. and Lemeshow, S. (1989) *Applied Logistic Regression*. Wiley and Sons, New York, New York.

Hough, R.A. (1974) Photorespiration and productivity in submersed aquatic vascular plants. *Limnology and Oceanography* 19, 912–927.

Houghton, J.T., Jenkins, G.J. and Ephraums, J.J. (1990) *Climate Change: The IPCC Scientific Assessment*. Cambridge University Press, Cambridge, UK.

Hudson, J. and Schindler, D. (2000) Phosphate concentrations in lakes. *Nature* 406, 54–56.

Huet, M. (1949) Aperçu des relations entre la pente et les populations piscicoles dans les eaux courantes. *Revue Suisse d'Hydrologie* 11, 332–351.

Huet, M. (1954) Biologie, priofils, en long et en travers des eaux courantes. *Bull. Fr. Pisc.* 175, 41–53.

Hughes, F.M.R. (1985) The Tana river floodplain forest, Kenya: ecology and the impact of development. PhD thesis, University of Cambridge, Cambridge, UK.

Hughes, F.M.R. (1988) The ecology of African floodplain forests in semi-arid zones: a review. *Journal of Biogeography* 15, 127–140.

Hughes, F.M.R. (1990) The influence of flooding regimes on forest distribution and composition in the Tana River floodplain, Kenya. *Journal of Applied Ecology* 27, 475–491.

Hughes, F.M.R. (1994) Environmental change, disturbance and regeneration in semi-arid floodplain forests. In: Millington, A.C. and Pye, K. (eds) *Environmental Change in Drylands*. John Wiley and Sons Ltd, Chichester, UK, pp. 321–345.

Hulme, M., Wigley, T.M.L. and Jones, P.D. (1990) Limitations of regional climate scenarios for impact future. In: Boer, M.M. and de Groot, R.S. (eds) *Landscape–Ecological Impact of Climatic Change*. IOS Press, Amsterdam, pp. 11–129.

Hulse, D. and Gregory, S. (2004). Integrating resilience into floodplain restoration. *Urban Ecosystems* 7, 295–314.

Hulshoff, R.M. (1995) Landscape indices describing a Dutch landscape. *Landscape Ecology* 10, 101–111.

Humboldt, A. (1845) *Entwurf einer physischen Weltheschreibung* (FHA von Humboldt). Stuttgart and Tubingen 5, 1845–62.

Humborg, C., Ittekkot, V., Cociasu, A. and von Bodungen, B. (1997) Effect of Danube River dam on Black Sea biogeochemistry and ecosystem structure. *Nature* 386, 385–388.

Humborg, C., Conley, D.J., Rahm, L., Wolff, F., Cociasu, A. and Ittekkot, V. (2000) Silicon retention in river basins: far reaching effects on biogeochemistry and aquatic food webs in coastal marine environments. *Ambio* 29, 45–50.

Humpesch, U. (1978) Preliminary notes on the effect of temperature and light-condition on the time of hatching in some Heptageniidae (Ephemeroptera). *Verhandlungen des Internationalen Vereins für Limnologie* 20, 2605–2611.

Hupfer, M., Gachter, R. and Giovanoli, R. (1995) Transformation of phosphorus species in settling seston and early sediment diagenesis. *Aquatic Science* 57, 305–324.

Hutchinson, G.E. (1959) Homage to Santa Rosalia or Why are there so many kinds of animals? *American Naturalist* 93, 145–159.

Hutchinson, G.E. (1975a) *A Treatise on Limnology. I. Geography, Physics, and Chemistry*. John Wiley, New York, New York, 540 pp.

Hutchinson, G.E. (1975b) *A Treatise on Limnology. III. Limnological Botany*. John Wiley, New York, New York, 660 pp.

Hutton, J. (1788) Theory of the Earth; or an investigation of the laws observed in the composition, dissolution, and restoration of land upon the globe. *Transactions* 1, 209–304.

Huxley, T.H. (1877) *Physiography*. MacMillan, London.

Hynes, H.B.N. (1970) *The Ecology of Running Waters*. Liverpool University Press, Liverpool, UK.

Ibbeken, H. and Schleyer, R. (1991) *Source and Sediment, A Case Study of Provenance and Mass Balance at an Active Plate Margin (Calabria, Southern Italy)*. Springer Verlag, Berlin, 286 pp.

Illies, J. (1961) Versuch einer allgemein biozonotischen Gliederung der Fließgewässer. *Internationale Revue der Gesamten Hydrobiologie und Hydrographie* 46, 205–213.

Illies, J. and Botosaneanu, L. (1963) Problèmes et methodes de la classification et de la zonation écologique des eaux courantes, considerées surtout du point de vue faunistique. *Mitteilungen der Internationalen Vereinigung für Limnologie* 12, 1–57.

Ingersoll, C.G. and Claussen, D.L. (1984) Temperature selection and critical thermal maxima of the fantail darter, *Etheostoma flabellare*, and johnny darter, *E. nigrum*, related to habitat and season. *Environmental Biology of Fish* 11, 131–138.

International Council for Science (ICSU) (2005) *Harnessing science, technology, and innovation for sustainable development*. A report from the ICSU-ISTS-TWAS Consortium Advisory Group.

Imboden, D.M. and Gachter, R. (1975) Modeling and control of lake eutrophication. In: *Proceedings of the 6th Triennial World Congress*, Boston, Massachusetts, August.

Imboden, D.M. and Gachter, R. (1978) A dynamic lake model for tropic state prediction. *Ecological Modeling* 4, 77–98.

IPCC (2007a) *Climate Change 2007 – The Physical Science Basis. Contribution of Working Group I to the Fourth Assessment Report of the IPCC*. Cambridge University Press, Cambridge, UK.

IPCC (2007b) *Climate Change 2007 – Impacts, Adaptation and Vulnerability. Contribution of Working Group II to the Fourth Assessment Report of the IPCC*. Cambridge University Press, Cambridge, UK.

IPCC (2007c) *Climate Change 2007 – Mitigation of Climate Change Contribution of Working Group III to the Fourth Assessment Report of the IPCC*. Cambridge University Press, Cambridge, UK.

Irfanullah, H.M. and Moss, B. (2004) Factors influencing the return of submerged plants to a clear-water, shallow temperate lake. *Aquatic Botany* 80, 177–191.

Irvine, J.R. (1985) Effects of successive flow perturbations on stream invertebrates. *Canadian Journal of Fisheries and Aquatic Science* 42, 1922–1927.

Irvine, J.R. and Henriques, P.R. (1984) A preliminary investigation on effects of fluctuating flows on invertebrates of the Hawea River, a large regulated river in New Zealand. *New Zealand Journal of Marine and Freshwater Research* 18, 283–290.

Irvine, J.R., Jowett, I.G. and Scott, D. (1987) A test of the instream flow incremental methodology for underyearling rainbow trout, *Salmo gairdneri*, in experimental New Zealand streams. *New Zealand Journal of Marine and Freshwater Research* 21, 35–40.

Ittekkot, V. and Zhang, S. (1989) Pattern of particulate nitrogen transport in world rivers. *Global Biogeochemical Cycles* 3, 383–391.

Ittekkot, V., Humborg, C. and Schaefer, P. (2000) Hydrological alterations and marine biogeochemistry: a silicate issue? *BioScience* 50, 776–782.

IUCN (2003) *Tana River, Kenya: Integrating Downstream Values into Hydropower Planning*. Integrating Wetland Economic Values into River Basin Management, Case Study No. 6. IUCN, Gland, Switzerland.

IUCN (2008) *Application of the Ecosystem Approach* (Agenda Item 3.6). Position paper for 9th meeting of the conference of the parties to the convention on biological diversity (COP9), Bonn, Germany. http://cmsdata.iucn.org/downloads/ecosystem_approach_cop9.pdf (Accessed 19–30 May 2008).

Ivlev, V.S. (1939) Transformation of energy by aquatic animals. *Internationale Revue der Gesamten Hydrobiologie und Hydrographie* 38, 449–458.

Jacobs, T.J. and Gilliam, J.W. (1985) Riparian losses of nitrate from agricultural drainage waters. *Journal of Environmental Quality* 14, 472–478.

Jacobsen, O.S. (1983) Biological processes in the ecosystem. In: Jörgensen, S.E. (ed.) *Application of Ecological Modeling in Environmental Management*. Elsevier Scientific Publishing Co., Amsterdam, pp. 55–103.

Jaetzold, R. and Schmidt, H. (1983) *Farm Management Handbook of Kenya: Natural Conditions and Farm Management Information*, Vols 1 and 2. Ministry of Agriculture, Nairobi.

Janauer, G.A. (1981) Die Zonierung submerser Wasserpflanzen und ihre Beziehung zur Gewässerbelastung am Beispiel der Fischa (Niederösterreich). *Verhandlungen der Zoologisch.-Botanischen Gesellschaft Österreichs* 120, 73–98.

Janauer, G.A. (1997) Macrophytes, hydrology, and aquatic ecotones: a GIS-supported ecological survey. *Aquatic Botany* 58, 379–391.

Janauer, G.A. (2006) Ecohydrological control of macrophytes in floodplain lakes. *Ecohydrology and Hydrobiology* 6, 19–24.

Janauer, G.A. and Kum, G. (1996) Macrophytes and flood plain water dynamics in the River Danube ecotone research region (Austria). *Hydrobiologia* 340, 137–140.

Janauer, G.A. and Pall, K. (1999) Makrophytenvegetation. In: Österreichische Elektrizitätswirtschafts-Aktiengesellschaft (ed.) *Giessgang Greifenstein*, Vol. 53. *Vegetation*. Österreichische Elektrizitätswirtschafts-Aktiengesellschaft, Wien, pp. 9–97.

Janauer, G.A. and Schmidt, B. (2004) *Vergleichende Studie zur Entwicklung der Makropyhtenvegetation an ausgewählten Abschnitten der Donau und in Augewässern des Machlandes. Forschung im Verbund 90*. AHP-Verbund, Vienna, 69 pp.

Janauer, G.A. and Wychera, U. (2000) Biodiversity, succession and the functional role of macrophytes in the New Danube (Vienna, Austria). *Large Rivers* 12, 71–74.

Janauer, G.A., Jolankai, G.G. and Exler, N. (2006) River restoration in the Tisza River Basin: conflicting interests and the future of the aquatic macrophyte vegetation. *Archive of Hydrobiology* 158, 525–540.

Jansen, W., Kappus, B., Böhmer, J., Beiter, T. and Rahmann, H. (1996) Fish community structure, distribution patterns and migration in the vicinity of different types of fishways on the Enz River (Germany). In: Leclerc, M., Capra, H., Valentin, S., Boudreault, A. and Coté, Y. (eds) *Ecohydraulics 2000. Proceedings of the 2nd International Symposium on Habitat Hydraulics*. INRS-Eau, Québec City, Canada, pp. 903–914.

Janssen, M. (1998) Use of complex adaptive systems for modeling global change. *Ecosystems* 1, 457–463.

Jansson, M., Leonardson, L. and Fejes, J. (1994) Denitrification and nitrogen retention in a farmland stream in southern Sweden. *Ambio* 23, 326–331.

Jansson, M.B. (1982) *Land Erosion by Water in Different Climates*. UNGI Rapport 57. Department of Physical Goegraphy, University of Uppsala, Uppsala, Sweden, 151 pp.

Jenkins, A., Ferrier, R.C., Harriman, R. and Ogunkoya, Y.O. (1994) A case study in catchment hydrochemistry: conflicting interpretations from hydrological and chemical observations. *Hydrological Processes* 8, 335–349.

Jenkins, R.E. and Freeman, C.A. (1972) Longitudinal distribution and habitat of the fishes of Mason Creek, an upper Roanoke River drainage tributary, Virginia. *Virginia Journal of Science* 23, 194–202.

Jenssen, P.D., Mæhlum, T. and Krogstad, T. (1993) Potential use of constructed wetlands for wastewater treatment in northern environments. *Water Science and Technology* 28, 149–157.

Jenssen, P.D., Mæhlum, T., Roseth, R., Braskerud, B., Syversen, N., Njs, A. and Krogstad, T. (1994) The potential of natural ecosystem self-purifying measures for controlling nutrient inputs. *Marine Pollution Bulletin* 29, 420–425.

Jeppesen, E., Sondergaard, Ma., Sondergaard, Mo. and Christoffersen, K. (eds) (1998a) *The Structuring Role of Submerged Macrophytes in Lakes. Ecological Studies*, Vol. 131. Springer, New York, New York, 423 pp.

Jeppesen, E., Lauridsen, T.L., Kairesalo, T. and Perrow, M.R. (1998b) Impact of submerged macrophytes on fish–zooplankton interaction in lakes. In: Jeppesen, E., Sondergaard, Ma., Sondergaard, Mo. and Christoffersen, K. (eds) *The Structuring Role of Submerged Macrophytes in Lakes. Ecological Studies*, Vol. 131, Springer, New York, New York, pp. 91–114.

JICA (1997) Feasibility Study on Mutonga/Grand Falls Hydropower Project. Report to Republic of Kenya, Ministry of Energy, Nairobi.

Johnson, B.H. (1982) *Development of a Numerical Modelling Capability for the Computation of Unsteady Flow on the Ohio River and its Major Tributaries*. Technical Report H-82-20. US Army Engineer Waterways Experiment Station, Vicksburg, Mississippi.

Johnson, B.H. (1983) *User's Guide for Branched Implicit River Model (BIRN) with Application to the Lower Mississippi River*. Hydraulics Laboratory, US Army Engineer Waterways Experiment Station, Vicksburg, Mississippi.

Johnson, L.B., Richards, C.H., Host, G.E. and Arthur, J.W. (1997) Landscape influence on water chemistry in Midwestern stream ecosystems. *Freshwater Biology* 37, 193–208.

Johnston, C.A. (1991) Sediment and nutrient retention by freshwater wetlands: effects on surface water quality. *Critical Reviews in Environmental Control* 21, 491–565.

Jolánkai, G. (1983) Modeling of non-point source pollution. In: Jörgensen, S.E. (ed.) *Application of Ecological Modeling in Environmental Management*. Elsevier Scientific Publishing Co., Amsterdam, pp. 283–379.

Jolánkai, G. (1986) SENSMOD. A simple experimental non-point source model system. In: *Proceedings of the International Conference on Water Quality Modelling in the Inland Natural Environment*, Bournmouth, UK, 10–13 June, pp. 77–91.

Jolánkai, G. (1992) *Hydrological, Chemical and Biological Processes of Contaminant Transformation and Transport in River and Lake Systems*. UNESCO Series, Technical Documents in Hydrology, WS-93/WS.15. UNESCO, Paris.

Jolánkai, G. and Bíró, I. (2000) GIS based integrated water management decision support model for the Zala River Basin, Hungary. In: *Proceedings of the International Conference on Water Resources Management in the 21st Century, with Particular Reference to Europe*, Budapest, 1–3 June, pp. 113–120.

Jolánkai, G. and Bíró, I. (2001) *Basic River and Lake Water Quality Models, Computer aided Learning Programme on Water Quality Modelling (WQMCAL Version 2) (with an Outlook to 'Ecohydrological' Applications). Software and Description.* UNESCO International Hydrology Programme Documents on CD-ROM Series No. 1. UNESCO, Paris.

Jolánkai, G. and Roberts, G. (eds) (1984) *Proceedings of the UNESCO/MAB-5 Workshop on Land Use Impacts and Aquatic Systems*, Budapest, 10–14 October.

Jolánkai, G. and Szöllősi-Nagy, A. (1978) A simple eutrophication model for the bay of Keszthely, Lake Balaton. In: *Proceedings of the IAHS–AISH Symposium on Modeling the Water Quality of the Hydrological Cycle*, Baden, Germany, September. IAHS–AISH Publication No. 1. Centre for Ecology and Hydrology, Wallingford, UK, pp. 137–149.

Jolánkai, G., Bíró, I. and Ajkay, R. (1993) *Analysis of the Balance of Point and Non-point Source Loading of Selected Chemicals in the Rhine River Basin.* Special Issue. Department for Surface Water Hydrology, the Water Resources Research Centre VITUKI, Budapest, p. 35.

Jolánkai, G., Bíró, I., Varga, Gy., Urbán, J., Gillyénné-Hofer, A. and Michels, G. (1994) A Balatoni Vízkészletgazdálkodási és Vízminőségszabályozási Döntéstámogató Rendszer Kidolgozásának Kezdeti Lépései (Initial steps of developing a decision support system for the management of water resources and water quality of Lake Balaton). In: *Magyar Hidrológiai Társaság XII*, Országos Vándorgyűlése, május 17–19 Siófok, Vol. I. Magyar Hidrológiai Társaság (Hungarian Hydrological Society) Budapest, pp. 129–139.

Jolánkai, G., Panuska, J. and Rast, W. (1999) Modelling of nonpoint source pollutant loads. In: Thornton, J.A., Rast, W., Holland, M.M., Jolánkai, G. and Ryding, S.O. (eds) *Assessment and Control of Non-point Source Pollution of Aquatic Systems: A Practical Approach. Man and the Biosphere Series*, Vol. 23. UNESCO, Paris and Parthenon Publishing, Carnforth, UK, pp. 291–338.

Jolánkai, G., *et al.* (~40 co-authors) (2005) Final Report of the Tisza River Project. Real-life scale integrated catchment models for supporting water- and environmental management decisions. www.vituki.hu (main report can be downloaded under menu item 'Projekts').

Jones, F.R.H. (1968) *Fish Migration.* St. Martins, New York, New York

Jones, J.B., Fisher, S.G. and Grimm, N.B. (1995) Nitrification in the hyporheic zone of a desert stream ecosystem. *Journal of the North American Benthological Society* 14, 249–258.

Jorga, W. and Weise, G. (1977) Biomasseentwicklung submerser Makrophyten in langsam fließender Gewässern in Beziehung zum Sauerstoffhaushalt. *Internationale Revue der Gesamten Hydrobiologie* 62, 209–234.

Jorgensen, B.B. and Richardson, K. (eds) (1996) *Eutrophication in Coastal Marine Ecosystems.* Coastal and Estuarine Studies No. 52. American Geophysical Union, Washington, DC, 273 pp.

Jörgensen, S.E. (1976) An eutrophication model for a lake. *Ecological Modeling* 2, 147–165.

Jörgensen, S.E. (ed.) (1983) *Application of Ecological Modelling in Environmental Management.* Elsevier Scientific Publishing Co., Amsterdam, 735 pp.

Jörgensen, S.E. (ed.) (1988) *Fundamentals of Ecological Modeling.* Elsevier Science Publishers BV, Amsterdam.

Jörgensen, S.E. and Harleman, D.R.F. (1977) *Summary Report of the IIASA Workshop on Geophysical and Ecological Modeling of Deep Lakes and Reservoirs*, Laxenburg, Austria, 12–15 December. International Institute for Applied Systems Analysis, Laxenburg, Austria.

Jorgensen, S.E. and Kay, J. (eds) (2000) *Thermodynamics and Ecological Modelling.* CRC Press, Inc., Boca Raton, Florida.

Jorgensen, S.E. and Svirezhev, Y.M. (2004) *Towards a Thermodynamic Theory for Ecological Systems.* Elsevier Science BV, Amsterdam.

Jowett, I.G. (1989) *River Hydraulic and Habitat Simulation, RHYHABSIM Computer Manual.* New Zealand Fisheries Miscellaneous Report No. 49. Ministry of Agriculture and Fisheries, Christchurch, New Zealand.

Jowett, I.G. (1998) Hydraulic geometry of New Zealand rivers and its use as a preliminary method of habitat assessment. *Regulated Rivers* 14, 451–466.

Jowett, I.G. and Biggs, B.J.F. (1997) Flood and velocity effects on periphyton and silt accumulation in two New Zealand rivers. *New Zealand Journal of Marine and Freshwater Research* 31, 287–300.

Jowett, I.G. and Duncan, M.J. (1990) Flow variability in New Zealand rivers and its relationship to in-stream habitat and biota. *New Zealand Journal of Marine and Freshwater Research* 24, 305–317.

Jowett, I.G. and Richardson, J. (1990) Microhabitat preferences of benthic invertebrates in a New Zealand river and the development of in-stream flow-habitat models for *Delatidium* sp. *New Zealand Journal of Marine and Freshwater Research* 24, 19–30.

Jowett, I.G. and Richardson, J. (1995) Habitat preferences of common, riverine New Zealand native fishes and implications for flow management. *New Zealand Journal of Marine and Freshwater Research* 29, 13–23.

Joyce, J.C. (1993) Practical uses of aquatic weeds. In: Pieterse, A.H. and Murphy, K.J. (eds) *Aquatic Weeds. The Ecology and Management of Nuisance Aquatic Vegetation.* Oxford University Press, Oxford, UK, pp. 274–291.

Juday, C. (1940) The annual budget of an inland lake. *Ecology* 21, 438–450.

Junk, W.J. (1982) Amazonian floodplains: their ecology, present and potential use. *Rev. Hydrobiol. Trap.* 15, 285–301.

Junk, W.J. (ed.) (1997) *The Central Amazon Floodplain: Ecology of a Pulsing System. Ecological Studies*, Vol. 126. Springer Verlag, Berlin, 515 pp.

Junk, W.J. (2000) Mechanisms for the development and maintenance of biodiversity in a neotropical floodplain. In: Gopal, B., Junk, W.J. and Davis, J.A. (eds) *Biodiversity in Wetlands: Assessment, Function and Conservation*, Vol. 1. Backhuys Publishers, Leiden, The Netherlands, pp. 119–139.

Junk, W.J. and Wantzen, K.M. (2003) The flood pulse concept: new aspects, approaches and appraisals – an update. Paper presented at the *Large Rivers Symposium (LARS2)*, Phnom Penh, Cambodia, 11–14 February; available at www.lars2.org

Junk, W.J., Bailey, P.B. and Sparks, R.E. (1989) The flood-pulse concept in river–floodplain systems. In: Dodge, D.P. (ed.) Proceedings of the International Large Rivers Symposium. *Canadian Special Publication of Fisheries and Aquatic Science* 106, 110–127.

Jury, M.R. (1996) Regional teleconnection patterns associated with summer rainfall over South Africa, Namibia and Zimbabwe. *International Journal of Climatology* 16, 135–153.

Jury, M.R., McQueen, C. and Levey, K. (1994) SOI and QBO signals in the African region. *Theoretical and Applied Climatology* 50, 103–115.

Justic, D., Rabalais, D.D., Turner, R.E. and Dortch, Q. (1995) Changes in nutrient structure of river-dominated coastal waters: stoichiometric nutrient balance and its consequences. *Estuarine, Coastal and Shelf Science* 40, 33 –356.

Kaczmarek, Z., Strzepek, K.M., Somlyódy, L. and Priazhinskaya, V. (eds) (1996) *Water Resources Management in the Face of Climatic/Hydrologic Uncertainties*. Kluwer, Dordrecht, The Netherlands and International Institute for Applied Systems Analysis, Laxenburg, Austria.

Kadomura, H. (1995) Palaeoecological and palaeohydrological changes in the humid tropics during the last 20 000 years with reference to equatorial Africa. In: Gregory, K.J., Starkel, L. and Baker, V.R. (eds) *Global Continental Palaeohydrology*. John Wiley and Sons Ltd, Chichester, UK, pp. 177–202.

Kaenel, B.R. and Uehlinger, U. (1998) Effects of plant cutting and dredging on habitat conditions in streams. *Archive of Hydrobiology* 143, 257–273.

Kaenel, B.R., Matthaei, C.D. and Uehlinger, U. (1998) Disturbance by aquatic plant management in streams: effects on benthic invertebrates. *Regulated Rivers: Research and Management* 14, 341–356.

Kalicki, T. (1991) The evolution of the Vistula river valley between Cracow and Niepołomice in Late Vistulian and Holocene times. In: Evolution of the Vistula River Valley During the Last 15 000 Years, Part IV. *Geographical Studies Special Issue* 6, 11–37.

Kamler, E. (1992) *Early Life History of Fish. An Energetics Approach*. Chapman and Hall, London, 267 pp.

Karasyov, I.F. (1975) *The Channel Processes under Transfer of Flow*. Hydrometeoizdat, Leningrad, Russia (in Russian).

Karim, M.F. and Kennedy, J.F. (1986) Velocity and sediment–concentration profiles in river flows. *Journal of Hydraulic Engineering* 113, 159–178.

Keller, R. (1962) *Gewässer und Wasserhaushalt des Festlandes*. Teubner, Leipzig, Austria.

Kelly, M.H, and Gore, J.A. (2008) Florida river flow patterns and the Atlantic Multidecadal Oscillation. *River Research and Applications* (in press).

Kelly, R.A. (1973) Conceptual ecological model of the Delaware Estuary (manuscript). Quality of the Environment Program, Resources for the Future Inc., Washington, DC.

Kelly, R.A. and Spofford, W.O. (1977) Application of an ecosystem model to water quality management: the Delaware Estuary. In: Hall, C.A.S. and Day, J.W. (eds) *Models as Ecological Tools: Theory and Case Histories*. Wiley Interscience Inc., New York, New York, pp. 420–443.

Kemmers, R.H. (1993) Staalkaarten voor een ecologische landevaluatie (Sample cards for an ecological land evaluation). *Landschap* 10, 5–22.

Kemp, J.L. and Harper, D.M. (1997) *River Deben Alleviation of Low Flows Scheme: Proposals for Channel Habitat Physical Rehabilitation in Association with Flow Augmentation*. Environment Agency, London.

Kemp, J.L., Harper, D.M. and Crosa, G.A. (1999) Use of 'functional habitats' to link ecology with morphology and hydrology in river rehabilitation. *Aquatic Conservation – Marine and Freshwater Ecosystems* 9, 159–178.

Kemp, J.L., Harper, D.M. and Crosa, G.A. (2000) The habitat scale ecohydraulics of rivers. *Ecological Engineering* 16, 17–29.

Kemp, J.L., Harper, D.M. and Crosa, G.A. (2002) A deeper understanding of river habitat-scale ecohydraulics: interpreting the relationship between habitat type, depth and velocity using knowledge of sediment dynamics and macrophyte growth. *International Journal of Ecohydrology and Hydrobiology* 2, 271–282.

Kerekes, J. (1983) Predicting tropic response to phosphorus addition in a Cape Breton island lake. *Proceedings of the Nova Scotia Institute of Science* 33, 7–18.

Kindler, J. and Matthews, G. (1997) The central Asian Aral Sea basin programme. In: Lundqvist, J. and Gleick, P. (eds) *Sustaining Our Waters into the 21st Century. Background Report to Comprehensive Assessment of the Freshwater Resources of the World*. Stockholm Environment Institute, Stockholm, pp. 29–32.

King, J.M. (1981) The distribution of invertebrate communities in a small South African river. *Hydrobiologia* 83, 43–65.

Kirchner, J.W. (1989) The Gaia hypothesis: can it be tested? *Reviews of Geophysics* 27, 223–235.

Kirnbauer, R., Pirkl, H., Haas, P. and Steidl, R. (1996) Abflußmechanismen – Beobachtung und Modellierung (Runoff mechanisms – observations and modelling). *Österreichische Wasser- und Abfallwirtschaft* 48, 15–26.

Kitchell, J.F., Stewart, D.J. and Weiniger, D. (1977) Applications of a bioenergetics model to yellow perch (*Perca flavescens*) and walleye (*Stizostedion vitreum vitreum*). *Journal of the Fisheries Research Board of Canada* 34, 1922–1935.

Klekowski, R.Z. and Duncan, A. (1975) Physiological approach to ecological energetics. In: Grodzinski, W., Klekowski, R.Z. and Duncan, A. (eds) *Methods for Ecological Bioenergetics*. IBP Handbook No. 24. Blackwell Scientific Publications, Oxford, UK, pp. 15–64.

Klijn, F., Groen, C.L.G. and Witte, J.P.M. (1996) Ecoseries for potential site mapping, an example from the Netherlands. *Landscape and Urban Planning* 35, 53–70.

Klijn, F., van Ek, R. and Witte, J.P.M. (1998) Modellentrein ontspoord. Met NTM-rijtuig achter een stoomlocomotief op smalspoor (Critical review of the NTM-model). *Landschap* 15, 235–241.

Klimanov, W.A. (1990) Quantitative characteristics of the Northern Eurasia in the Lateglacial. *Izvestiya Akademii Nauk SSSR, Seriya Geograficheskaya* 4, 116–126 (in Russian).

Klimek, K. and Rotnicki, K. (1977) The palaeogeography of the Tot-Nuurin-khotnor Basin in the southern piedmont area of the Khangai Mts. *Bulletin Academie Polish Science, Series Sciences de la Terre* 25, 119–124.

Klimek, K. and Starkel, L. (1974) History and actual tendency of flood-plain development at the border of the Polish Carpathians. In: *Report of the Commission on Present-day Processes of the IGU*. Nachrichten Akademie, Göttingen, Germany, pp. 185–196.

Klopatek, J.M. (1978) Nutrient dynamics of freshwater riverine marshes and the role of emergent macrophytes. In: Good, R.E., Whigham, D.F. and Simpson, R.L. (eds) *Freshwater Wetlands: Ecological Processes and Management Potential*. Academic Press, New York, New York, pp. 195–216.

Klosowski, S. (1993) The shore vegetation in selected lakeland areas in north-eastern Poland. *Hydrobiologia* 251, 227–237.

Klosowski, S. and Tomaszewicz, H. (1993) Analysis of distribution, structure and habitat conditions of riparian vegetation in the river–lake system of the Upper Szeszupa (Suwalski Landscape Park, North-Eastern Poland). *Ekologia polska* 41, 125–156.

Kloss, M. (1993) Differentiation and development of peatlands in hollows without run-off on young glacial terrain. *Polish Ecological Studies* 19, 115–220.

Kloss, M. and Wilpiszewska, I. (2002) Diversity, disturbance and spatial structure of wetland vegetation along a lake shore: Jorka river system (Masurian Lakeland, Poland). *Polish Journal of Ecology* 50, 489–514.

Knight, B. (1985) Energetics and fish farming. In: Tytler, P. and Calow, P. (eds) *Fish Energetics: New Perspectives*. Croom Helm, London, pp. 309–340.

Knoblauch, A. (1977) Mathematische Simulation von Stoffkreislaufen stehender Gewasser, aufgezeigt am Phosphorkreislauf der Wahnbachtalsperre. *Vom Wasser* 49, 55–70.

Knox, J.C. (1977) Human impact on Wisconsin stream channels. *Annals of the Association of American Geographers* 76, 323–342.

Knox, J.C. (1983) Responses of river systems to Holocene climates. In: Wright, H.E. (ed.) *Late Quaternary Environment of the United States.* Vol. 2. *The Holocene.* University of Minnesota Press, Minneapolis, Minnesota, pp. 26–41.

Knox, J.C. (1995) Fluvial systems since 20 000 years BP. In: Gregory, K.J., Starkel, L. and Baker, V.R. (eds) *Global Continental Palaeohydrology.* John Wiley and Sons Ltd, Chichester, UK, pp. 87–108.

Kohler, A. (1978) Methoden der Kartierung von Flora und Vegetation von Süßwasserbiotopen. *Landschaft + Stadt* 10, 73–85.

Kohler, A. and Janauer, G.A. (1995) Zur Methodik der Untersuchung von Fließgewässern. In: Steinberg, Ch., Bernhardt, H. and Klapper, H. (eds) *Handbuch für Angewandte Limnologie,* Vol. VIII-1.1.3. Ecomed, Landsberg-Lech, Germany, pp. 1–22.

Kohler, A., Vollrath, H. and Beisl, E. (1971) Zur Verbreitung, Vergesellschaftung und Ökologie der Gefäßmakrophyten im Fließgewässersystem Moosach (Münchner Ebene). *Archive of Hydrobiology* 69, 333–365.

Koltunova, A.S. (1951) Admissible concentrations of copper under outfalls to water bodies of industrial sewage. In: *Sanitary Characteristics of Water Bodies.* Stroiizdat, Moscow, pp. 88–118 (in Russian).

Kondracki, J. (1988) *Geografia fizyczna Polski (Physical Geography of Poland).* PWN, Warsaw, 454 pp.

Kornijow, R. and Kairesalo, T. (1994) *Elodea canadensis* sustains rich environment for macroinvertebrates. *Verhandlungen des Internationalen Vereins für Limnologie* 25, 4098–4111.

Kotarba, A, (1992) High-energy geomorphic events in the Polish Tatra Mountains. *Geografiska Annaler* 48A, 123–131.

Koutaniemi, L. (1991) Glacio-isostatically adjusted palaeohydrology, the river Ivalojoki and Oulankajoki, Northern Finland. In: Starkel, L., Gregory, K.J. and Thornes, J.B. (eds) *Temperate Palaeohydrology.* John Wiley and Sons Ltd, Chichester, UK, pp. 65–78.

Kouwen, N. and Fahti-Moghadam, M. (1996) Friction factors for vegetation. In: Leclerc, M., Capra, H., Valentin, S., Boudreault, A. and Coté, Y. (eds) *Ecohydraulics 2000. Proceedings of the 2nd International Symposium on Habitat Hydraulics.* INRS-Eau, Québec City, Canada, pp. 251–257.

Kouwen, N. and Li, R.M. (1980) Biomechanics of vegetative channel linings. *Journal of the Hydraulics Division: Proceedings of the American Society of Civil Engineers* 106, 1085–1103.

Kovalak, W.P. (1976) Seasonal and diel changes in the positioning of *Glossosoma nigrior* Banks (Trichoptera: Glossosomatidae) on artificial substrates. *Canadian Journal of Zoology* 54, 1585–1594.

Kovalak, W.P. (1978) Relationships between size of stream insects and current velocity. *Canadian Journal of Zoology* 56, 178–186.

Kovalenko, E.P. and Fashchevsky, B.V. (1986) *Water, Nature, Man.* Ouradjay, Minsk, Belarus (in Russian).

Kővári, I., Kovács, Zs., Nádorné, I. and Rákli, M.Zs. (1997) A Zala folyó vízgyűjtőterületének vízminőségi állapotfeltárása (State of water quality in the Zala river catchment) (manuscript). Szombathely, Hungary.

Kozák, M. and Rátky, I. (1999) Effect of the width of the flood plain and its built-in area on the flood water levels. *Vizügyi Közlemenyek* 2, 311–317 (in Hungarian).

Kozarski, S. (1991) Warta – a case study of a lowland river. In: Starkel, L., Gregory, K.J. and Thornes, J.B. (eds) *Temperate Palaeohydrology.* John Wiley and Sons Ltd, Chichester, UK, pp. 189–215.

Kozarski, S. and Rotnicki, K. (1977) Valley floors and changes of river channel pattern in the North Polish Plain during the Late-Würm and Holocene. *Quaestiones Geographicae* 4, 51–93.

Krapiec, M. (1992) Skale dendrochronologiczne późnego holocenu poludniowej i centralnej Polski. *Kwartalnik AGH Geologia* 18, 37–119.

Krauze, K. and Wagner, I. (2008) An ecohydrological approach for the protection and enhancement of ecosystem services. In: Petrosillo, I., Jones, B., Muller, F., Zurlini, G., Krauze, K. and Victorov, S. (eds) *Use of Landscape Sciences for the Assessment of Environmental Security*. Springer-Verlag, Berlin, pp. 177–207.

Krohn, M.M. and Boisclair, D. (1994) Use of a stereo-video system to estimate the energy expenditure of free-swimming fish. *Canadian Journal of Fisheries and Aquatic Science* 51, 1119–1127.

Kronvang, B., Grant, R. and Lambel, A.L. (1997) Sediment and phosphorus export from a lowland catchment: quantification of sources. *Water, Soil and Air Pollution* 99, 465–476.

Kros, J., Reinds, G.J., de Vries, W., Latour, J.B. and Bollen, M. (1995) *Modelling of Soil Acidity and Nitrogen Availability in Natural Ecosystems in Response to Changes in Acid Deposition and Hydrology*. Winand Staring Centre, Wageningen, The Netherlands.

Kruchkova, N.M. (1972) *Zooplankton as Agent of Reservoir Selfpurification. Theory and Practice of Biological Selfpurification of Polluted Waters*. Moscow University Publisher, Moscow, pp. 58–61 (in Russian).

Kruk, M. (1987) The influence of the mire properties in a drainingless catchment area on the trophic state of mire waters. *Ekologia polska* 35, 679–698.

Kruk, M. (1990) The processing of elements by mires in agricultural landscape: mass balance based on sub-surface hydrology. *Ekologia polska* 38, 73–117.

Kruk, M. (1991) The significance of peatlands in cycling of mineral elements in landscape. *Wiadomości ekologiczne* 37, 79–96 (in Polish with English summary).

Kruk, M. (1996) Biogeochemical consequences of watershed inflow of mineral elements into fens in agricultural landscape. *Polish Ecological Studies* 22, 105–127.

Kruk, M. (1997) Effect of draining on nitrogen inflow through mires in agricultural landscape. *Polish Journal of Ecology* 45, 441–460.

Kufel, L. (1990) Nutrient sedimentation at the river inflow to a lake. *Verhandlungen des Internationalen Vereins für Limnologie* 24, 1772–1774.

Kulik, B.H. (1990) A method to refine the New England aquatic base flow policy. *Rivers* 1, 8–22.

Kundzewicz, Z.W. (2002a) Ecohydrology for sustainable development and management of water resources. *Ecohydrology and Hydrobiology* 2, 49–58.

Kundzewicz, Z.W. (2002b) Ecohydrology — seeking consensus on interpretation of the notion. *Hydrological Sciences Journal* 47, 799–804.

Kutas, T. and Herodek, D. (1986) A complex model for simulating the Lake Balaton ecosystem. In: Somlyódy, L. and van Straten, G. (eds) *Modeling and Managing Lake Eutrophication with Application to Lake Balaton*. Springer Verlag, Berlin, pp. 309–323.

Kutzbach, J.E. (1980) Estimates of past climate of paleolake Chad, North Africa, based on a hydrological and energy-balance model. *Quaternary Research* 14, 210–223.

Kutzbach, J.E. (1983) Modelling of Holocene climates. In: Wright, H.E. (ed.) *Late Quaternary Environment of the United States*. Vol. 2. *The Holocene*. University of Minnesota Press, Minneapolis, Minnesota, pp. 271–277.

Kutzbach, J.E. (1992) Modelling earth system changes of the past. In: Ojima, D. (ed.) *Modelling the Earth System*. UCAR, Boulder, Colorado, pp. 377–404.

Kvasov, D.D. (1976) *Paleohydrology of Eastern Europe during Valdai Period*, Problemy Paleagidrologii, Moskva, pp. 260–266.

Kyakk, V.A. and Lartsina, L.E. (1974) About temperature regime of waterbodies, using for cooling of thermal electric power station. *Proceedings of the Coordination Meetings in Hydraulic Engineering, Leningrad* 96, 35–78 (in Russian).

Laë, R. (1997) Effects of climatic changes and developments on continental fishing in West Africa: the examples of the central delta of the Niger in Mali and coastal lagoons in Togo. In: Remane, K. (ed.) *African Inland Fisheries, Aquaculture and the Environment*. Fishing New Books, Oxford, UK, pp. 66–86.

Lake, P.S., Barmuta, L.A., Boulton, A.J., Campbell, I.C. and St. Clair, R.M. (1986) Australian streams and Northern Hemisphere stream ecology: comparisons and problems. *Proceedings of the Ecological Society of Australia* 14, 61–82.

Lamb, B.L. (1998) Protection of instream uses of water in the US. In: Blazkova, S., Stalnaker, C. and Novicky, O. (eds) *Hydroecological Modeling: Research, Practice, Legislation, and Decision-making*. T.G. Masaryk Water Research Institute, Prague, Czech Republic and MESC, Fort Collins, Colorado, pp. 55–56.

Lamb, B.L. and Doerksen, H.R. (1990) Instream water use in the United States – water laws and methods for determining flow requirements. In: Chase, E.B., Paulson, R.W. and Moody, D.W. (eds) *National Water Summary 1987 – Hydrologic Events and Water Supply and Use*. Water Supply Paper No. 2350. US Geological Survey, Washington, DC, pp. 109–116.

Lamouroux, N. (1997) Hydraulique statistique et prediction de caracteristique du peuplement pisciocole: modeles pour l'ecosysteme fluvial. PhD thesis, Universite Claude Bernard, Lyon, France.

Lamouroux, N. (1998) Depth probability distributions in stream reaches. *ASCE Journal of Hydraulic Engineering*.

Lamouroux, N., Souchon, Y. and Herouin, E. (1995) Predicting velocity frequency distributions in stream reaches. *Water Resources Research* 31, 2367–2375.

Lamouroux, N., Capra, H. and Pouilly, M. (1996) A perspective to predict habitat suitability of lotic fish: linking statistical hydraulic with multivariate habitat use models. In: Leclerc, M., Capra, H., Valentin, S., Boudreault, A. and Coté, Y. (eds) *Ecohydraulics 2000. Proceedings of the 2nd International Symposium on Habitat Hydraulics*. INRS-Eau, Québec City, Canada, pp. B96–B117.

Lancaster, J., Hildrew, A.G. and Gjerlov, C. (1996) Invertebrate drift and longitudinal transport processes in streams. *Canadian Journal of Fisheries and Aquatic Science* 53, 572–582

Lapshev, M.N. (1977) *Computations of Sewage Outfalls*. Stroiizdat, Moscow (in Russian).

Laraque, C. and Olivry, J.C. (1996) Évolution de l'hydrologie du Congo-Zaire et de ses affluents rive droite et dynamique des transports solides et dissous. In: *L'hydrologie tropicale: géoscience et outil pour le développement, Proceedings of the Paris Conference*, May 1995. IAHS Publication No. 238. Centre for Ecology and Hydrology, Wallingford, UK, pp. 271–288.

Larsen, D.P., Mercier, H.T. and Malueg, K.W. (1974) Modeling algal growth dynamics in Shagawa Lake, Minnesota, with comments concerning projected restoration of the lake. In: Middlebrooks, E.J., Falkenborg, D.H. and Maloney, T.E. (eds) *Modeling the Eutrophication Process*. Ann Arbor Science Publishers, Ann Arbor, Michigan, pp. 15–31.

Larson, H.N. (1981) *New England Flow Policy*. Memorandum, Interim Regional Policy for New England Stream Flow Recommendations. US Fish and Wildlife Service, Region 5, Boston, Massachussets.

Latour, J.B., Reiling, T. and Wiertz, J. (1993) MOVE: a multiple stress model for vegetation. In: Hooghart, J.C. and Posthumus, C.W.S. (eds) *The Use of Hydro-ecological*

Models in The Netherlands. TNO Committee on Hydrological Research, The Hague, The Netherlands, pp. 53–66.

Latour, J.B., Reiling, R. and Wiertz, J. (1994) A flexible multiple stress model: who needs a priori classification? In: Klijn, F. (ed.) *Ecosystem Classification for Environmental Management.* Kluwer, Dordrecht, The Netherlands, pp. 183–198.

Layzer, J.B. and Madison, L.M. (1995) Microhabitat use by freshwater mussels and recommendations for determining their instream flow needs. *Regulated Rivers* 10, 329–345.

Leclerc, M., Boudrealut, A., Bechara, J. and Corfa, G. (1995) Two-dimensional hydrodynamic modeling: a neglected tool in the instream flow incremental methodology. *Transactions of the American Fisheries Society* 124, 645–662.

Leclerc, M., Capra, H., Valentin, S., Boudreault, A. and Coté, Y. (1996) (eds) *Ecohydraulics 1996. Proceedings of the 2nd International Symposium on Habitat Hydraulics.* INRS-Eau, Québec City, Canada, 991 pp.

Leftwich, K.N., Angermeier, P.L. and Dolloff, C.A. (1997) Factors influencing behavior and transferability of habitat models for a benthic stream fish. *Transactions of the American Fisheries Society* 126, 725–734.

Leighley, J.B. (1934) Turbulence and the transportation of rick debris by streams. *Geographical Review* 24, 453–464.

Lenton, T.M. (1998) Gaia and natural selection. *Nature* 394, 439–447.

Lenton, T.M. and Lovelock, J.E. (2000) Daisyworld is Darwinian: constraints on adaptation are important for planetary self-regulation. *Journal of Theoretical Biology* 205, 109–114.

Leonard, P.M. and Orth, D.J. (1988) Use of habitat guilds to determine instream flow requirements. *North American Journal of Fisheries Management* 8, 399–409.

Leopold, L.B and Langbein, W.B. (1962) *The concept of entropy in Lanscape evolution.* Untited States Geological survey, Professional Paper 500-A.

Leopold, L.B., Wolman, M.G. and Miller, J.P. (1964) *Fluvial Processes in Geomorphology.* W.H. Freeman, San Francisco, California.

Lethimaki, J., Sivonen, K., Luukkainen, R. and Niemela, S.I. (1994) The effects of incubation time, temperature, light, salinity and phosphorus on growth and hepatotoxin production by *Nodularia* strains. *Archiv für Hydrobiologie* 130, 269–282.

Lewis, S. and Nir, A. (1978) A study of parameter estimation procedures of a model for lake phosphorus dynamics. *Ecological Modeling* 4, 99–117.

Li, X., Jongman, R.H.G., Hu, Y., Bu, R., Harms, B., Bregt, A.K. and He, H.S. (2005) Relationship between landscape structure metrics and wetland nutrient retention function: a case study of Liaohe Delta, China. *Ecological Indicators* 5, 339–349.

Lillie, R. and Budd, J. (1992) Habitat architecture of *Myriophyllum spicatum* L. as an index to habitat quality for fish and macroinvertebrates. *Journal of Freshwater Ecology* 7, 113–125.

Lincoln, R.J., Boxhall, G.A. and Clark, P.F (1982) *A Dictionary of Ecology, Evolution and Systematics.* Cambridge University Press, Cambridge, UK.

Lindeman, R.L. (1942) The trophic–dynamic aspect of ecology. *Ecology* 23, 399–418.

Lock, M.A. and John, P.H. (1979) The effect of flow patterns on uptake of phosphorus by river periphyton. *Limnology and Oceanography* 24, 376–383.

Lockwood, J.G. (1979) Water balance of Britain, 50 000 yr BP to the present day, *Quaternary Research* 12, 297–310.

Lodge, M.D., Cronin, G., van Donk, E. and Froelich, A.J. (1998) Impact of herbivory on plant standing crop: comparisons among biomes, between vascular and nonvascular plants, and among freshwater herbivore taxa. In: Jeppesen, E., Sondergaard, Ma., Sondergaard, Mo. and Christoffersen, K. (eds) *The Structuring Role of Submerged*

Macrophytes in Lakes. Ecological Studies, Vol. 131. Springer, New York, New York, pp. 149–174.

Loizeau, J.-L. and Dominik, J. (2000) Evolution of the Upper Rhone discharge and suspended sediment load during the late 80 years and some implications for Lake Geneva. *Aquatic Sciences* 62, 54–67.

Lorenzen, M.W. (1974) Predicting the effects of nutrient diversion on lake recovery. In: Middlebrooks, E.J., Falkenborg, D.H. and Maloney, T.E. (eds) *Modeling the Eutrophication Process*. Ann Arbor Science Publishers, Ann Arbor, Michigan, pp. 205–211.

Lotka, A. (1925) *Elements of Physical Biology*. Williams and Wilkins, Baltimore, Maryland. Republished as *Elements in Mathematical Biology*, Dover, New York, New York, 1956.

Lotter, A.F. and Zbinden, H. (1989) Late-Glacial palynology, oxygen isotope record and radiocarbon stratigraphy from Rotsee (Lucerne), Central Swiss Plateau. *Ecologae Geologica Helvetica* 82, 87–98.

Lovelock. J.E. (1972) Gaia as seen through the atmosphere. *Atmospheric Environment* 6, 579–580.

Lovelock, J. (1979) *Gaia: A New Look at Life on Planet Earth*. Oxford University Press, Oxford, UK.

Lovelock. J.E. (1988) *The ages of Gaia*. W.W. Norton and Company, New York.

Lovelock.J.E. and Margulis L. (1974) Atmospheric homeostais by and for the bioshere: the gaia hyphothesis. *Tellus* 26, 2–9.

Lucas, M.C. (1994) Heart rate as an indicator of metabolic rate and activity in adult Atlantic salmon, *Salmo salar*. *Journal of Fish Biology* 44, 889–903.

Lucassen, E.C.H.E.T., Smolders, A.J.P., Lamers, L.P.M. and Roelofs, J.G.M. (2005) Water table fluctuations and groundwater supply are important in preventing phosphate-eutrophication in sulphate-rich fens: consequences for wetland restoration. *Plant and Soil* 269, 109–115.

Lukyanenko, V.I. (1987) *Ecological Aspects of Ichthyology*, Agropromizdat, Moscow (in Russian).

Lung, W.S., Canale, R.P. and Freedman, P.L. (1976) Phosphorus models for eutrophic lakes. *Water Research* 10, 1101–1114.

Luther, H. (1949) Vorschlag zu einer ökologischen Grundeinteilung der Hydrophyten. *Acta Botanica Fennica* 44, 1–15.

Ľvovič, M.I. (1974) *World Water Resources and Their Future*. Translated by Nace, R., American Geophysical Union, Washington, DC, 1979.

McGarrigle, M.L. (1993) Aspects of river eutrophication in Ireland. *Annals of Limnology* 29, 355–364.

Machata-Wenninger, C. and Janauer, G.A. (1991) The measurement of current velocities in macrophyte beds. *Aquatic Botany* 39, 221–230.

McIntyre, M.E. (1993) The Quasi-Biennial oscillation (QBO): some points about the terrestrial QBO and the possibility of related phenomena in the solar interior. In: *Proceedings of NATO Advanced Research Workshop ARW 920946*, Paris, October.

Mackay, H.M. (1999) *Water Resources Policy Protection Implementation: Resource Directed Measures for Protection of Water Resources: Integrated Manual, v 1.0.* Report N/28/29. Department of Water Affairs and Forestry, Pretoria, South Africa.

Margalef, R. (1968) *Perspectives in Ecological Theory*. University of Chicago Press, Chicago.

McNaughton, S.J. (1979) Grazing as an optimzation process: grass–ungulate relationships in the Serengeti. *American Naturalist* 113, 691–699.

McNaughton, S.J. (1985) Ecology of a grazing ecosystem: the Serengeti. *Ecological Monographs* 55, 259–294.

McNaughton, S.J. (1988) Mineral nutrition and spatial concentrations of African ungulates. *Nature* 334, 343–345.

McNaughton, S.J. (1990) Mineral nutrition and seasonal movements of African migratory ungulates. *Nature* 345, 613–615.

McNaughton, S.J., Ruess, R.W. and Eagle, S.W. (1988) Large mammals and process dynamics in African ecosystems. *BioScience* 38, 794–800.

Madsen, T.V. and Warncke, E. (1983) Velocities of currents around and within submerged aquatic vegetation. *Archive of Hydrobiology* 97, 389–394.

Madison, C.E., Blevins, R.L., Frye, W.W. and Barfield, B.J. (1992) Tillage and grass filter strip effects upon sediment and chemical losses. *Agronomy Abstracts* 331.

Magny, M. (1993) Holocene fluctuations of lake levels in the French Jura and sub-Alpine ranges and their implications for past general circulation pattern. *The Holocene* 3, 306–313.

Mahamah, D.S. and Bhagat, S.K. (1983) Use and abuse of empirical phosphorus models in lake management. In: Lauenroth, W.K., Skogerobe, G.V. and Flug, M. (eds) *Analysis of Ecological Systems: State-of-the-Art in Ecological Modeling*. Elsevier Scientific Publishing Co., Amsterdam, pp. 593–599.

Mainstone, C., Gulson, J. and Parr, W. (1994) *Phosphates in Freshwater. Standards for Nature Conservation*. Research Report No. 873. English Nature, London.

Maizels, J.K. (1983) Channel changes, paleohydrology and deglaciation: evidence from some lateglacial sandur deposits, northeast Scotland. Proceedings of the Holocene Symposium. *Quaternary Studies in Poland (Poznań)* 4, 171–187.

Maizels, J.K. (1989) Sedimentology, paleoflow dynamics and flood history of jökul-hlaup deposits: paleohydrology of Holocene sediment sequences in Southern Iceland sandur deposits. *Journal of Sedimentary Petrology* 59, 204–223.

Maizels, J.K. (1995) Palaeohydrology of polar and subpolar regions over the past 20 000 years. In: Gregory, K.J., Starkel, L. and Baker, V.R. (eds) *Global Continental Palaeohydrology*. John Wiley and Sons Ltd, Chichester, UK, pp. 259–299.

Malati, M.A. and Fox, I. (1985) A review of the role of bed deposits in the phosphate eutrophication of lakes and rivers. *International Journal of Environmental Studies* 26, 43–54.

Mande, U., Kull, A., Tamm, V., Kuusemets, V. and Karjus, R. (1998) Impact of climatic conditions and land use change on runoff and nutrient losses in rural landscape. *Landscape and Urban Planning* 41, 229–238.

Mann, K.H. (1978) Estimating the food consumption of fish in nature. In: Gerking, S.D. (ed.) *Ecology of Freshwater Fish Production*. Blackwell Scientific Publications, Oxford, UK, pp. 250–273.

Manning, R. (1891) On the flow of water in open channels and pipes. *Transactions of the Institute of Civil Engineering of Ireland* 20, 161.

Mansikkaniemi, H. (1991) Development of river valleys originated by emergence, S-Finland. Case study of the river Kyrönjoki. In: Starkel, L., Gregory, K.J. and Thornes, J.B. (eds) *Temperate Palaeohydrology*. John Wiley and Sons Ltd, Chichester, UK, pp. 79–104.

Margalef, D.R. (1963) On certain unifying principles in ecology. *American Naturalist* 97, 357–374.

Margalef, R. (1960) Ideas for a synthetic approach to the ecology of running waters. *Internationale Revue der Gesamten Hydrobiologie* 45, 133–153.

Margulis, L. (1998) *The Symbiotic Planet: A New Look at Evolution*. Weidenfield and Nicholson, London, 146 pp.

Marshall, E.J.P. and Westlake, D.F. (1990) Water velocities around water plants in chalk streams. *Folia Geobotanica et Phytotaxonomica* 25, 279–289.

Mason, Y. (1995) Natural manganese oxides associated with Metallogenium-like particles as scavengers of metals in lakes. PhD dissertation no. 11393, Swiss Federal Institute of Technology, Zurich, Switzerland, 171 pp.

Mathews, R.C. Jr and Bao, Y. (1991) The Texas method of preliminary instream flow assessment. *Rivers* 2, 295–310.

Mathur, D., Bason, W.H., Purdy, E.J. Jr and Silver, C.A. (1985) A critique of the instream flow incremental methodology. *Canadian Journal of Fisheries and Aquatic Science* 42, 825–831.

Matthews, W.J. and Maness, J.D. (1979) Critical thermal maxima, oxygen tolerances and success of cyprinid fishes in a southwestern river. *American Midland Naturalist* 102, 374–377.

Maude, S.H. and Williams, D.D. (1983) Behavior of crayfish in water currents: hydrodynamics of eight species with reference to their distribution patterns in southern Ontario. *Canadian Journal of Fisheries and Aquatic Science* 40, 68–77.

May, L., House, W.A., Bowes, M. and McEvoy, J. (2001) Seasonal export of phosphorus from a lowland catchment: upper River Cherwell in Oxfordshire. England. *Science of the Total Environment* 269, 117–130.

Mayer, T.D. and Jarrell, W.M. (1996) Assessing colloidal forms of phosphorus and iron in the Tualatin River Basin. *Journal of Environmental Quality* 24, 1117–1124.

Meadows, D., Meadows, D. and Randers, J. (1992) *Beyond the Limits*. Earthscan Publications, London.

Medley, K.E. and Hughes, F.M.R. (1996) Riverine forests. In: McClanahan, T.R. and Young, T.P. (eds) *East African Ecosystems and Their Conservation*. Oxford University Press, New York, New York, pp. 361–383.

Melillo, J.M., Callaghan, T.V., Woodward, F.I., Salati, E. and Sinha, S.K. (1990) Effect on Ecosystems. In: Houghton, J.T., Jenkins, G.J. and Ephraums, J.J. (eds) *Climate Change: The IPCC Scientific Assessment,* Intergovernmental Panel on Climate Change (IPCC). Cambridge University Press, Cambridge, UK, pp. 281–310.

Membratu, D. (1998) Sustainability and sustainable development: historiacal and conceptual review. *Environmental Impact Assessment Review* 18, 493–520.

Mercier, Y. (1996) Wetlands conservation along the St.-Lawrence River: an examplary partnership. In: Leclerc, M., Capra, H., Valentin, S., Boudreault, A. and Coté, Y. (eds) *Ecohydraulics 2000. Proceedings of the 2nd International Symposium on Habitat Hydraulics*. INRS-Eau, Québec City, Canada, pp. 865–870.

Messer. J.J., Grenney, W.J. and Ho, J. (1983) Simulation of phytoplankton growth in a small fast-flushing reservoir – eutrophication management implications. In: Lauenroth, W.K., Skogerobe, G.V. and Flug, M. (eds) *Analysis of Ecological Systems: State-of-the-Art in Ecological Modeling*. Elsevier Scientific Publishing Co., Amsterdam, pp. 787–806.

Meuleman, A.F., Kloosterman, R.A., Koerselman, W., den Besten, M. and Jansen, A.J.M. (1996) NICHE: een nieuw instrument voor hydro-ecologische effectvoorspelling (NICHE: a new tool for hydro-ecological effect predictions). *H$_2$O* 5, 37–39.

Meybeck, M. (1998) The IGBP water group: a response to a growing global concern. *Global Change Newsletters* 36, 8–12.

Meybeck, M. (2003) Global analysis of river systems: from Earth system controls to Anthropocene syndromes. *Philosophical Transactions of the Royal Society of London B* 358, 1935–1955.

Meybeck, M., Chapman, D.V. and Helmer, R. (1989) *Global Freshwater Quality: A First Assessment*. World Health Organization and UN Environment Programme. Basil Blackwell Ltd, Oxford, UK.

Meyer, G., Masliev, I. and Somlyódy, L. (1994) *Impact of Climatic Change on Global Sensivity of Lake Stratification*. International Institute for Applied Systems Analysis, Laxenburg, Austria.

Milhous, R.T., Bartholow, J.M., Updike, M.A. and Moos, A.R. (1990) *Reference MAnual for Generation and Analysis of Habitat Time Series – Version II.* Biological Report 90(16). US Fish and Wildlife Service, Washington, DC.

Millennium Ecosystem Assessment (2005) *Millennium Ecosystem Assessment: Ecosystems and Human Well-Being – Biodiversity Synthesis.* UN Millennium Ecosystem Assessment Series. Island Press, Washington, DC.

Milliman, J.D. (1991) Flux and fate of fluvial sediment and water in coastal seas. In: Mantoura, R.F.C., Martin, J.M. and Wollast, R. (eds) *Ocean Margin Processes in Global Change.* John Wiley and Sons Ltd, Chichester, UK, pp. 69–89.

Milliman, J.D. (1997) Blessed dams or dammed dams? *Nature* 386, 325–327.

Milliman, J.D. and Meade, R.H. (1983) World-wide delivery of river sediment to the oceans. *Journal of Geology* 91, 1–21.

Milliman, J.D., Xie, Q. and Yang, Z. (1984) Transfer of particulate organic carbon and nitrogen from the Yangtze river to the ocean. *American Journal of Science* 284, 824–834.

Milner, N.J., Scullion, J., Carling, P.A. and Crisp, D.T. (1981) The effects of discharge on sediment dynamics and consequent effects on invertebrates and salmonids in Upland Rivers. *Advances in Applied Biology* 6, 153–220.

Minarik, W. (1990) Standortfaktoren von Höheren Wasserpflanzen. Entwicklung der Expertensystem-Vorstufe ‚Makrophyt'. Master thesis, University of Vienna, Vienna.

Minshall, G.W., Cummins, K.W., Petersen, R.C., Cushing, C.E. Bruns, D.A., Sedell, J.R. and Vannote, R.L. (1985) Development in stream ecosystem theory. *Canadian Journal of Fisheries and Aquatic Science* 42, 1045–1055.

Miranda, L.E. (2002) A review of guidance and criteria for managing reservoirs and associated riverine environments to benefit fish and fisheries. In: Marmulla, G. (ed.) *Dams, Fish and Fisheries: Opportunities, Challenges, Conflict Resolution.* FAO Fisheries Technical Paper No. 419. FAO, Rome, pp. 93–141.

Mitsch, W. and Jørgensen, S.E. (1989) *Ecological engineering: an introduction to ecotechnology.* John Wiley and Sons, New York.

Mitsch, W. and Jørgensen, S.E. (2004) *Ecological Engineering and Ecosystem Restoration.* John Wiley and Sons Inc., New York, New York, 411 pp.

Mitsch, W.J. and Gosselink, J.G. (2000) *Wetlands,* 3rd ed. John Wiley, New York, New York.

Mitsch, W.J., Dorge, C.L. and Wiemhoff, J.R. (1979) Ecosystem dynamics and a phosphorus budget of an alluvial cypress swamp in southern Illinois. *Ecology* 60, 1116–1124.

Mitsch, W.J, Cronx, J.K., Wu, X., Nairn, R.W. and Hey, D.L. (1995) Phosphorus retention in constructed freshwater riparian marshes. *Ecological Applications* 5, 830–895.

Mitsch, W.J., Zhang, L., Anderson, C.J., Altor, A.E. and Hernandez, M.E. (2005) Creating riverine wetlands: ecological succession, nutrient retention, and pulsing effects. *Ecological Engineering* 25, 510–527.

Mitchell, St. F. and Perrow, M.R. (1998) Interactions between grazing birds and macrophytes. In: Jeppesen, E., Sondergaard, Ma., Sondergaard, Mo. and Christoffersen, K. (eds) *The Structuring Role of Submerged Macrophytes in Lakes. Ecological Studies,* Vol. 131. Springer, New York, New York, pp. 175–196.

Mogaka, H., Gichere, S., Davis, R. and Hirji, R. (2002) *Impacts and Costs of Climate Variability and Water Resources Degradation in Kenya.* World Bank Report. World Bank, New York, New York, 88 pp.

Mörner, N.A. (1976) The Pleistocene/Holocene boundary: a proposed boundary – stratotype in Gothenburg, Sweden. *Boreas* 5, 193–275.

Moklyak, V.I. (1976) Definition of within-year distribution flow in years of different probability. *Amelioration and Water Management* 36, 39–43 (in Russian).

Moldan, B. and Cerny, J. (eds) (1994) *Biogeochemistry of Small Catchments – A Tool for Environmental Research*. SCOPE 51. John Wiley and Sons Ltd, New York, New York, 460 pp.

Moog, O. (1993) Quantification of daily peak hydropower effects on aquatic fauna and management to minimize environmental impacts. *Regulated Rivers: Research and Management* 8, 5–14.

Moore, V.M. and Folt, C.L. (1993) Zooplankton body size and community structure: effect of thermal and toxicant stress. *TREE* 8, 178–183.

Moore, V.M., Folt, C.L. and Stemberger, R.S. (1996) Consequences of elevated temperatures for zooplankton assembledges in temperate lakes. *Archive of Hydrobiology* 135, 289–319.

Moss, B., McGowan, S. and Carvalho, L. (1994) Determination of phytoplankton crops by top-down and bottom-up mechanisms in a group of English lakes, the West Midland meres. *Limnology and Oceanography* 39, 1020–1029.

Mounier, F. (1986) The Senegal River Scheme: development for whom? In: Goldsmith, E. and Hildyard, N. (eds) *The Social and Environmental Effects of Large Dams*. Vol. 2. *Case Studies*. Wadebridge Ecological Centre, UK.

Mozgawa, J. (1993) Photointerpretation analysis of landscape structure in lake watersheds of Suwalski Landscape Park (North-Eastern Poland). *Ekologia polska* 41, 53–74.

Mtahiko, M.G.G., Gereta, E., Kajuni, A., Chiombola, E., Ng'umbi, G.Z., Coppolillo, P. and Wolanski, E. (2006) Towards an ecohydrology-based restoration of the Usangu wetlands and the Great Ruaha River, Tanzania. *Wetlands Ecology and Management* **14**, 489–503.

Mulholland, P.J., Marzolf, E.R., Webster, J.R., Hart, D.R. and Hendricks, S.P. (1997) Evidence that hyporheic zones increase heterotrophic metabolism and phosphorus uptake in forest streams. *Limnology and Oceanography* 42, 443–451.

Munson, A.B. and Delfino, J.J. (2007) Minimum wet-season flows and levels in Southwest Florida Rivers. *Journal of the American Water Resources Association* 43, 522–532.

Munson, A.B., Delfino, J.J. and Leeper, D.A. (2005) Determining minimum flows and levels: the Florida experience. *Journal of the American Water Resources Association* 41, 1–10.

Murphy, K.J. and Pieterse, A.H. (1993) Present status and prospects of integrated control of aquatic weeds. In: Pieterse, A.H. and Murphy, K.J. (eds) *Aquatic Weeds. The Ecology and Management of Nuisance Aquatic Vegetation*. Oxford University Press, Oxford, UK, pp. 222–227.

Muscutt, A.D. and Withers, P.J.A. (1996) The phosphorus content of rivers in England and Wales. *Water Research* 30, 1258–1268.

Nachtigall, W. and Bilo, D. (1965) Die Strömungsmechanik des Dytiscus-Rumpfes. *Zetischrift für vergleichende Physiologie* 50, 371–401.

Nagy, I. (1997) Role of protection-forests of the Tisza river flood-plain in reducing flood-load on the levees. *Vizügyi Köslemenyek* 1, 5–23 (in Hungarian).

Naiman, R.J. (1976) Production of a herbivorous pupfish population (*Cyprinodon nevadensis*) in a warm desert stream. *Journal of Fish Biology* 9, 125–137.

Naiman, R.J. and Decamps, H. (1990) *The ecology and management of aquatic–terrestrial ecotones* . UNESCO MAB series. The Parthenon Publishing Group/UNESCO. Paris, pp. 316.

Naiman, R.J and Decamps, H. (1991) *The Ecology and Management of Aquatic–Terrestrial Ecotones. Man and the Biosphere Series*, Vol. 4. UNESCO, Paris and Parthenon Publishing, Carnforth, UK, 316 pp.

Naiman, R.J. and Decamps, H. (1997) The ecology of interfaces: riparian zones. *Annual Review of Ecology and Systematics* 28, 621–658.

Naiman, R.J., Melillo, J.M. and Hobbie, J.E. (1986) Ecosystem alteration of boreal forest streams by beaver (*Castor canadensis*). *Ecology* 67, 1254–1269.

Naiman, R.J., Décamps, H., Pastor, J. and Johnston, C.A. (1988) The potential importance of boundaries to fluvial ecosystems. *Journal of the North American Benthological Society* 7, 289–306.

Naiman, R.J., Decamps, H. and Fournier, F. (eds) (1989) *Role of Land/Inland Water Ecotones in Landscape Management and Restoration, Proposals for Collaborative Research*. UNESCO, Vendome, France.

Naiman, R.J., Beechie, T.J., Benda, L.E., Berg, D.R., Bisson, P.A., MacDonald, L.H.M., O'Connel, M.D., Olson, P.L. and Steel, E.A. (1992) Fundamental elements of ecologically healthy watersheds in the Pacific Northwest Coastal Ecoregion. In: Naiman, R.J. (ed.) *Watershed Management*. Springer-Verlag, New York, New York, pp. 127–188.

Naiman, R.J., Bilby, R.E., Schindler, D.E. and Helfield, J.M. (2002) Pacific salmon, nutrients, and the dynamics of freshwater and riparian ecosystems. *Ecosystems* 5, 399–417.

Nelson, F.A. (1980) Evaluation of selected instream flow methods in Montana. *Proceedings of the Western Association of Fish and Wildlife Agencies* 4, 412–432.

Nesje, A. and Johannessen, T. (1992) What were the primary forcing mechanims of high frequency Holocene glacier and climatic variations? *The Holocene* 2, 70–84.

Nestler, J.M., Milhous, R.T. and Layzer, J.B. (1989) Instream habitat modeling techniques. In: Gore, J.A. and Petts, G.E. (eds) *Alternatives in Regulated River Management*. CRC Press, Inc., Boca Raton, Florida, pp. 295–315.

Newbold, D.J., Elwood, J.W., O'Neill, R.V. and Sheldon, A.L. (1983) Phosphorus dynamics in a woodland stream ecosystem: a study of nutrient spiralling. *Ecology* 64, 1249–1265.

Newbury, R.W. (1984) Hydrologic determinants of aquatic insect habitats. In: Resh, V.M. and Rosenberg, D.M. (eds) *Ecology of Aquatic Insects*. Praeger, New York, New York, pp. 323–357.

Ney, J.J. (1990) Trophic economics in fisheries: assessment of demand–supply relationships between predators and prey. *Reviews in Aquatic Sciences* 2, 55–81.

NGRP (1974) Instream Needs Subgroup Report. Working Group C: Water. Northern Great Plains Resource Program, Billings, Montana.

Nicholson, S.E. and Kim, E. (1997) The relationship of the El-Niño Southern Oscillation to African rainfall. *International Journal of Climatology* 17, 117–135.

Niemann, E. (1963) Beziehungen zwischen Vegetation und Grundwasser (Relations between vegetation and groundwater). *Archive für Naturschutz und Landschaftsforschung* 3, 3–36.

Niemela, S.L. (1989) The influence of peaking hydroelectric discharges on habitat selection and movement patterns of rainbow trout (*Oncorhynchus mykiss*). M.Sc. thesis, Tennessee Technical University, Cookeville, Tennessee.

Niemi, J. (1979) *Application of an Ecological Simulation Model to Lake Paijanne*. National Board of Waters, Helsinki, p. 39.

Noest, V. (1994) A hydrology–vegetation interaction model for predicting the occurrence of plant species in dune slacks. *Journal of Environmental Management* 40, 119–128.

Nortier, I.W. and van der Velde, H. (1968) *Hydrologie voor waterbouwkundigen*. Stam, Culemborg, The Netherlands.

Nottage, A.S., Carpenter, K.E. and Chesher, T.J. (1996) Hydraulic and ecological considerations for coastal realignment. In: Leclerc, M., Capra, H., Valentin, S., Boudreault, A. and Coté, Y. (eds) *Ecohydraulics 2000. Proceedings of the 2nd International Symposium on Habitat Hydraulics*. INRS-Eau, Québec City, Canada, pp. 513–524.

Novotny, V. and Chesters, G. (1981) *Handbook of Nonpoint Source Pollution: Sources and Management.* Van Nostrand Reinhold Co., New York, New York, 545 pp.

Nowakowski, C., Smolenski, A. and Soszka, H. (1996) Underground waters and their role in supply and the transport of matter to lakes. *Zeszyty Naukowe Komitetu Polskiej Akademii Nauk „Człowiek i Srodowisko"* 13, 71–96 (in Polish with English summary).

Nowell, A.R.M. and Jumars, P.A. (1984) Flow environments of aquatic benthos. *Annual Review of Ecology and Systematics* 15, 303–328.

Nwokedi, G.I.C. and Obodo, G.A. (1993) Pollution of the River Niger and its main tributaries. *Bulletin of Environmental Contamination and Toxicology* 51, 282–288.

Odum, E.P. (1953) *Fundamentals of Ecology.* W.B. Saunders, Philadelphia, Pennsylvania.

Odum, E.P. (1968) The watershed as an ecological unit. Reprinted in *Ecological Vignettes: Ecological Approaches to Dealing with Human Predicaments.* Harwood Academic Publishers, Amsterdam, 1998, pp. 127–132.

Odum, E.P. (1969) The strategy of ecosystem development: an understanding of ecological sucession provides a basis for resolving man's conflict with nature. *Science* 164, 262–269.

Odum, E.P. (1985) Trends expected in stressed ecosystems. *BioScience* 35, 419–422.

Odum, E.P. and Odum, H.T. (1959) *Fundamentals of Ecology, Second edition.* W.B. Saunders, Philadelphia.

Odum, E.P. and Odum, H.T. (1971) *Fundamentals of Ecology, Third edition.*W.B. Saunders, Philadelphia.

Odum, H.T. (1957) Trophic structure and productivity of Silver Springs, Florida. *Ecological Monographs* 27, 55–112.

Odum, H.T. (1971) *Environment Power and Society.* John Wiley, New York.

Odum, H.T. (1983) *Systems Ecology: An Introduction.* John Wiley, New York.

OECD (1982) *Eutrophication of Waters, Monitoring and Assessment.* OECD Publications Office, Paris.

Officer, C.B. and Ryther, J.H. (1980) The possible importance of silicon in marine eutrophication. *Marine Ecology Progress Series* 3, 83–91.

Ohl, C., Krauze, K. and Grünbühel, C. (2007) Towards an understanding of long-term ecosystem dynamics by merging socio-economic and environmental research criteria for long-term socio-ecological research sites selection. *Ecological Economics* 63, 383–391.

Ohno, A., Marui, A., Castro, E.S., Benitez, A.A., Elio-Calvo, D., Kasitani, H., Ishii, Y. and Yamaguchi, K. (1997) Enteropathogenic bacteria in the La Paz river of Bolivia. *American Journal of Tropical Medicine and Hygiene* 57, 438–444.

Okabe, T., Yuuki, T. and Kojima, M. (1997) Bed-load rate on movable beds covered by vegetation. In: Holly, F.M. and Alsaffar, A. (eds) *Proceedings of the 27th Congress of the International Association for Hydraulic Research*, San Francisco, California, 10–15 August, pp. 1396–1401.

O'Keeffe, J.A. and Davies, B.R. (1991) Conservation and management of the rivers of Kruger National Park: suggested methods for calculating instream flow needs. *Aquatic Conservation* 1, 55–71.

Olivry, J.C. (1987) Les conséquences durables de la sécheresse actuelle sur l'écoulement du fleuve Sénégal et l'hypersalinisation de la Basse-Casamance. In: *The Influence of Climate Change and Climatic Variability on the Hydrologic Regime and Water Resources (Proceedings of the Vancouver Symposium, August 1987).* IAHS Publication No. 168. Centre for Ecology and Hydrology, Wallingford, UK, pp. 501–512.

Olson, D.M., Dinerstein, E., Canevari, P., Davidson, I., Castro, G., Morisset, V., Abell, R. and Toledo, E. (eds) (1998) *Freshwater Biodiversity of Latin America and the*

Caribbean: A Conservation Assessment. America Verde Publications, Biodiversity Support Programme, Washington, DC, 70 pp.

Orcutt, J.D., Porter, K.G. (1983) Diel vertical migration by zooplankton: constant and fluctuating temperature effects on life history parameters of *Daphnia*. *Limnology and Oceanography* 28, 720–730.

Orlob, G.T. (1977) *Mathematical Modeling of Surface Water Impoundments.* US Department of the Interior Project T-0006. Resource Management Associates Inc., Lafayette, California, p. 119.

Orloci, I. (1966) Geometric models in ecology. I. The theory and application of some ordination methods. *Journal of Ecology* 54, 193–215.

Orth, D.J. (1987) Ecological considerations in the development and application of instream flow–habitat models. *Regulated Rivers* 1, 171–181.

Pachur, H.J. and Kröpelin, S. (1987) Wadi Howar: paleoclimatic evidence from an extinct system in the southeastern Sahara. *Science* 237, 298–300.

Pacini, N. (1994) Coupling of land and water: phosphorus fluxes in the Upper Tana River catchment. PhD thesis, University of Leicester, Leicester, UK.

Pacini, N. and Gächter, R. (1999) Speciation of riverine particulate phosphorus during rain events. *Biogeochemistry* 47, 87–109.

Pacini, N. and Harper, D.M. (2000) River conservation in Central and Tropical Africa. In: Boon, P.J., Davies, B.R. and Petts, G.E. (eds) *Global Perspectives on River Conservation: Science, Policy and Practice.* John Wiley and Sons Ltd, Chichester, UK, pp. 155–178.

Pacini, N. and Harper, D.M. (2008) Aquatic, semi-aquatic and riparian vertebrates. In: Dudgeon, D. (ed.) *Tropical Stream Ecology.* Academic Press, New York, New York, pp. 24–43.

Pacini, N., Harper, D.M. and Mavuti, K.M. (1993) A sediment-dominated tropical impoundment: Masinga Dam, Kenya. *Verhandlungen der internationale Vereinigung theoretische und angewandte Limnologie* 25, 1275–1279.

Pacini, N., Harper, D.M. and Mavuti, K.M. (1998) Hydrological and ecological considerations in the management of a catchment controlled by a reservoir cascade: the Tana river, Kenya. In: Harper, D.M. and Brown, A.G. (eds) *The Sustainable Management of Tropical Catchments.* Wiley Interscience, New York, New York, pp. 239–258.

Palczynski, A. (1996) Paleophytosociological research on peatlands in the drainage basin of the rivers Dajna and Jorka (Masurian Lakeland, Poland). *Zeszyty Naukowe Komitetu Polskiej Akademii Nauk „Człowiek i Srodowisko"* 13, 399–421 (in Polish with English summary).

Pall, K. and Janauer, G.A. (1995) Die Makrophytenvegetation von Flußstauen am Beispiel der Donau zwischen Fluß-km 2552,0 und 2511,8 in der Bundesrepublik Deutschland. *Archive für Hydrobiologie* 9(Suppl. Large Rivers), 91–109.

Pall, K. and Janauer, G.A. (1998) Makrophyteninventar der Donau. Totalinventarisierung der Makrophytenvegetation des österreichischen Abschnittes. *Schriftenreihe Forschung im Verbund* 38, 116.

Pall, K., Rath, B. and Janauer, G.A. (1996) Die Makrophyten in dynamischen und abgedämmten Gewässersystemen der Kleinen Schüttinsel (Donau-Fluß-km 1848 bis 1806) in Ungarn. *Limnologica* 26, 105–115.

Parde, A. (1955) *Fleuves et rivers.* Colin, Paris.

Pardo, I. and Armitage, P.D. (1997) Species assemblages as descriptors of mesohabitats. *Hydrobiologia* 344, 111–128.

Park, R.A., Groden, T.W. and Desormeau, C.J. (1979) Modification to model CLEANER, requiring further research. In: Scavia, D. and Robertson, A. (eds) *Perspectives on*

Lake Ecosystem Modelling. Ann Arbor Science Publishers, Ann Arbor, Michigan, pp. 87–108.

Park-Lee, Y-O. (1986) Beitrag zur Erfassung der jahreszeitlichen Veränderungen der Entrobacteriaceen auf submersen Makrophyten und im freien Wasser. PhD thesis, University of Vienna, Vienna.

Pastoors, M.J.H., Lieste, R. and Kovar, K. (1993) Toepassing van landelijk grondwater model voor milieutoekomstverkenningen (Application of a nation-wide groundwater model for environmental enquiries of the future). In: Hooghart, J.C. (ed.) *Gebruik van GIS bij grondwatermodellering*. TNO Committee on Hydrological Research, Delft, The Netherlands, pp. 93–116.

Patten, B.D. (1979) Summary report of module B – instream fishery ecosystem. *Instream Flow Criteria and Modeling Workshop*. Exp. Stn. IS No. 40. Colorado State University, Fort Collins, Colorado.

Patten B.C. and Odum, E.P. (1981) *The cybernetic nature of ecosystems*. The American Naturalist 118, 886–895.

Paugy, D., Traoré, K. and Diouf, P.S. (1994) Faune ichtyologique des eaux douces d'Afrique. In: Teugels, G.G., Guégan, J.F. and Albaret, J.J. (eds) Proceedings of a Symposium on Biological diversity of African fresh- and brackish water fishes. *Annales du Musée Royal de l'Afrique Centrale, Zoologie* 275, 35–66.

Paul, J.F. (1976) Modeling the hydrodynamic effects of large man-made modification to lakes. In: Ott, W.R. (ed.) *Proceedings of the EPA Conference on Environmental Modeling and Simulation*, Cincinnati, 19–22 April. EPA 600/9-76-016. US Environmental Protection Agency, Washington, DC.

Pavese, M.P., Banzon, V., Colacino, M., Gregori, G.P. and Pasqua, M. (1992) Three historical data series on floods and anomalous climatic events in Italy. In: Bradley, R.S. and Jones, P.D. (eds) *Climate Since AD 1500*. Routledge, London, pp. 155–170.

Penczak, T. (1985) Phosphorus, nitrogen, and carbon cycling by fish populations in two small lowland rivers in Poland. *Hydrobiologia* 120, 159–165.

Penczak, T. (1992) Contribution to energy transformation by fish populations ina small tropical river, North Venezuela. *Comparative Biochemistry and Physiology* 101A, 791–798.

Penczak, T. (1999) Fish production and food consumption in the Warta River (Poland): continued post-impoundment study (1990–1994). *Hydrobiologia* 416, 107–123.

Penczak, T., Zalewski, M. and Pfeifer, K. (1977) The ecology of roach, *Rutilus rutilus* (L.), in the barbel region of the polluted Pilica River. IV. Production and food consumption. *Ekologia polska* 25, 241–255.

Penczak, T., Galicka, W., Molinski, M., Kusto, E. and Zalewski, M. (1982) The enrichment of a mesotrophic lake by carbon, phosphorus and nitrogen from the cage aquaculture of rainbow trout, *Salmo gairdneri*. *Journal of Applied Ecology* 19, 371–393.

Penczak, T., Kusto, E., Krzyanowska D., Molinski, M. and Suszycka, E. (1984) Food consumption and energy transformations by fish populations in two small lowland rivers in Poland. *Hydrobiologia* 108, 135–144.

Penczak, T., Agostinho, A.A., Hahn, N.S., Fugi, R. and Gomes, L.C. (1999) Energy budgets of fish populations in two tributaries of the Paraná River, Paraná, Brazil. *Journal of Tropical Ecology* 15, 159–177.

Penczak, T., Galicka, W., Głowacki, Ł. and Koszaliński, H. (2001) The importance of fish growth and consumption on the nutrient budget of the impounded Warta River. *Archive of Hydrobiology* 139(Suppl. 1), 117–138.

Pennak, R.W. (1971) Toward a classification of lotic habitats. *Hydrobiologia* 38, 321–334.

Pennington, W. (1981) Record of a lake's life in time: the sediments. *Hydrobiologia* 79, 197–215.

Persson, L. (1988) Asymmetry in competitive and predatory interactions in fish populations. In: Ebenman, B. and Persson, L. (eds) *Size-structurised Populations – Ecology and Evolution*. Springer-Verlag, Heidelberg, Germany, pp. 203–218.

Persson, L. and Eklöv, P. (1995) Prey refuges affecting interactions between piscivorous perch and juvenile perch and roach. *Ecology* 76, 70–81.

Peters, D.L., Buttle, J.M., Taylor, C.H. and La Zerte, B.D (1995) Runoff production in a forested, shallow soil, Canadian Shield Basin. *Water Resources Research* 31, 1291–1304.

Peters, E.J., Holland, R.S., Callam, M.A. and Bunnell, D.L. (1989) *Platte River Suitability Criteria . . . Habitat Utilization, Preference and Suitability Index Criteria for Fish and Aquatic Invertebrates of the lower Platte River*. Nebraska Technical Series No. 17. Nebraska Game and Parks Commission, Lincoln, Nebraska.

Petersen, R.C., Petersen, L.B.M. and Lacoursiere, J. (1992) A building-block model for stream restoration. In: Boon, P.J., Calow, P. and Petts, G.E. (eds) *River Conservation and Management*. John Wiley and Sons Ltd., Chichester, Ltd, pp. 293–309.

Peterson, C.G. (1986) Effects of discharge reduction on diatom colonization below a large hydroelectric dam. *Journal of the North American Benthological Society* 5, 278–289.

Peterson, C.G. (1987) Influences of flow regime on development and desiccation response of lotic diatom communities. *Ecology* 68, 946–954.

Petts, G.E. (1984) *Impounded Rivers, Perspectives for Ecological Management*. John Wiley and Sons Ltd, Chichester, UK.

Petts, G. and Amoros, C. (eds) (1996a) *Fluvial Hydrosystems*. Chapman and Hall, London.

Petts, G.E. and Amoros, C. (1996b) Fluvial hydrosystems: a management perspective. In: Petts, G.E. and Amoros, C. (eds) *Fluvial Hydrosystems*. Chapman and Hall, London, UK, pp. 263–278.

Pfeifer, R.F. and McDiffett, W.F. (1975) Some factors affecting primary productivity of stream riffle communities. *Archiv für Hydrobiologie* 75, 306–317.

Pfister, C. (1992) Monthly temperate and precipitation in central Europe 1525–1979: quantifying documentary evidence on weather and its effects. In: Bradley, R.S. and Jones, P.D. (eds) *Climate Since 1500 AD*. Routledge, London, pp. 118–142.

PHARE (1998) *Water Management Development Project Ráckeve–Soroksár Danube Branch, Final Report*. OSS No. HU9407-03-L001, VITUKI, Budaest, p. 191.

Phillips, G.L., Eminson, D. and Moss, B. (1978) A mechanism to account for macrophyte decline in progressively eutrophicated freshwaters. *Aquatic Botany* 4, 103–126.

Pieterse, A.H. (1993a) Introduction. In: Pieterse, A.H. and Murphy, K.J. (eds) *Aquatic Weeds. The Ecology and Management of Nuisance Aquatic Vegetation*. Oxford University Press, Oxford, UK, pp. 3–16.

Pieterse, A.H. (1993b) Biological control of aquatic weeds. In: Pieterse, A.H. and Murphy, K.J. (eds) *Aquatic Weeds. The Ecology and Management of Nuisance Aquatic Vegetation*. Oxford University Press, Oxford, UK, pp. 174–177.

Pieterse, A.H. and Murphy, K.J. (eds) (1993) *Aquatic Weeds. The Ecology and Management of Nuisance Aquatic Vegetation*. Oxford University Press, Oxford, UK, 593 pp.

Pionke, H.B., Gburek, W.S., Schnabel, R.R., Sharpley, A.N. and Elwinger, G.F. (1999) Seasonal flow, nutrient concentrations and loading patterns in stream flow draining an agricultural hill–land watershed. *Journal of Hydrology* 220, 62–73.

Pitlo, R.H. (1982) Flow resistance of aquatic vegetation. In: *Proceedings of the 6th EWRS Symposium on Aquatic Weeds*. European Weed Research Society, Doorwerth, The Netherlands, pp. 225–234.

Pitlo, R.H. (1986) Towards a larger capacity of vegetated channels. In: *Proceedings of the 7th EWRS/AAB Symposium on Aquatic Weeds*. European Weed Research Society, Doorwerth, The Netherlands, pp. 245–250.

Pitlo, R.H. and Dawson, F.H. (1993) Flow-resistance of aquatic weeds. In: Pieterse, A.H. and Murphy, K.J. (eds) *Aquatic Weeds. The Ecology and Management of Nuisance Aquatic Vegetation.* Oxford University Press, Oxford, UK, pp. 74–84.

Platania, S.P. (1991) Fishes of the Rio Chama and upper Rio Grande, New Mexico, with preliminary comments on their longitudinal distribution. *Southwestern Naturalist* 36, 186–193.

Platts, W.S. (1979) Relationships among stream order, fish populations, and aquatic geomorphology in an Idaho river drainage. *Fisheries* 4, 5–9.

Poff, N.L. and Ward, J.V. (1989) Implications of streamflow variability and predictability for lotic community structure: a regional analysis of streamflow patterns. *Canadian Journal of Fisheries and Aquatic Science* 46, 1805–1818.

Polprasert, C., Edwards, P., Rajput, V.S. and Pacharrakiti, C. (1986) Integrated biogas technology in the tropics. I. Performance of small-scale digesters. *Solid Waste Management Research* 4, 197–213.

Popa, A. (1993) Liquid and sediment inputs of the Danube river into the north-western Black Sea. *Mitteilungen aus dem Geologisch-Paleontologischen Intitut der Universität Hamburg* 74, 137–149.

Prandtl, L. (1904) Über flüssigkeitsbewegung bei sehr kleiner Reibung. In: *Verhandlungen III Internationalen Mathematiker Kongress Heidelberg*, pp. 484–491.

Preobrajensky, V.S., Muhina, L.I. and Kazanskaya, N.C. (1975) Methodical indications of natural conditions of recreational region. In: Preobrajensky, V.S. (ed.) *Geographical Problems of Tourism and Rest Organisation.* Naouka, Moscow, pp. 50–112 (in Russian).

Puckridge, J.T., Sheldon, F., Walker, K.F. and Boulton, A.J. (1998) Flow variability and the ecology of large rivers. *Marine and Freshwater Research* 49, 55–72.

Pyrozhnikov, P.L. (1932) *Investigation and Utilisation of Siberian Water Bodies.* Selhozizdat, Moscow (in Russian).

Quinn, W.H. and Neal, V.T. (1992) The historical record of El Nino events. In: Bradley, R.S. and Jones, P.D. (eds) *Climate Since AD 1500.* Routledge, London, pp. 623–649.

Rabotnov, F.V. (1984) *The Meadow Cultivation.* Moscow University Publisher, Moscow (in Russian).

Rahm, L., Conley, D., Sandén, P., Wulff, F. and Stålnacke, P. (1996) A time series analysis of nutrient inputs to the Baltic Sea and changing DSi/N ratios. *Marine Ecology Progress Series* 130, 221–228.

Rapp, A. (1987) Desertification. In: Gregory, K.J. and Walling, D.E. (eds) *Human Activity and Environmental Changes.* John Wiley and Sons Ltd, Chichester, UK, pp. 425–443.

Raven, P.J., Holmes, N.T.H., Dawson, F.H. and Everard, M. (1998) Quality assessment using River Habitat Survey data. *Aquatic Conservation – Marine and Freshwater Ecosystems* 8, 477–499.

Razavian, D. (1990) Hydrological responses of an agricultural watershed to various hydrological and management conditions. *Water Resources Bulletin* 26, 777–785.

Rechkow, K.H. (1979) *Quantitative Techniques for the Assessment of Lake Quality.* EPA-440/5-79-015. US Environmental Protection Agency, Washington, DC, p. 146.

Reddy, K.R., Kadlec, R.H., Flaig, E. and Gale, P.M. (1999) Phosphorus retention in streams and wetlands: a review. *Critical Reviews in Environmental Science and Technology* 29, 83–146.

Redman, C.L., Grove, M.J. and, Kuby, L.H. (2004) Integrating social science into the Long-Term Ecological Research (LTER) network: social dimensions of ecological change and ecological dimensions of social change. *Ecosystems* 7, 161–171.

Regier, H.A. and Meisner, J.D. (1990) Anticipated effects of climate change on freshwater fish and their habitat. *Fisheries* 15, 10–15.

Rekolainen, S. (1989) Phosphorus and nitrogen load from forest and agricultural areas in Finland. *Aqua Fennica* 12, 95–10.

Resh, V.H., Brown, A.V., Covich, A.P., Gurtz, M.E., Li, H.W., Minshall, G.W., Reice, S.R., Sheldon, A.L., Wallace, J.B. and Wissmar, R. (1988) The role of disturbance in stream ecology. *Journal of the North American Benthological Society* 7, 433–455.

Reynolds, C.S. (1984) *The Ecology of Freshwater Phytoplankton*. Freshwater Biological Association and University Press, Cambridge, UK.

Rezabek, H. and Schügerl, W.-D. (1999) The New Danube – the quantitative effect on the left bank hinterland. In: Weber, G., Wimmer, W., Mager, G., Klotz, A. and Neckar, H. (eds) *Perspektiven, 2-3/1999*. Gewässergüteforschung Neue Donau, Vienna, pp. 62–74 (English version p. 95).

Ribbink, A.J. (1994) Biodiversity and speciation of freshwater fishes with particular reference to African cichlids. In: Giller, P.S., Hildrew, A.G. and Raffaelli, D.G. (eds) *Aquatic Ecology: Scale, Pattern and Process*. Blackwell Scientific Publications, Oxford, UK, pp. 261–268.

Richter, B.D., Baumgartner, J.V., Powell, J. and Braun, D.P. (1996) A method for assessing hydrologic alteration within ecosystems. *Conservation Biology* 10, 1163–1174.

Richter, B.D., Baumgartner, J.V., Wiggington, R. and Braun, D.P. (1997) How much water does a river need? *Freshwater Biology* 37, 231-249.

Ripl, W. (2003) Water – the bloodstream of the biosphere. In: Falkenmark, M. and Folke, C. Freshwater and Welfare Fragility – Syndromes, Vulnerability and Challenges. Freshwater Special Issue. *Philosophical Transactions of the Royal Society of London B* 358, 1921–1934.

Ripl, W. and Wolter, KD (2002) Ecosystem function and degradation. In: Williams, J. le B., Thomas, D.N. and Reynolds, C.S. (eds) *Phytoplankton Productivity – Carbon Assimilation in Marine and Freshwater Ecosystems*. Blackwell Science Ltd, Oxford, UK, pp. 291–317.

Ripl, W., Pokorny, J. and Scheer, H. (unpublished report) Memorandum on Climate Change. The necessary reforms of society to stabilize the climate and solve the energy issues.

Robert, J.L., Anctil, F. and Ouellet, Y. (1996) Two layer finite element numerical model for simulating salmon river estuaries. In: Leclerc, M., Capra, H., Valentin, S., Boudreault, A. and Coté, Y. (eds) *Ecohydraulics 2000. Proceedings of the 2nd International Symposium on Habitat Hydraulics*. INRS-Eau, Québec City, Canada, pp. 155–166.

Robertson, K.M. and Augspurger, C.K. (1999) Geomorphic processes and spatial patterns of primary forest succession on the Bogue Chitto River, USA. *Journal of Ecology* 87, 1052–1063.

Rodriguez-Iturbe, I. (2000) Ecohydrology: a hydrological perspective of climate–soil–vegetation dynamics. *Water Resource Research* 36, 3–9.

Roggeri, H. (ed) (1995) *Tropical Freshwater Wetlands – A Guide to Current Knowledge and Sustainable Management*. Kluwer Academic Publishers, Dordrecht, The Netherlands, 349 pp.

Romanenko, V.D., Oksiyuk, O.P. and Zhukinsky V.N. (1990) *Ecological Assessment of the Impact Hydrotechnical Structures on the Water Bodies*. Naukova Dumka, Kyiv (in Russian).

Romanowska-Duda, Z., Lupa, D., Skóra, M. and Knypl, J.S. (1998) Possible ecophysiological significance of phosphatase activity in Sulejów Aquene in connection with the blue-green algae (Cyanobacteria) blooms. *Acta Physiologia Plantarum* Suppl. 1, 67.

Rossi, G. and Premazzi, G. (1991) Delay in lake recovery caused by internal loading. *Water Research* 25, 567–575.

Rotnicki, K. (1991) Retrodiction of palaeodischarges of meandering and sinuous alluvial rivers and its palaeohydroclimatic implications, In: Starkel, L., Gregory, K.J. and

Thornes, J.B. (eds) *Temperate Palaeohydrology*. John Wiley and Sons Ltd, Chichester, UK, pp. 431–471.

Rowntree, K.M. and Wadeson, R.A. (1996) Translating channel morphology into hydraulic habitat: application of the hydraulic biotope concept to an assessment of discharge related habitat changes. In: Leclerc, M., Capra, H., Valentin, S., Boudreault, A. and Coté, Y. (eds) *Ecohydraulics 2000. Proceedings of the 2nd International Symposium on Habitat Hydraulics*. INRS-Eau, Québec City, Canada, pp. 281–292.

Różanski, K., Goslar, T., Dulinski, M., Kuc, T., Pazdur, M.F. and Walanus, A. (1992) The late-Glacial–Holocene transition in laminated sediments from Lake Gosciaz (central Poland), In: Bard, E. and Broecker, W. (eds) *The Last Deglaciation: Absolute and Radiocarbon Chronologies*. NATO Advanced Research Workshop, Erice, Sicily, December 1990. Springer, Berlin, pp. 69–80.

Rtshiladze, S.I. (1951) Pollution and selfpurification of the Kura River. In: *Sanitary Characteristics of Water Bodies*. Stroiizdat, Moscow, pp. 89–98 (in Russian).

Rudoy, A.N. and Baker, V.R. (1993) Sedimentary effects of cataclysmic late Pleistocene glacial outburst flooding, Altay Mountains, Siberia. *Sedimentary Geology* 85, 53–63.

Ruhlov, F.N. (1973) About river period of Siberian salmon life. *Proceedings of the Pacific Research Institute* 91, 25–30 (in Russian).

Ruhlyadev, U.P. (1964) Dependence of biomass and zooplankton flow from basic factors of environment. *Zoological Journal* 13, 3–16.

Rulon, J.J., Rodwy, R. and Freeze, R.A. (1985) The development of multiple seepage faces on layered slopes. *Water Resources Research* 21, 1625–1636.

Runhaar, J. (1999) Impact of hydrological changes on nature conservation areas in the Netherlands. PhD thesis, University of Leiden, Leiden, The Netherlands.

Runhaar, J. and Udo de Haes, H.A. (1994) Site factors as classification characteristics. In: Klijn, F. (ed.) *Ecosystem Classification for Environmental Management*. Kluwer, Dordrecht, The Netherlands, pp. 139–172.

Runhaar, J., Groen, C.L.G., van der Meijden, R. and Stevers, R.A.M. (1987) Een nieuwe indeling in ecologische groepen binnen de Nederlandse flora (A new division in ecological groups in the flora of the Netherlands). *Gorteria* 13, 276–359.

Runhaar, J., Witte, J.P.M. and Jongman, R.H.G. (1994) Ellenberg-indicatiewaarden: verbetering met reciprocal averaging? (Ellenberg's indicator values: improvement by reciprocal avaraging?) *Landschap* 11, 41–47.

Runhaar, J., van Gool, C.R. and Groen, C.L.G. (1996a) Impact of hydrological changes on nature conservation areas in the Netherlands. *Biological Conservation* 76, 269–276.

Runhaar, J., Witte, J.P.M. and van der Linden, M. (1996b) Waterbeheer en natuur. Effectvoorspelling met het landelijke model DEMNAT (Water management and nature. Predicting effects on wet and moist ecosystems). *Landschap* 13, 65–77.

Runhaar, J., Witte, J.P.M. and Verburg, P.H. (1997a) Ground-water level, moisture supply, and vegetation in the Netherlands. *Wetlands* 17, 28–38.

Runhaar, J., van Ek, R., Bos, H.B. and van't Zelfde, M. (1997b) *Dosis-effectmodule DEMNAT versie 2.1 (Dose–Effect Module of DEMNAT Version 2.1)*. RIZA, Lelystad, The Netherlands.

Ruttenberg, K. and Berner, R.A. (1993) Authigenic apatite formation and burial in sediments from non-upwelling continental margin environments. *Geochimica et Cosmochimica Acta* 576, 991–1007.

Rybak, J. (2000) Long-term and seasonal dynamics of nutrient export rates from lake watersheds of diversified land cover pattern. *Verhandlungen des Internationalen Vereins für Limnologie* 27, 3132–3136.

Rybak, J. (2002) Seasonal and long-term export rate of nutrients with surface runoff in the river Jorka catchment basin (Masurian Lakeland, Poland). *Polish Journal of Ecology* 50, 439–458.

Ryszkowski, L. (1992) Energy and material flows across boundaries in agricultural land-scapes. In: Hansen, J.A. and di Castri, F. (eds) *Landscape Boundaries: Consequences for Biotic Diversity and Ecological Flows. Ecological Studies*, Vol. 92. Springer-Verlag, Berlin, pp. 270–284.

Ryszkowski, L., Bartoszewicz, A. and Kędziora, A. (1997) The potential role of mid-field forests as buffer zones. In: Haycock, N.E., Burt, T.P., Goulding, K.W.T. and Pinay, G. (eds) *Buffer Zones: Their Processes and Potential in Water Protection*. Quest Environ-mental, Harpenden, UK, pp. 171–191.

Rytelewski, J., Solarski, H., Korybut-Daszkiewicz, S. and Mirowski, Z. (1985) Hydrologi-cal and pedological studies in the Jorka River watershed. *Polish Ecological Studies* 11, 187–200.

Rzepecki, M. (2000) Wetlands in lake protection: nutrient dynamics and removal in ecotones of a river-lake system (Masurian Lakeland, Poland). *Verhandlungen des Internationalen Vereins für Limnologie* 27, 1685–1689.

Rzepecki, M. (2002) Wetland zones along lake shores as barrier systems: field and ex-perimental research on nutrient retention and dynamics. *Polish Journal of Ecology* 50, 527–542.

Saint-Hilaire, A., Caissie, D., Morin, G. and El-Jab, N. (1996) Importance of hydrological models in habitat issues: Comparison between a stochastic and a deterministic at Catamaran Brook (New-Brunswick). In: Leclerc, M., Capra, H., Valentin, S., Boudreault, A. and Coté, Y. (eds) *Ecohydraulics 2000. Proceedings of the 2nd International Sym-posium on Habitat Hydraulics*. INRS-Eau, Québec City, Canada, pp. 409–420.

Salas, H.J. and Martino, P. (1991) A simplified phosphorus trophic state model for warm-water tropical lakes. *Water Research* 25, 341–350.

Sale, M.J. (1985) Aquatic ecosystem response to flow modification: an overview of the issues. In: Olson, F.W., White, R.G. and Hamre, R.H. (eds) *Proceedings of the Sym-posium on Small Hydropower and Fisheries*. American Fisheries Society, Bethesda, Maryland, pp. 22–31.

Salem-Murdock, M. (1996) Social science inputs to water management and wetland con-servation in the Senegal River Valley. In: Acreman, M.C. and Hollis, G.E. (eds) *Water Management and Wetlands in sub-Saharan Africa*. IUCN, Gland, Switzerland, pp. 125–144.

Sandén, P., Rahm, L. and Wulff, F. (1991) Non-parametric trend test of Baltic Sea data. *Environmetrics* 2, 263–278.

Sand-Jensen, K. and Gordon, D.M. (1984) Differential ability of marine and freshwater macrophytes to utilize HCO_3 and CO_2. *Marine Biology* 80, 247–523.

Salvucci, G.D. and Entekhabi, D. (1995) Hillslope and climatic controls on hydrologic fluxes. *Water Resources Research* 31, 1725–1739.

Sas, H. (1989) *Lake Restoration by Reduction of Nutrient Loading: Expectations, Experi-ences, Extrapolations*. Academia Verlag, Sankt Augustin, Germany.

Sasi, M.N. (1994) A relationship between equatorial lower stratospheric QBO and El Niño. *Journal Atmospheric and Terrestrial Physics* 56, 1563–1570.

Scarnecchia, D.L. (1988) The importance of streamlining in influencing fish community structure in channelized and unchannelized reaches of a prairie stream. *Regulated Rivers* 5, 155–166.

Scarsbrook, M.R. and Townsend, C.R. (1993) Stream community structure in relation to spatial and temporal variation: a habitat templet study of two contrasting New Zealand streams. *Freshwater Biology* 29, 395–410.

Schick, A.P. (1988) Hydrologic aspects of floods in extreme arid environments. In: Baker, V.R., Kochel, R.C. and Patton, P.C. (eds) *Flood Geomorphology*. John Wiley and Sons Ltd, Chichester, UK, pp. 189–203.

Schiel, W. (1999) The New Danube: the entire area – the whole problem. In: Weber, G., Wimmer, W., Mager, G., Klotz, A. and Neckar, H. (eds) *Perspektiven, 2-3/1999*. Gewässergüteforschung Neue Donau, Vienna, p. 1826 (English version pp. 84–85).

Schiemer, F., Spindler, T., Wintersberger, H., Schneider, A. and Chovanec, A. (1991) Fish fry associations: important indicators for the ecological status of large rivers. *Verhandlungen des Internationalen Vereins für Limnologie* 24, 2497–2500.

Schiemer, F., Keckeis, H. and Flore, L. (2001) Ecotones and hydrology: key conditions for fish in large river. *Ecohydrology and Hydrobiology* 1, 49–55.

Schirmer, W. (1983) Die Talentwicklung an Main und Regnitz seit dem Hochwürm. *Geologisches Jahrbuch* A71, 11–43.

Schlögel, G. (1997) Die Verbreitung und quantitative Erfassung der Gewässervegetation in der Lobau. M.Sc. thesis, University of Vienna, Vienna.

Schlosser, I.J. (1985) Flow regime, juvenile abundance, and the assemblage structure of stream fishes. *Ecology* 66, 1484–1490.

Schmidt, B., Straif, M., Waidbacher, H. and Janauer, G.A. (2006) Man-made near-natural structures offer new habitats to macrophytes, as well as fish, in the Austrian Danube (Vienna, hydro-power plant Freudenau). In: *36th IAD Conference (CD)*. Austrian Committee Danube Research/International Association Danube Research, Vienna, pp. 112–116

Schneider, S.H. and Boston, P.J (1991) *Scientists on Gaia*. MIT Press, Cambridge.

Schrödinger, E. (1944) *What is life?* Cambridge University Press, Cambridge.

Schumm, S.A. (1965) Quaternary palaeohydrology. In: Wright, H.E. and Frey, D.G. (eds) *The Quaternary of the United States*. Princeton University Press, Princeton, New Jersey, pp. 783–794.

Schumm, S.A. (1977) The Fluvial System. John Wiley and Sons Ltd, Chichester, UK

Schumm, S.A. (1981) Evolution and response of the fluvial system, sedimentologic implications. *Society of Economic Paleontologists and Mineralogists Special Publication* 31, 19–29.

Schwencke-Hofmann, J. (1987) Jahreszeitliche Schwankungen in der Zusammensetzung des Phytoplanktons und Phytobenthos in Altwässern der Lobau bei Wien. *Archive für Hydrobiologie Supplement* 68, 269–308.

Schwertner, R. (1995) *Lemna minor*, ein geeigneter Organismus für Ökotoxicitätstests. Diploma thesis, University of Vienna, Vienna.

Schwöppe, W. (1994) Die lanschaftökologischen Veränderungen im Bereich des Nationalparkes Djoudj (Senegal). PhD thesis, University of Hamburg, Hamburg, Germany.

Scott, D. and Shirvell, C.S. (1987) A critique of the instream flow incremental methodology and observations on flow determination in New Zealand. In: Craig, J.F. and Kemper, J.B. (eds) *Regulated Streams: Advances in Ecology*. Plenum Press, New York, New York, pp. 27–43.

Sculthorpe, C.D. (1967) *The Biology of Aquatic Vascular Plants*. E. Arnold, London. Reprinted by Koeltz Scientific, Königstein, Germany, 1985.

Seager, J. (ed.) (1990) *The State of the World Atlas*. Simon and Schuster, Inc., New York, New York.

Sedell, J.R., Richey, J.E. and Swanson, F.J. (1989) The river continuum concept: a basis for the expected ecosystem behaviour of very large rivers? In: Dodge, D.P. (ed.) Proceedings of the International Large River Symposium. *Canadian Special Publication of Fisheries and Aquatic Science* 106, 49–55.

Selby, M.J. (1987) Slopes and weathering. In: Gregory, K.J. and Walling, D.E. (eds) *Human Activity and Environmental Processes*. John Wiley and Sons Ltd, Chichester, UK, pp. 183–205.

Semazzi, F.H.M., Mehta, V. and Sud, Y.C. (1988) An investigation of the relationship between sub-Saharan rainfall and global sea surface temperatures. *Atmosphere–Ocean* 26, 118–138.

Shanahan, P. and Harleman, D.R.F. (1986) Lake eutrophication model: coupled hydro-physical–ecological model. In: Somlyódy, L. and van Straten, G. (eds) *Modeling and Managing Shallow Lake Eutrophication*. Springer Verlag, Berlin, pp. 256–285.

Shanahan, P. and Somlyody, L. (1995) *Modelling the Impact of Diffuse Pollution on Receiving Water Quality*. IIASA Working Paper, WP-95-2. International Institute for Applied Systems Analysis, Laxenburg, Austria, 53 pp.

Sharpley, A.N. and Smith, S.J. (1990) Phosphorus transport in agricultural runoff: the role of soil erosion. In: Boardman, J., Foster, I.D.L. and Dearing, J.A. (eds) *The Selective Erosion of Plant Nutrients in Runoff*. John Wiley and Sons Ltd, Chichester, UK, pp. 351–366.

Sharpley, A.N., Chapra, S.C., Wedepohl, R., Sims, J.T., Daniel, T.C. and Reddy, K.R. (1994) Managing agricultural phosphorus for protection of surface waters: issues and options. *Journal of Environmental Quality* 23, 437–451.

Sheldon, A.L. (1968) Species diversity and longitudinal succession in stream fishes. *Ecology* 49, 193–198.

Shiklomanov, I.A. (1990) Global water resources. *Nature and Resources* 26(3), 34–43.

Shrag, V.I. (1969) *Floodplain Soils, Their Amelioration and Agriculture Utilization*. Agro-promizdat, Moscow, 269 pp. (in Russian).

Shrestha P. (1997a) Ecological study on the aquatic macrophyte vegetation of Lake Phewa and Lake Rupa, Nepal. PhD thesis, University of Vienna, Vienna.

Shrestha, P. (1997b) Selected bibliography of ethnobotanical literatures of Nepal. In: Shresta, K.K., Iha, P.K., Sahengji, P., Rastogi, A., Rajbhandary, S. and Joshi, M. (eds) *Ethnobotany for Conservation and Community Development. Proceedings of the National Training Workshop in Nepal*, Kathmandu, 6–13 January. ESON (Ethnobotanical Society of Nepal), Kathmandu, pp. 121–144.

Shrestha, P. and Janauer, G.A. (2001) Management of aquatic macrophyte resource: a case of Phewa Lake, Nepal. In: Jha, S.R., Baral., S.R., Marmacharya, S.B., Lekhak, H.D., Lacoul., P. and Baniya, C.B. (eds) *Environment and Agriculture: Biodiversity, Agriculture and Pollution in South-Asia*. Ecological Society (ECOS), Kathmandu, pp. 99–107.

Shuler, S.W. and Nehring, R.B. (1993) Using the physical habitat simulation model to evaluate a stream habitat enhancement project. *Rivers* 4, 175–193.

Sidle, R.C., Tsuboyama, Y., Noguchi, S., Hosoda, I., Fujieda, M. and Shimizu, T. (2000) Stormflow generation in steep forested head waters: a linked hydrogeomorphic paradigm. *Hydrological Processes* 14, 369–385.

Sidorchuk, A., Panin, A. and Borisova, O. (2003) The Lateglacial and Holocene palaeo-hydrology of Northern Eurasia. In: Gregory, K.J. and Benito, G. (eds) *Palaeohydrology, Understanding Global Change*. John Wiley and Sons Ltd, Chichester, UK.

Sigg, L. (1987) Surface chemical aspects of the distribution and fate of metal ions in lakes. In: Stumm, W. (ed.) *Aquatic Surface Chemistry: Chemical Processes at the Particle–Water Interface*. Wiley Interscience, New York, New York, Chapter 12.

Simon, A. and Darby, S.E. (eds) (1999) *Incised River Channels: Processes, Forms, Engineering and Management*. John Wiley and Sons Ltd, Chichester, UK.

Simon, A. and Hupp, C.R. (1986) Channel evolution in modified Tennessee channels. In: *Proceedings of the 4th Federal Interagency Sedimentation Conference*, Las Vegas, Nevada. US Government Printing Office, Washington, DC, pp. 5.71–5.82.

Sinclair, A.R. and Arcese, P. (1995) *Serengeti II: Dynamics, Management and Conserva-tion of an Ecosystem*. University of Chicago Press, Chicago, Illinois, 666 pp.

Sivonen K. (1990) Effects of light, temperature, nitrate, orthophosphate and bacteria on growth of and hepatotoxin production by *Oscillatoria* strains. *Applied and Environ-mental Microbiology* 56, 2658–2666.

SIWI (2001) Water security – opportunity for development and cooperation in the Aral Sea area. Proceedings of a SIWI/RSAS/UNIFEM Seminar.

Sloane, W.T., Ewen, J., Kilsby, C.G., Fallows, C.S. and O'Connell, P.E. (1997) A physically based model for large river basins. In: Holly, F.M. and Alsaffar, A. (eds.) *Proceedings of the 27th Congress of the International Association for Hydraulic Research*, San Francisco, California, 10–15 August, pp. 853–858.

Smayda, T.J. (1990) Novel and nuisance phytoplankton blooms in the sea: evidence for a global epidemic. In: Graneli, E., *et al.* (eds) *Toxic Marine Phytoplankton*. Elsevier, Amsterdam, pp 29–40.

SMEC (1977) *Hydrologic Model of the Upper White Nile Basin*. Prepared for the World Meteorological Organization, Geneva. Snowy Mountains Engineering Corporation, Cooma, New Soth Wales, Australia.

Smirnova, N.N. (1986) *Accumulating Capacity of Higher Aquatic Plants in Mouth Re-gions of Rivers. Black Sea North-western Hydrobiology Danube and Limanov of Black Sea*. Naukova Dumka, Kiev, pp. 135–151 (in Russian)

Smith, N.J.H., Serrão, E.A.S., Alvim, P.T. and Falesi, I.C. (1995) *Amazonia: Resiliency and Dynamism of the Land and Its People*. United Nations University Press, New York, New York.

Smolska, E. (1993) Dynamics of contemporary morphogenetic processes in the drainage basin of the Upper Szeszupa River (Suwalski Landscape Park). *Ekologia polska* 41, 27–42.

Smolska, E., Mazurek, Z. and Wojcik, J. (1995) Dynamics of geomorphological processes on slopes as a factor of habitat formation in lakeland landscape. *Zeszyty Naukowe Komitetu Polskiej Akademii Nauk „Człowiek i Srodowisko"* 12, 205–220.

Smayda, T.J. (1990) Novel and nuisance phytoplankton blooms in the sea: evidence for a global epidemic. In: Graneli, E., Sundström, B., Edler, L. and Anderson, D.M. (eds) *Toxic Marine Phytoplankton*. Elsevier, Amsterdam, pp. 20–40.

Soja, R. (1977) Deepening of channel in the light of the cross profile analysis (Carpathian river as example). *Studia Geomorphologica Carpatho-Balcanica* XI, 127–138.

Soja, R. (1994) Quantitative palaeohydrology. *Przegląd Geograficzny* 66, 159–167 (in Polish).

Solanes, M. and Gonzalez-Villarreal, F. (1999) *The dublin principles for water as reflected in a comparative assessment of institutional and legal arrangements for integrated water resources management*. Global Water Partnership/Swedish International Development Cooperation Agency, S105-25 Stockholm, Sweden.

Solovkina, L.M. (1975) *Fish Resources of Komi Republic*. Sever, Syktyvkar, Russia (in Russian).

Somlyódy, L. and van Straten, G. (eds) (1986) *Modeling and Managing Shallow Lake Eutrophication; With Application to Lake Balaton*. Springer Verlag, Berlin, 386 pp.

Soltan, M.E., Awadallah, R.M. and Moalla, S.M.N. (1996) Speciation of major, minor, and trace elements in River Nile mud. *International Journal of Environment and Pollution* 6, 300–305.

Sondergaard, M. and Moss, B. (1998) Impact of submerged macrophytes on phytoplankton in shallow freshwater lakes. In: Jeppesen, E., Sondergaard, Ma., Sondergaard, Mo. and Christoffersen, K. (eds) *The Structuring Role of Submerged Macrophytes in Lakes. Ecological Studies*, Vol. 131. Springer, New York, New York, pp. 115–132.

Sondergaard, M. and Sand-Jensen, K. (1979) Carbon uptake by leaves and roots of *Littorella uniflora* (L) Aschers. *Aquatic Botany* 6, 1–12.

Sonntag, E., Pozzi, D., Penska, K., Zeltner, G.-H., Björk, S. and Kohler, A. (1999) Makrophyten-Vegetation und Standorte im eutrophen Kävlinge-Fluß (Skane, Südschweden). *Berichte des Instituts für Landschafts- und Pflanzenökologie. Universität Hohenheim Beihefte* 9, 113.

Soranno, P.A., Hubber, L., Carpentier, S.P. and Lethrop, R.C. (1996) Phosphorus load to surface waters: a simple model to account for spatial pattern of land use. *Ecological Applications* 6, 865–878.

Soulsby, C. (1995) Contrasts in storm event hydrochemistry in an acidic afforested catchment in upland Wales. *Journal of Hydrology* 170, 159–179.

Southwood, T.R.E. (1977) Habitat, the templet for ecological strategies? *Journal of Animal Ecology* 46, 337–365.

Southwood. T.R.E. (1988) Tactics, strategies and templets. *Oikos* 52, 3–18.

Spencer, H. (1844) Remarks upon the theory of reciprocal dependence in the animal and vegetable creations, as regards its bearing upon paleontology. *The London, Edinburgh and Dublin Philosophical Magazine and Journal of Science* 24, 90–94 (reprinted in Cloud, P. (ed) Adventures in Earth History (1970) 207–209, W.H. freeman, New York.)

Spencer, W. and Bowes, G. (1985) *Limnophila* and *Hygrophila*: a review and physiological assessment of their weedy potential in Florida. *Journal of Aquatic Plant Management* 23, 7–16.

Spencer, W. and Bowes, G. (1993) Ecophysiology of the world's most troublesome aquatic weeds. In: Pieterse, A.H. and Murphy, K.J. (eds) *Aquatic Weeds. The Ecology and Management of Nuisance Aquatic Vegetation.* Oxford University Press, Oxford, UK, pp. 39–73.

Stabel, H.-H. and Geiger, M. (1985) Phosphorus adsorption to riverine suspended matter. *Water Research* 19, 1347–1352.

Stachurski, A. and Zimka, J. (1994) Transfer of elements in the watersheds along the increase of areal of wetland. *Ekologia polska* 42, 34–58.

Stalnaker, C.B., Milhous, R.T. and Bovee, K.D. (1989) Hydrology and hydraulics applied to fishery management in large rivers. In: Dodge, D.P. (ed.) Proceedings of the International Large Rivers Symposium. *Canadian Special Publication of Fisheries and Aquatic Science* 106, 13–30.

Stalnaker, C., Lamb, B.L., Henrikson, J., Bovee, K. and Bartholow, J. (1995) *The Instream Flow Incremental Methodology. A Primer for IFIM.* Biological Report 29. National Biological Service, Washington, DC.

Stanford, J.A., Ward, J.V., Liss, W.J., Frissell, C.A., Williams, R.N., Lichatowich, J.A. and Coutant, C.C. (1996) A general protocol for restoration of regulated rivers. *Regulated Rivers* 12, 391–413.

Starkel, L. (1972) The role of catastrophic rainfall in the shaping of the relief of the Lower Himalaya (Darjeeling Hills). *Geographia Polonica* 21, 103–147.

Starkel, L. (1976) The role of extreme (catastrophic) meteorological events in the contemporaneous evolution of slopes. In: Derbyshire, E. (ed.) *Geomorphology and Climate.* John Wiley and Sons Ltd, Chichester, UK, pp. 203–246.

Starkel, L. (1979) Typology of river valleys in the temperate zone during the last 15,000 years. *Acta Universitatis Ouluensis, Series A* 82, 9–18.

Starkel, L. (1983) The reflection of hydrologic changes in the fluvial environment of the temperate zone during the last 15,000 years. In: Gregory, K.J. (ed.) *Background to Palaeohydrology. A Perspective.* John Wiley and Sons Ltd, Chichester, UK, pp. 213–235.

Starkel, L. (1985) The reflection of Holocene climatic variations in the slope and fluvial deposits and forms in the European mountains. *Ecologia Mediterranea* 11, 91–98.

Starkel, L. (1987) Man as a cause of sedimentological changes in the Holocene. *Anthropogenic Sedimentological Changes during the Holocene Striae* 26, 5-12.

Starkel, L. (1989) Global paleohydrology. *Quaternary International* 2, 25–33.

Starkel, L. (1990a) In: Evolution of the Vistula River Valley During the Last 15 000 Years, Part III. *Geographical Studies Special Issue* 5, 220.

Starkel, L. (1990b) Fluvial environment as an expression of geoecological changes, *Zeitschrift für Geomorphologie* 79(Suppl.), 133–152.

Starkel, L. (1993) Late Quaternary continental paleohydrology as related to future environmental change. *Global and Planetary Change* 7, 95–108.

Starkel, L. (1994) Reflection of the glacial–interglacial cycle in the evolution of the Vistula river basin, Poland. *Terra Nova* 6, 1–9.

Starkel, L. (1995a) Palaeohydrology of the temperate zone. In: Gregory, K.J., Starkel, L. and Baker, V.R. (eds) *Global Continental Palaeohydrology*. John Wiley and Sons Ltd, Chichester, UK, pp. 233–257.

Starkel, L. (1995b) Changes of river channels in Europe during the Holocene. In: Gurnell, A. and Petts, G. (eds) *Changing River Channels*. John Wiley and Sons Ltd, Chichester, UK, pp. 29–42.

Starkel, L. (1996a) Geomorphic role of extreme rainfalls in the Polish Carpathians. *Studia Geomorphologica Carpatho-Balcanica* 30, 21–38.

Starkel, L. (1996b) Palaeohydrological reconstructions: advantages and disadvantages. In: Branson, J., Brown, A.G. and Gregory, K.J. (eds) *Global Continental Changes, The Context of Palaeohydrology*. Special Publication No. 115. Geological Society, London, pp. 9–17.

Starkel, L. (1999) Geomorphic and sedimentologic effects of river floods in space and time. In: *Proceedings of 1996 Conference La difeza dalle alluvioni*, Florence, Italy, pp. 21–45.

Starkel, L. (2003) Short-term hydrological changes, In: Gregory, K.J. and Benito, G. (eds) *Palaeohydrology, Understanding Global Change*. John Wiley and Sons Ltd, Chichester, UK, pp. 337–356.

Starkel, L. and Sarkar, S. (2002) Different frequency of threshold rainfalls transforming the margin of Sikkimese and Bhutanese Himalaya. *Studia Geomorphologica Carpatho-Balcanica* 36, 51–67.

Starkel, L., Gębica, P., Niedziałkowska, E. and Podgórska-Tkacz, A. (1991) Evolution of both the Vistula floodplain and lateglacial–early Holocene palaeochannel systems in the Grobla Forest (Sandomierz Basin). In: Evolution of the Vistula River Valley During the Last 15 000 Years, Part VI. *Geographical Studies Special Issue* 87–99.

Starkel, L., Kalicki, T., Krapiec, M., Soja, R., Gębica, P. and Czyżowska, E. (1996) Hydrological changes of valley floors in upper Vistula basin during late Vistulian and Holocene. In: Evolution of the Vistula River Valley During the Last 15 000 Years. *Geographical Studies Special Issue* 9, 7–128.

Statzner, B. (1981) The relation between 'hydraulic stress' and microdistribution of benthic macroinvertebrates in a lowland running water system, the Schierenseebrooks (North Germany). *Archive für Hydrobiologie* 91, 192–218.

Statzner, B. (1987a) Ökologische Bedeutung der sohlennahen Strömungsgeschwindigkeit für benthische Wirbellose in Fließgewässern. Habilitation, University of Karlsruhe, Karhlsruhe, Germany.

Statzner, B. (1987b) Characteristics of lotic ecosystems and consequences for future research directions. In: Schulze, E.D. and Zwölfer, H. (eds) *Potentials and Limitations of Ecosystem Analysis. Ecological Studies*, Vol. 61. Springer Verlag, Berlin, pp. 365–390.

Statzner, B. (1988) Growth and Reynolds number of lotic macroinvertebrates: a problem for adaptation of shape to drag. *Oikos* 51, 84–87.

Statzner, B. and Higler, B. (1985) Questions and comments on the river continuum concept. *Canadian Journal of Fisheries and Aquatic Science* 42, 1038–1044.

Statzner, B. and Higler, B. (1986) Stream hydraulics as a major determinant of benthic invertebrate zonation patterns. *Freshwater Biology* 16, 127–139.

Statzner, B. and Holm, T.F. (1982) Morphological adaptations of benthic invertebrates to stream flow – an old question studied by means of a new technique (laser Doppler anemometry). *Oecologia* 53, 290–292.

Statzner, B. and Holm, T.F. (1989) Morphological adaptation of shape to flow: microcurrents around lotic macroinvertebrates with known Reynolds numbers at quasi-natural flow conditions. *Oecologia* 78, 145–157.

Statzner, B., Gore, J.A. and Resh, V.H. (1988) Hydraulic stream ecology: observed patterns and potential applications. *Journal of the North American Benthological Society* 7, 307–360.

Statzner, B., Gore, J.A. and Resh, V.H. (1998) Monte Carlo simulation of benthic macroinvertebrate populations: estimates using random, stratified, and gradient sampling. *Journal of the North American Benthological Society* 17, 324–337.

Stephan, U. and Wychera, U. (1996) Analysis in flow velocity fluctuations in different macrophyte banks in a natural open channel. In: Leclerc, M., Capra, H., Valentin, S., Boudreault, A. and Coté, Y. (eds) *Ecohydraulics 2000. Proceedings of the 2nd International Symposium on Habitat Hydraulics*. INRS-Eau, Québec City, Canada, pp. 191–202.

Stockton, C.W. (1990) Climatic hydrologic and water supply inferences from tree rings. *Civil Engineering Practice* Spring, 37–52.

Straskraba, M, (1994) Vltava cascade as teaching grounds for reservoir limnology. *Water Scienec and Technology* 30, 289–297.

Street-Perrott, F.A. and Harrison, S.P. (1985) Lake levels and climatic reconstruction. In: Hecht, A.D. (ed.) *Paleoclimate Analysis and Modeling*, John Wiley and Sons Ltd, Chichester, UK, pp. 291–340.

Stromberg, J.C. (1993) Instream flow models for mixed deciduous riparian vegetation within a semiarid region. *Regulated Rivers: Research and Management* 8, 225–235.

Stromberg, J.C. and Patten, D.T. (1990) Riparian vegetation instream flow requirements: a case study from a diverted stream in the Eastern Sierra Nevada, California, USA. *Environmental Management* 14, 185–194.

Stromberg, J.C. and Patten, D.T. (1991) Instream flow requirements for cottonwoods at Bishop Creek, Inyo County, California. *Rivers* 2, 1–11.

Stromberg, J.C. and Patten, D.T. (1996) Instream flow and cottonwood growth in the Eastern Sierra Nevada of California, USA. *Regulated Rivers: Research and Management* 12, 1–12.

Stumm, W. and Morgan, J.J. (1981) *Aquatic Chemistry*. John Wiley and Sons, New York, New York.

Sweeney, B.W. and Schnack, J.A. (1977) Egg development, growth, and metabolism of *Sigara alternata* (Say) (Hemiptera: Corixidae) in fluctuating thermal environments. *Ecology* 58, 265–277.

Sweeting, R.A. (1994) River pollution. In: Calow, P. and Petts, G.E. (eds) *The Rivers Handbook: Hydrological and Ecological Principles*. Blackwell Scientific, Oxford, UK.

Szilas, C.P., Borggaard, O.K., Hansen, H.C.B. and Rauer, J. (1998) Potential iron and phosphate mobilisation during flooding of soil material. *Water, Air and Soil Pollution* 106, 97–109.

Szumański, A. (1983) Paleochannels of large meanders in the river valleys of the Polish Lowland. *Quaternary Studies in Poland* 4, 207–216.

Tachet, H., Pierrot, J.P., Roux, C. and Bournaud, M. (1992) Net-building behaviour of six *Hydrophsyche* species (Trichoptera) in relation to current velocity and distribution along the Rhône river. *Journal of the North American Benthological Society* 11, 350–365.

Tansley, A. (1935) The use and abuse of vegetational concepts and terms. *Ecology* 16, 284–307.

Tarczyńska, M., Wagner, I., Romanowska-Duda, Z. and Zalewski, M. (1999) Identification of factors determining the occurrence of toxic cyanobacterial blooms as a tool to effective restoration and management in lowland reservoir, Poland. *Sustainable Lake Management* I, s.S2A-6.

Tarczynska, M., Izydorczyk, K. and Zalewski, M. (2001) Optimization of monitoring strategy for eutrophic reservoirs with toxic cyanobacterial blooms. In: *Proceedings of 9th International Conference on the Conservation and Management of Lakes*, Otsu City, Shiga Prefecture, Japan, 11–16 November. 3C/D-P71: pp. 530–533.

Tarlock, A.D., Corbridge, J.N. Jr and Getches, D.H. (1993) *Water Resource Management. A Casebook in Law and Public Policy.* Foundation Press, Westbury, New York.

Telitchenko, M.M. (1972) *About Possibility of Selfpurification Management by Biological Methods. Theory and Practice of Biological Selfpurification of Polluted Waters.* Moscow University Publishing, Moscow, pp. 58-61 (in Russian).

Teller, J.T. (1990) Meltwater and precipitation runoff to the North Atlantic, Arctic and Gulf of Mexico from the Laurentide ice sheet and adjacent regions during the Younger Dryas. *Paleoceanography* 5, 897–905.

Tennant, D.L. (1976) Instream flow regimens for fish, wildlife, recreation, and related environmental resources. *Fisheries* 1, 6–10.

Ter Braak, C.J.F. and Gremmen, N.J.M. (1987) Ecological amplitudes of plant species and the internal consistency of Ellenberg's indicator values for moisture. *Vegetatio* 69, 79–87.

Thomann, R.V., Di Toro, D.M. and O'Connor, D.J. (1974) Preliminary model of Potomac Estuary phytoplankton. *Journal of Environmental Engineering Division, American Society of Civil Engineers* 100, 699–715.

Thomas, C.D., Cameron, A., Green, R.E., Bakkenes, M., Beaumont, L.J, Collingham, Y.C., Erasmus, B.F.N., Ferreira De Siqueira, M., Grainger, A., Hannah, L., Hughes, L., Huntley, B., Van Jaarsveld, A.S., Midgley, G.F., Miles, L., Ortega-Huerta, M.A., Peterson, A.T., Phillips, O.L. and Williams, S.E. (2004) Feeling the heat: climate change and biodiversity loss. *Nature* 427, 145–148.

Thomas, J.A. and Bovee, K.D. (1993) Application and testing of a procedure to evaluate transferability of habitat suitability criteria. *Regulated Rivers* 8, 285–294.

Thompson, J.R. (1996) Africa's floodplains: a hydrological overview. In: Acreman, M.C. and Hollis, G.E. (eds) *Water Management and Wetlands in sub-Saharan Africa.* IUCN, Gland, Switzerland, pp. 5–20.

Thorarinsson, S. (1957) The jökulhlaups in Iceland. *Miscellaneous Papers, Museum of Natural History, Reykjavik* 18, 21–25.

Thornes, J.B. (1976) *Semi-arid Erosional Systems: Case Studies from Spain.* Paper No. 7. Geography Department, London School of Economics, London, 96 pp.

Thornton, J.A., Rast, W., Holland, M.M., Jolánkai, G. and Ryding, S.O. (eds) (1999) *Assessment and Control of Non-point Source Pollution of Aquatic Systems: A Practical Approach. Man and the Biosphere Series*, Vol. 23. UNESCO, Paris and Parthenon Publishing, Carnforth, UK, 466 pp.

Thorp, J.H. and Delong, M.D. (1994) The riverine productivity model: an heuristic view of carbon sources and organic processing in large river ecosystems. *Oikos* 70, 305–308.

Tiessen, H. (1995) In: Tiessen, H. (ed.) *Phosphorus in the Global Environment.* SCOPE 54. John Wiley and Sons Ltd, Chichester, UK.

Timchenko, V.M. (1990) *Ecohydrological Investigations on the Water Bodies of the North-Western Black Sea Region.* Naukova Dumka, Kyiv (in Russian).

Timchenko, V.M. (1996) External water exchange of floodplain water bodies of the Dnieper-River mouth zone as a controlling function for their ecosystems. *Hydrobiological Journal* 32, 90–102.

Timchenko, V. and Oksiyuk, O. (2002) Ecosystem condition and water quality control at impounded sections of rivers by the regulated hydrobiological regime. *Ecohydrology and Hydrobiology* 2, 259–264.

Timchenko, V., Oksiyuk, O. and Gore, J. (2000) A model for ecosystem condition and water quality management in the Dnieper River delta. *Ecological Engineering* 16, 119–125.

Timofeeva-Resovskaya, E.N. (1963) Distribution of radioisotopes on basic components of freshwater bodies. *Proceedings of Biological Institute of Ural Branch of the Academy of Sciences USSR* 30, 59–63 (in Russian).

Tockner, K. and Standford, J.A. (2002) Riverine flood plains: present state and future trends. *Environmental Conservation* 29, 308–330.

Tourre, Y.M. and White, W.B. (1995) ENSO signals in global upper-ocean temperature. *Journal of Physical Oceanography* 25, 1317–1332.

Townsend, C.R. (1980) *The Ecology of Streams and Rivers.* Edward Arnold, London.

Townsend, C.R. and Hildrew, A.G (1994) Species traits in relation to a habitat templet for river systems. *Freshwater Biology* 31, 265–275.

Tréguer, P.D., Nelson, M., van Bennekom, A.J., DeMaster, D.J., Leynaert, A. and Queguiner, B. (1995) The silica balance in the world ocean: a re-estimate. *Science* 268, 375–379.

Tricart, J., *et al.* (1962) Mechanismes normaux and phenomens catastro-phiques dans l'evolution des versants du bassin du Guil (Hautes Alpes, France). *Zeitschrift für Geomorphologie* 5, 277–301.

Triska, F.J. (1984) Role of wood debris in modifying channel geomorphology and riparian areas of a large lowland river under pristine conditions: a historical case study. *Verhandlungen des Internationalen Vereins für Limnologie* 22, 1876–1892.

Triska, F.J., Kennedy, V.C., Avanzino, R.J., Zellweger, G.W. and Bencala, K. (1989) Retention and transport of nutrients in a third order stream: channel processes. *Ecology* 70, 1877–1892.

Triska, F.J., Duff, J.H. and Avanzino, R.J. (1993) Patterns of hydrological exchange and nutrient transformation in the hyporheic zone of a gravel-bottom stream: examining terrestrial–aquatic linkages. *Freshwater Biology* 29, 259–274.

Trojanowska, A., Tarczyńska, M., Wagner, I., Romanowka-Duda, Z. and Zalewski, M. (2001) The importance pf phosphatase activity as compensatory mechanism for phytoplankton primary production in lowland reservoir (Poland). In: *Proceedings of 9th International Conference on the Conservation and Management of Lakes,* Otsu City, Shiga Prefecture, Japan, 11–16 November. 3C/D-P83: pp. 572–575.

Trump, C.L. and Leggett, W.C. (1980) Optimum swimming speeds in fish: the problem of currents. *Canadian Journal of Fisheries and Aquatic Science* 37, 1086.

Tsujimoto, T. (1996) Fish habitat and micro structure of flow in gravel bed stream. In: Leclerc, M., Capra, H., Valentin, S., Boudreault, A. and Coté, Y. (eds) *Ecohydraulics 2000. Proceedings of the 2nd International Symposium on Habitat Hydraulics.* INRS-Eau, Québec City, Canada, pp. 293–304.

Turkowska, K. (1988) Evolution des valle'as fluviatiles sur le Plateau de Lódz au cours du Quaternaire Tardif. *Acta Geographica Lodziensia* 57 (in Polish with a French summary), pp. 1–124.

Turner, J.V. and Macpherson, D.K. (1990) Mechanisms affecting streamflow and stream water quality: an approach via stable isotope, hydrochemical and time series analysis. *Water Resources Research* 26, 3005–3019.

Turner, R.E. and Rabalais, N.N. (1994) Coastal eutrophication near the Mississippi delta. *Nature* 368, 619–621.

Turner, R.E., Qureshi, N., Rabalais, N.N., Dortch, Q., Justic, D., Shaw, R.F. and Cope, J. (1998) Fluctuating silicate: nitrate ratios and coastal plankton food webs. *Proceedings of the National Academy of Sciiences of the USA* 95, 13048–13051.

Tüxen, R. (1954) Pflanzengesellschaften und Grundwasser-Ganglinien (Plant communities and groundwater duration lines). *Angewandte Pflanzensoziologie* 8, 64–97.

Tytherleigh, A. (1997) The establishment of buffer zones – The Habitat Scheme Water Fringe Option, UK. In: Haycock, N.E., Burt, T.P., Goulding, K.W.T. and Pinay, G. (eds) *Buffer Zones: Their Processes and Potential in Water Protection*. Quest Environmental, Harpenden, UK, pp. 255–264.

USACE (1982) *HEC-2 Water Surface Profiles Program*. United States Army Corps of Engineers, Hydrologic Engineering Center, Davis, California.

USEPA (1998) *Guidelines for Ecological Risk Assessment*. EPA/630/R-95%002F. US Environmental Protection Agency, Washington, DC, 124 pp.

UNESCO (1996) *Hydrology and Water Resources Development in a Vulnerable Environment. Detailed Plan of the Fifth Phase (1996–2000) of the IHP*. UNESCO, Paris, 54 pp.

UNESCO (2006) *Water Dependencies: Systems Under Stress and Societal Responses. Draft Strategic Plan for the 7th Phase of the IHP, 2008–2013*. SC-2006/CONF.203/CLD.22; IHP/IC-XVII/INF.9. Intergovernmental Council of the International Hydrology Programme, Paris, 51 pp.

UN Foundation (2007) Confronting Climate Change: Avoiding the Unmanageable and Managing the Unavoidable. Executive Summary. Scientific Expert Group Report on Climate Change and Sustainable Development. Prepared for the 15th Session of the Commission on Sustainable Development. United Nations Foundation and Sigma Xi; available at http://www.unfoundation.org/files/pdf/2007/SEG_ExecSumm.pdf

United Nations (2005) *The Millenium Development Goals Repor, 2005*. United Nations Department of Public Information, New York.

US NAP (2000) *Global Change Ecosystem Research*. Ecosystem Panel, Oversight Group for Ecosystem Panel, National Research Council. National Academy Press, Washington, DC.

Uusi-Kämppä, J., Turtula, E., Hartikanen, H. and Yläranta, T. (1997) The interactions of buffer zones and phosphorus runoff. In: Haycock, N.E., Burt, T.P., Goulding, K.W.T. and Pinay, G. (eds) *Buffer Zones: Their Processes and Potential in Water Protection*. Quest Environmental, Harpenden, UK, pp. 43–53.

Valett, H.M., Morrice, J.A., Dahm, C.N. and Campana, M.E. (1996) Parent lithology, surface–groundwater exchange and nitrate retention in headwater streams. *Limnology and Oceanography* 41, 333–345.

Van Dam, J.C., Huygen, J., Wesseling, J.G., Feddes, R.A., Kabat, P., van Walsum, P.E.V., Groenendijk, P. and van Diepen, C.A. (1997) *Theory of SWAP version 2.0. Simulation of Water Flow, Solute Transport and Plant Growth in the Soil–Water–Atmosphere–Plant Environment*. Department of Water Resources, Wageningen Agricultural University, Wageningen, The Netherlands.

Van de Griend, A.A., Seyhan, E., Engelen, G.B. and Geirnaert, W. (1986) Hydrological characteristics of an Alpine glacial valley in the North Italian Dolomites. *Journal of Hydrology* 88, 275–299.

Van den Bosch, R., Leigh, T.F., Falcon, L.A., Stern, V.M., Gonzales, D. and Hagen, K.S. (1971) The development program of integrated control of cotton pests in California. In: Huffacker, C.B. (ed.) *Biological Control*. Plenum, New York, New York, pp. 377–394.

Van der Meijden, R., Van Duuren, L., Weedam E.J. and Plate, C.L. (1991) Standaardlijst van de Nederlandse flora 1990 (Standard list of the flora of the Netherlands 1990). *Gorteria* 17, 75–127.

Van der Meijden, R., Groen, C.L.G., Vermeulen, J.J., Peterbroers, T., van't Zelfde, M. and Witte, J.P.M. (1996) *De landelijke flora-databank FLORBASE-1. Eindrapport (The National Databank FLORBASE-1. Final Report)*. National Herbarium, Leiden, The Netherlands.

Van Donk, E. (1998) Switches between clear and turbid water stated in a biomanipulated lake (1986–1996): the role of herbivory on macrophytes. In: Jeppesen, E., Sondergaard, Ma., Sondergaard, Mo. and Christoffersen, K. (eds) *The Structuring Role of Submerged Macrophytes in Lakes. Ecological Studies*, Vol. 131. Springer, New York, New York, pp. 290–297.

Van Ek, R., Witte, J.P.M., Runhaar, J. and Klijn, F. (2000) Ecological effects of water management in the Netherlands: the model DEMNAT. *Ecological Engineering* 16, 127–141.

Vannote, R.L., Minshall, G.W., Cummins, K.W., Sedell, J.R. and Cushing, C.E. (1980) The river continuum concept. *Canadian Journal of Fisheries and Aquatic Sciences* 37, 130–137.

van Oene, H., Berendse, F. and de Kovel, C.G.F. (1999) Model analysis of the effects of historic CO_2 levels and nitrogen inputs on vegetation succession. *Ecological Applications* 9, 920–935.

van Valen, L. (1973) A New Evolutionary Law. *Evolution Theory* 1, 1–30.

van Wirdum, G. (1986) Water related impacts on nature protection sites. In: Hooghart, J.C. (ed.) *Water Management in Relation to Nature, Forestry and Landscape Management*. TNO Committee on Hydrological Research, The Hague, The Netherlands, pp. 25–57.

van Wirdum, G. (1991) Vegetation and hydrology of floating rich-fens. PhD thesis, University of Amsterdam, Amsterdam.

Varis, O. and Somlyódy, L. (1994) *Potential Impacts of Climatic Change on Lake and Reservoir Water Quality*. International Institute for Applied Systems Analysis, Laxenburg, Austria.

Varlygin, D.L. and Bazilevich, N.I. (1992) Production linkages of zonal world plant formations with some climate parameters. *Izvestiya Akademii Nauk SSSR, Seriya Geograficheskaya* 1, 36–64.

Vasiliev, U.S. and Kukushkina, V.A. (1988) *Use of Lakes and Rivers with Goal of Recreation*. Hydrometeoizdat, Leningrad, Russia (in Russian).

Vehanen, T., Hyvärinen, P. and Mäki-Petays, A. (1996) Fish migration from two regulated lakes outcoming rivers monitored by hydroacustics. In: Leclerc, M., Capra, H., Valentin, S., Boudreault, A. and Coté, Y. (eds) *Ecohydraulics 2000. Proceedings of the 2nd International Symposium on Habitat Hydraulics*. INRS-Eau, Québec City, Canada, pp. 967–978.

Velz, C.J. (1984) *Applied Stream Sanitation*. John Wiley and Sons, New York, New York.

Verdonschot, P.F.M., Driessen, J.M.C., Mosterdijk, H.G. and Schot, J.A. (1998) The 5-S model: an integrated approach for stream rehabilitation. In: Hansen, H.O. and Madsen, B.L. (eds) *River Restoration '96*. National Environmental Research Institute, Copenhagen, pp. 36–44.

Vermulst, J.A.P.H. and De Lange, W.J. (1999) An analytic-based approach for coupling models for unsaturated and saturated groundwater flow at different scales. *Journal of Hydrology* 226, 262–273.

Vermulst, J.A.P.H., Hoogeveen, J., de Lange, W.J., Bos, H.B. and Pakes, U. (1996) MONA, an interface for GIS-based coupled saturated and unsaturated groundwater modelling in the Netherlands. In: Holzmann, H. and Nachtnebel, H.P. (eds) *Application* of Geographical Information Systems in Hydrology and Water Resources Management. Vienna, pp. 358–365.

Vincke, P.P. and Thiaw, I. (1995) Protected areas and dams: the case of the Senegal River Delta. *Parks* 5, 32–38.

Vincon-Leite, B. and Tassin, B. (1990) Modelisation de la qualité des lacs profonds: modéle thermique et biogéochimique du lac du Baurget. *La Huille Blanche* 3/4, 321–236.

Viner, A.B. (1987) Nutrients transported on silt in rivers. *Archivium Hydrobiologiae Beihefte (Ergebnisse der Limnologie)* 28, 63–71.

Vipper, P.B., Dorofyeuk, N.P., Metelcova, E.P. and Sokolowskaya, W.T. (1987) Landscape–climatic changes in central Mongolia during the Holocene. In: Khotinsky, N.A. (ed.) *Paleoklimaty pozdnielednikovia i golocena*. Nauka, Moscow, pp. 160–167 (in Russian)

Vita-Finzi, C. (1969) *The Mediterraneacn Valleys*. Cambridge University Press, Cambridge. UK.

VITUKI plc (1991) *Relationship between the Structure of Flood-plain Vegetation and the Water Regime*. VITUKI Research Report No. 7611/2/2001 (theme leader: J. Gaspar). VITUKI, Budapest (in Hungarian).

VITUKI plc (1993) *Investigation of Channel Changes of Small Streams and Creeks*. VITUKI Research Report No. 731/2/2716 (theme leader: Gy. Szepessy). VITUKI, Budapest (in Hungarian).

VITUKI (1998) *Model Development and Application to the Zala River Catchment*. Contribution to the Final Report of EU-Project INCAMOD: Development of Integrated Catchment Models for Supporting Water Management Decisions. VITUKI, Budapest.

Volk, T. (1998) *Gaia's Body: toward a physiology of Earth*. Copernicus/Springer-Verlag, New York.

Vollenweider, R.A. (1968) *Les bases scientifiques de l'eutrophisation des lacs et des eaux courantes sous l'aspect particulier du phosphor et d l'azote comme facteurs d'eutrophisation*. OECD Technical Papers. OECD, Paris, 159 pp.

Vollenweider, R.A. (1969) Möglichkeiten und Grenzen elementarer Modelle der Stoffbilanz von Seen. *Archiv für Hydrobiologie* 66, 1–36.

Vollenweider R.A. (1970) *Scientific fundamentals of the eutrophication of lakes and flowing waters, with particular reference to nitrogen and phosphorus as factors in eutrophication*. OECD Technical Papers. OECD, Paris.

Vollenweider, R.A. (1976) Advances in defining the critical loading level for phosphorus in lake eutrophication. Mem.Ist.Ital.Idrobiol.Dott. Marco di Marchi 33, 53–83.

Vollenweider, R.A. (1989) Global problems of eutrophication and its control. *Symposia biologicae Hungaricae* 38, 19–41.

Vollenweider, R.A. and Kerekes, J. (1981) Background and summary results of the OECD Cooperative Programme on Eutrophication. In: *Proceedings of the International Symposium on Inland Waters and Lake Restoration*, Portland, Maine, 8–12 September. EPA 440/5-81-110. US Environmental Protection Agency, Washington, DC, pp. 25–36.

Vollenweider, R.A. and Kerekes, J. (1982) *Eutrophication of Waters. Monitoring, Assessment and Control*. OECD Cooperative Programme on Monitoring of Inland Waters (Eutrophication Control). Environment Directorate, OECD, Paris, 154 pp.

Vozhzennikova, J.F. (1958) Algae of the Katun River and its tributaries in region of Chermal resort. *Proceedings of Siberian Branch of the USSR Academy of Sciences* 8, 28–31 (in Russian).

Waddle, T. (1998) Development of 2-dimensional habitat models. In: Blazkova, S., Stalnaker, C. and Novicky, O. (eds) *Hydroecological Modeling: Research, Practice,*

Legislation, and Decision-making. T.G. Masaryk Water Research Institute, Prague, Czech Republic and MESC, Fort Collins, Colorado, pp. 19–22.

Wade, P.M. (1993) Physical control of aquatic weeds. In: Pieterse, A.H. and Murphy, K.J. (eds) *Aquatic Weeds. The Ecology and Management of Nuisance Aquatic Vegetation*. Oxford University Press, Oxford, UK, pp. 93–135.

Wagner, I. (2001) Influence of the selected climatic, hydrological and biological factors on eutrophication processes and symptoms in the Sulejow Reservoir. PhD thesis, University of Lodz, Lodz, Poland.

Wagner, I. and Zalewski, M. (1997) Potential effect of global climate changes on ecohydrological processes. In: Wiśniewski, R. and Zalewski, M. (eds) Application of Ecosystem Technologies Toward Freshwater Quality Improvement. *Zeszyty Naukowe "Człowiek i Środowisko"* 18, 37–50 (in Polish).

Wagner, I. and Zalewski, M. (2000) Effect of hydrological patterns of tributaries on biotic processes in a lowland reservoir – consequences for restoration. *Ecological Engineering* 16, 79–90.

Wagner, I., Marshalek, J. and Breil, P. (eds) (2008) *Aquatic Habitats in Sustainable Urban Water Management: Science, Policy and Practice*. Taylor and Francis/Balkema, Leiden, The Netherlands, 229 pp.

Wagner-Lotkowska, I. (2002) Influence of the selected climatic, hydrological and biological factors on eutrophication processes and symptoms in the Sulejów Reservoir. PhD thesis, University of Lodz, Lodz, Poland.

Wagner-Lotkowska, I., Bocian, J., Pypaert, P., Santiago-Fandino, V. and Zalewski, M. (2004) Environment and economy – dual benefit of ecohydrology and phytotechnology in water resources management: Pilica River Demonstration Project under the auspices of UNESCO and UNEP. Special Issue: Ecohydrology from Theory to Action. *Ecohydrology and Hydrobiology* 3, 345–352.

Wahby, S.D. and Bishara, N.F. (1980) The effect of the River Nile on Mediterranean water, before and after the construction of the High Dam at Aswan. In: Martin, J.M., Burton, J.D. and Eisma, D. (eds) *River Inputs to Ocean Systems*. UNEP/IOC/SCOR/United Nations, New York, New York, pp. 311–318.

Walling, D.E. (1987) Hydrological processes. In: Gregory, K.J. and Walling, D.E. (eds) *Human Activity and Environmental Processes*. John Wiley and Sons Ltd, Chichester, UK, pp. 53–86.

Walling, D.E., Russell, M.A. and Webb, B.W. (2000) Controls on the nutrient content of suspended sediment transported by British rivers. *Science of the Total Environment* 266, 113–123.

Wang, M. and Harleman, D.R.F. (1983) Modeling phytoplankton concentration in a stratified lake. In: Lauenroth, W.K., Skogerobe, G.V. and Flug, M. (eds) *Analysis of Ecological Systems: State-of-the-Art in Ecological Modeling*. Elsevier Scientific Publishing Co., Amsterdam, pp. 807–823.

Ward, J.V. (1990) Riverine–wetlands interactions. In: Sharitz, R.R. and Gibbons, J.W. (eds) *Freshwater Wetlands and Wildlife: Perspectives on Natural, Managed and Degraded Ecosystems*. USDOE CONF-8603101. Office of Scientific and Technical Information, US Department of Energy, Oak Ridge, Tennessee, pp. 385–400.

Ward, J.V. (1992) *Aquatic Insect Ecology*. Vol. 1. *Biology and Habitat*. Wiley, New York, New York.

Ward, J.V. and Stanford, J.A. (1982) Thermal responses in the evolutionary ecology of aquatic insects. *Annual Review of Entomology* 27, 97–117.

Ward, J.V. and Stanford, J.A. (1983) The serial discontinuity concept of lotic ecosystems. In: Fontaine, T.D. III and Bartell, S.M. (eds) *Dynamics of Lotic Ecosystems*. Ann Arbor Science Publishers, Ann Arbor, Michigan, pp. 29–42.

Ward, J.V. and Stanford, J.A. (1995) Ecological connectivity in alluvial river ecosystems and its disruption by flow regulations. *Regulated Rivers: Research and Management* 11, 105–119.

Ward, J.V. and Voelz, N.J. (1990) Gradient analysis of interstitial meiofauna along a longitudinal stream profile. *Stygologia* 5, 93–99.

Ward, J.V. and Wiens, J.A. (2001) Ecotones of riverine ecosystems: role and typology, spatio-temporal dynamics, and river regulation. *Ecohydrology and Hydrobiology* 1–2, 25–36.

Ward, J.V., Tockner, K., Uehlinger, U. and Malard, F. (2001) Understanding natural patterns and processes in river corridors as the basis for effective river restoration. *Regulated Rivers: Research and Management* 17, 311–323.

Waringer, J.A. (1989) Resistance of cased caddis larva to accidental entry into the drift: the contribution of active and passive elements. *Freshwater Biology* 21, 411–420.

Warren, C.E. and Davis, G.E. (1967) Laboratory studies on the feeding, bioenergetics, and growth of fish. In: Gerking, S.D. (ed.) *The Biological Basis of Freshwater Fish Production*. Blackwell Scientific Publications, Oxford, UK, pp. 175–214.

Wasmund, N., Breuel, G., Edler, L., Kuosa, H., Olsonen, R., Schultz, H., Pys-Wolska, M. and Wrzolek, L. (1996) Pelagic biology. In: *Baltic Sea Environ. Proc. 64B*. Helsinki Commission, Helsinki, pp. 89–100.

Wasylikowa, K., Starkel, L., Niedziałkowska, E., Skiba, S. and Stworzewicz, E. (1985) Environmental changes in the Vistula valley at Pleszów caused by neolithic man. *Przegląd Archeologiczny* 33, 19–55.

Watson, A.J. and Lovelock, J.E. (1983) Biological homeostasis of the global environment: the parable of Daisyworld. *Tellus* 35B, 284–289.

Watson, C.C., Harvey, M.D. and Garbrecht, J. (1986) Geomorphic–hydraulic simulation of channel evolution. In: *Proceedings of the 4th Federal Interagency Sedimentation Conference*, Las Vegas, Nevada. US Government Printing Office, Washington DC, pp. 5.21–5.30.

Watson, V.J. (1983) Application of a seasonal phosphorus dynamics model to lake ecosystem: assessing lake loading tolerance. In: Lauenroth, W.K., Skogerobe, G.V. and Flug, M. (eds) *Analysis of Ecological Systems: State-of-the-Art in Ecological Modeling*. Elsevier Scientific Publishing Co., Amsterdam, pp. 575–592.

Watt, S.B. (1981) Peripheral problems in the Senegal Valley. In: Saha, S.K. and Barrow, C.J. (eds) *River Basin Planning*. Wiley-Interscience, Chichester, UK.

Watts, D.M., Phillips, I., Callahan, J.D., Griebenow, W., Hyams, K.C. and Hayes, C.G. (1997) Oropouche virus transmission in the Amazon river basin of Peru. *American Journal of Tropical Medicine and Hygiene* 56, 148–152.

Waylen, P. (1995) Global hydrology in relation to palaeohydrological change. In: Gregory, K.J., Starkel, L. and Baker, V.R. (eds) *Global Continental Palaeohydrology*. John Wiley and Sons Ltd, Chichester, UK, pp. 61–86.

Webb, P.W. (1975) Hydrodynamics and energetics of fish propulsion. *Fisheries Research Board of Canada Bulletin* 190, 158–165.

Webster, J.R. and Pattern, B.C. (1979) Effects of watershed perturbation on stream potassium and calcium dynamics. *Ecological Monograph* 49, 51–72.

Weilguni, H., Janauer, G.A. and Wychera, U. (1999) The New Danube in 1998 – partial aspects of the trophic state and the aquatic plants. In: Weber, G., Wimmer, W., Mager, G., Klotz, A. and Neckar, H. (eds) *Perspektiven, 2-3/1999*. Gewässergüteforschung Neue Donau. Vienna, pp. 75–80 (English version pp. 95–96).

Welcomme, R.L. (1985) *River Fisheries*. FAO Fisheries Technical Paper No. 262. FAO, Rome, 330 pp.

Wesche, T.A. and Rechard, P.A. (1980) *A Summary of Instream Flow Methods for Fisheries and Related Research Needs.* Eisenhower Consortium Bulletin No. 9. University of Wyoming, Water Resources Research Institute, Laramie, Wyoming.

Westlake, D.F. (1974) Macrophytes. In: Vollenweider, R.A. (ed.) *A Manual on Methods for Measuring Primary Production in Aquatic Environments.* IBP Handbook No. 12. Blackwell Scientific Publications, Oxford, UK, pp. 32–42.

Wetzel, R.G. (1975) *Limnology.* W.B. Saunders, Philadelphia, Pennsylvania, 742 pp.

Whigham, D.F., Chitterling, C. and Palmer, B. (1988) Impacts of freshwater wetlands on water quality: a landscape perspective. *Environmental Management* 12, 663–671.

Whitton, B.A. (ed.) (1984) Ecology of European Rivers. Blackwell Scientific Publications, Oxford, UK, 644 pp.

Wiertz, J., van Dijk, J. and Latour, J.B. (1992) *De MOVE-vegetatiemodule. De kans op voorkomen van 700 plantesoorten als functie van vocht, pH, nutriënten en zout* (The MOVE Vegetation Module. The Occurrence Probability of 700 Plant Species as a Function of Moisture, pH and Salt). Institute of Forestry and Nature Research, Wageningen, The Netherlands.

Wiley, M.J. and Kohler, S.L. (1980) Positioning changes of mayfly nymphs due to behavioral regulation of oxygen consumption. *Canadian Journal of Zoology* 58, 618–622.

Williams, M.R. and Melack, J.M. (1997) Solute export from forested and partially deforested catchments in the central Amazon. *Biogeochemistry* 38, 67–102.

Williams, M.R., Fisher, T.R. and Melack, J.M. (1997) Solute dynamics in soil water and groundwater in a central Amazon catchment undergoing deforestation. *Biogeochemistry* 38, 303–335.

Williams, W.T. and Lambert, J.M. (1966) Multivariate methods in plant ecology. V. Similarity analysis and information analysis. *Journal of Ecology* 54, 427–445.

Willms, W., Colwell, D.D. and Kenzie, O. (1994) Water from dugouts can reduce livestock performance. *Weekly Letter* No. 3099. Agriculture and Agri-Food Research Centre, Lethbridge, Alberta, Canada.

Willms, W., Kenzie, O., Quinton, D. and Wallis, P. (1996) The water source as a factor affecting livestock production. In: Rode, L.M. (ed.) *Animal Science Research and Development: Meeting Future Challenges. Proceedings of the Canadian Society of Animal Science Annual Meeting 1996.* Canadian Society of Animal Science, Lethbridge, Alberta, Canada, pp. 41–46.

Wilpiszewska, I. (1990) Productivity and chemical valorization of mire vegetation in postglacial agriculture landscape. *Ekologia polska* 38, 3–72.

Wilpiszewska, I. and Kloss, M. (2002) Wetland patches (potholes) in a mosaic landscape (Masurian Lakeland, Poland): floristic diversity and disturbance. *Polish Journal of Ecology* 50, 515–526.

Wilson, E.O. (1988) The current state of biological diversity. In: Wilson, E.O. (ed.) *Biodiversity.* National Academy Press, Washington, DC, pp. 3–18.

Winberg, G.G. (1956) *Rate of Metabolism and Food Requirements of Fishes.* Nauchnyye Trudy Belorusskogo Gosudarstvennogo Universiteta, Minsk, 253 pp. Translated by *Journal of the Fisheries Research Board of Canada, Translation Series* 194, 1960.

Winberg, G.G. (1965) Biotic balance of matter and energy and the biological productivity of water basins. *Gidrobiologiczeskij Zhurnal* 1, 25–32 (in Russian with English summary).

Wischmeier, W.H. and Smith, D.D. (1978) *Predicting Rainfall Erosion Losses. A Guide to Conservation Planning.* USDA Handbook No. 537. US Department of Agriculture, Washington, DC.

Wishart, M.J., Gagneur, J. and El-Zanfaly, H.T. (2000) River conservation in North Africa and the Middle East. In: Boon, P.J., Davies, B.R. and Petts, G.E. (eds) *Global Perspec-*

tives on River Conservation: Science, Policy and Practice. John Wiley and Sons Ltd, Chichester, UK, 127–154.

Witte, J.P.M. (1996) De waarde van natuur. Zeldzaamheid en de botanische waardering van gebieden (The value of nature. Rarity and the botanical valuation of areas). *Landschap* 13, 79–95.

Witte, J.P.M. (1998) National water management and the value of nature. PhD thesis, Wageningen Agricultural University, Wageningen, The Netherlands.

Witte, J.P.M. (2002) The descriptive capacity of ecological plant species groups. *Plant Ecology* 162, 199–213.

Witte, J.P.M. and van der Meijden, R. (1990) *Natte en vochtige ecosystemen in Nederland (Wet and Moist Ecosystems in The Netherlands)*. KNNV, Utrecht, The Netherlands.

Witte, J.P.M. and van der Meijden, R. (1995) Verspreidingskaarten van de botanische kwaliteit in Nederland uit FLORBASE (Distribution maps of the botanical quality in the Netherlands from FLORBASE). *Gorteria* 21, 3–59.

Witte, J.P.M. and van der Meijden, R. (2000) Mapping ecosystem types by means of ecological species groups. *Ecological Engineering* 16, 143–152.

Witte, J.P.M., Groen, C.L.G., van der Meijden, R. and Nienhuis, J.G. (1993) A national model for the effects of water management on the vegetation. In: Hooghart, J.C. and Posthumus, C.W.S. (eds) *The Use of Hydro-ecological Models in The Netherlands*. TNO Committee on Hydrological Research, Delft, The Netherlands, pp. 31–51.

Wolanski, E. and Gereta, E. (2000) Oxygen cycle in a hippo pool, Serengeti National Park, Tanzania. *African Journal of Ecology* 37, 419–423.

Wolanski, E. and Gereta, E. (2001) Water quantity and quality as the factors driving the Serengeti ecosystem. *Hydrobiologia* 458, 169–180.

Wolanski, E., Boorman, L.A., Chicharo, L., Langlois-Saliou, E., Lara, R., Plater, A.J., Uncles, R.J. and Zalewski, M. (2005) Ecohydrology as a new tool for sustainable management of estuaries and coastal waters. *Wetlands Ecology and Management* 12, 235–276.

Wollast, R. and Mackenzie, F.T. (1983) The global cycle of silica. In: Aston, S.R. (ed.) *Silicon Geochemistry and Biochemistry*. Academic Press, London, pp. 39–77.

Wollheim, W.M. and Lovvorn, J.R. (1996) Effects of macrophyte growth forms on invertebrate communities in saline lakes of the Wyoming High Plains. *Hydrobiologia* 323, 83–96.

Wolman, M.G. (1967) A cycle of sedimentation and erosion in urban river channels. *Geografiska Annaler* 49A, 385–395.

Woo, M.K. (1986) Permafrost hydrology in North America. *Atmosphere–Ocean* 24, 201–234.

World Bank (1993) *Water Resources Management*. A World Bank Policy Paper. World Bank, Washington, DC.

World Bank (1999) *Global Development Finance 1999*. World Bank, Washington, DC.

World Resources (1998) *1998–1999 World Resources: A Guide to the Global Environment*. Oxford University Press, New York, New York.

Wu, P.I., Braden, J.B. and Johnson, G.V. (1989) Efficient control of cropland sediment: storm event versus annual average loads. *Water Resources Research* 25, 161–168.

Wulff, F. and Rahm, L. (1988) Long-term, seasonal and spatial variations of nitrogen, phosphorus and silicate in the Baltic: an overview. *Marine Environmental Research* 26, 19–37.

Wulff, F. and Stigebrandt, A. (1989) A time dependent budget model for nutrients in the Baltic. *Global Biogeochemical Cycles* 3, 63–78.

Wulff, F., Rahm, L., Hallin, A.-K. and Sandberg, J. (2000) A nutrient budget model of the Baltic Sea. In: Wulff, F., Rahm, L. and Larsson, P. (eds) *A Systems Analysis of the Changing Baltic Sea*. Springer Verlag, Berlin.

Würzbach, R., Zeltner, G.-H. and Kohler, A. (1997) Die Makrophyten-Vegetation des Fließgewässersystems der Moosach (Münchener Ebene). Ihre Entwicklung und Veränderung von 1970–1996. *Berichte des Instituts für Landschafts- und Pflanzenökologie. Universität Hohenheim* 4, 244–312.

Yamskich, A.F. (1993) *Deposition and Terrace Formation in the River Valleys of Southern Siberia.* Pedagogicheskiy Institut, Krasnoyarsk, Russia, 22 pp. (in Russian).

Yeasted, J.G. and Morel, F.M.M. (1978) Empirical insights into lake response to nutrient loading, with application to models of phosphorus in lakes. *Environmental Science and Technology* 12, 195–201.

Yelenevsky, R.A. (1936) *Questions of Studying and Mastering of Floodplains.* Selhozizdat, Moscow (in Russian).

Zalewski, M. (1988) Ecological basis for environment protection and structuring. In: Olaczek, R. (ed.) *Environment Protection and Structuring.* University of Lodz, Lodz, Poland, pp. 35–59 (in Polish).

Zalewski, M. (1999) Minimising the risk and amplifying the opportunities for restoration of shallow reservoirs. In: Harper, D.M., Brierley, B., Ferguson, A.J.D. and Phillips, G. (eds) The Ecological Bases for Lake and Reservoir Management. *Hydrobiologia* 395/396, 107–114.

Zalewski, M. (2000) Ecohydrology – the scientific background to use ecosystem properties as management tools towards sustainability of water resources. *Ecological Engineering* 16, 1–8.

Zalewski, M. (2002) Ecohydrology – the use of ecological and hydrological processes for sustainable management of water resources. *Hydrological Sciences Journal* 47, 823–832.

Zalewski, M. (2004) Ecohydrology as a system approach for sustainable water biodiversity and ecosystem services. *Ecohydrology and Hydrobiology* 4, 229–236.

Zalewski, M. (2006) Ecohydrology – an interdisciplinary tool for integrated protection and management of water bodies. *Archive of Hydrobiology* 158(Suppl. 4), 613–622.

Zalewski, M. and Harper, D.M. (2001) Ecohydrology – the use of ecosystem properties as management tools for enhancement of the absorbing capacity of ecosytems against human impact. *Ecohydrology and Hydrobiology* 1, 1–2.

Zalewski, M. and Naiman, R. (1985) The regulaton of riverine fish communities by a continuation of abiotic-biotic factors. In: Alabaster, J.S. (ed) *Habitat Modification and Freshwater Fisheries.* Butterworths Scientific, UK, pp 3–9.

Zalewski, M. and Wagner, I. (1998) Temperature and nutrients dynamic in freshwater eutrophic ecosystems. *Geographia Polonica* 71, 79–92.

Zalewski, M. and Wagner, I. (2005) Ecohydrology – the use of water and ecosystem processes for healthy urban environments. Special Issue: Aquatic Habitats in Integrated Urban Water Management. *Ecohydrology and Hydrobiology* 5, 263–268.

Zalewski M. and Wagner-Lotkowska, I. (2004). *Integrated Watershed Management - Ecohydrology and Phytotechnology Manual.* UNESCO, Regional Bureau for Science in Europe (ROSTE) Palazzo Zorzi, Castello 4930 – 30122 Venice, Italy.

Zalewski, M., Brewinska-Zaras, B., Frankiewicz, P. and Kalinowski, S. (1990) The potential for biomanipulation using fry communities in a lowland reservoir: concordance between water quality and optimal recruitment. *Hydrobiologia* 200/201, 549–556.

Zalewski, M., Janauer, G.A. and Jolankai, G. (1997) *Ecohydrology. A New Paradigm for the Sustainable Use of Aquatic Resources.* Technical Documents in Hydrology No. 7. UNESCO, Paris.

Zalewski, M., Bis, B., Łapińska, M., Frankiewicz, P. and Puchalski, W. (1998) The importance of the riparian ecotone and river hydraulics for sustainable basin-scale restoration scenarios. *Aquatic Conservation: Marine and Freshwater Ecosystems* 8, 287–307.

Zalewski, M., Tarczyńska, M. and Wagner-Łotkowska, I. (2000) Ecohydrological approaches to the elimination of toxic algal blooms in a lowland reservoir. *Verhandlungen des Internationalen Vereins für Limnologie* 27, 3178–3183.

Zalewski, M., Thorpe, J.E. and Scheimer, F. (2001) Guest Editorial. *Ecohydrology and Hydrobiology* 1, 5–7.

Zalewski, M., Santiago-Fandino, V. and Neate, J. (2003) Energy, water, plant interactions: 'green feedback' as a mechanism for environmental management and control through the application of phytotechnology and ecohydrology. *Hydrological Processes* 17, 2753–2767.

Zalewski, M., Harper D.M. and Roberts, R.D. (2004) Ecohydrology from Theory to Action. *Ecohydrology and Hydrobiology* 4(3), 227–352.

Zebidi, H. (1998) *Water: a looming crisis.* UNESCO IHP-V Technical Documents in Hydrology No 18.

Zhang, P. and Wu, X. (1990) Regional response to global warming: a case study in China. In: Li, Y.Y. (ed.) *World Laboratory and CCAST Workshop Series*, Vol. 5. Gordon and Breach Scientific Publishers, Reading, UK, pp. 26–41.

Index